Current Topics in Microbiology and Immunology

253

Editors

R.W. Compans, Atlanta/Georgia
M. Cooper, Birmingham/Alabama · Y. Ito, Kyoto
H. Koprowski, Philadelphia/Pennsylvania · F. Melchers, Basel
M. Oldstone, La Jolla/California · S. Olsnes, Oslo
M. Potter, Bethesda/Maryland
P.K. Vogt, La Jolla/California · H. Wagner, Munich

Springer
Berlin
Heidelberg
New York
Barcelona
Hong Kong
London
Milan
Paris
Singapore
Tokyo

The Mechanisms of Neuronal Damage in Virus Infections of the Nervous System

Edited by G. Gosztonyi

With 28 Figures and 4 Tables

Springer

Professor GEORG GOSZTONYI, M.D., Ph.D.
FU Berlin, Fachbereich Humanmedizin
Universitätsklinikum Benjamin Franklin
Abt. für Neuropathologie
Hindenburgdamm 30
D-12200 Berlin

Cover Illustration: Distribution of p24 antigen of Borna disease virus (BDV) in the hippocampal formation of the rat. Only the dentate gyrus and the CA3 subfield of the hippocampus are infected; CA1 neurons remain resistant to infection. The viral protein is carried anterogradely along the axons of the CA3 neurons, filling them in their entire length and resulting in a laminar distribution pattern of the antigen in the CA1 subfield. The permissiveness of dentate gyrus and CA3 neurons is explained by the presence of the kainate 1 (KA-1) receptor on their surface, and the resistance of CA1 neurons by the lack of expression of this receptor protein, which seems to represent the BDV receptor. The illustration demonstrates that expression of a viral receptor on the cell surface renders the neuron vulnerable to, and absence of the receptor protects the neuron from, infection. For further details see the chapter by Gosztonyi and Ludwig, this volume.

ISSN 0070-217X
ISBN 3-540-67617-1 Springer-Verlag Berlin Heidelberg New York

This work is subject to copyright. All rights are reserved, whether the whole or part of the material is concerned, specifically the rights of translation, reprinting, reuse of illustrations, recitation, broadcasting, reproduction on microfilm or in any other way, and storage in data banks. Duplication of this publication or parts thereof is permitted only under the provisions of the German Copyright Law of September 9, 1965, in its current version, and permission for use must always be obtained from Springer-Verlag. Violations are liable for prosecution under the German Copyright Law.

Springer-Verlag Berlin Heidelberg New York
a member of BertelsmannSpringer Science + Business Media GmbH

http://www.springer.de

© Springer-Verlag Berlin Heidelberg 2001
Library of Congress Catalog Card Number 15-12910
Printed in Germany

The use of general descriptive names, registered names, trademarks, etc. in this publication does not imply, even in the absence of a specific statement, that such names are exempt from the relevant protective laws and regulations and therefore free for general use.

Product liability: The publishers cannot guarantee the accuracy of any information about dosage and application contained in this book. In every individual case the user must check such information by consulting other relevant literature.

Cover Design: *design & production GmbH*, Heidelberg
Typesetting: Scientific Publishing Services (P) Ltd, Madras
Printed on acid-free paper   SPIN: 10675019   27/3136/M – 5 4 3 2 1 0

Preface

Neuronal damage in virus infections of the nervous system is not a simple story any more. Not one single, but several different mechanisms may result in the demise of neurons, and our knowledge of these various mechanisms, and their interrelationships, is progressively increasing. Therefore, it seemed to be justified to compile this recent knowledge into one *Current Topics in Microbiology and Immunology* volume.

   Neurons are the most cherished cell type of the brain, since they are highly differentiated and irreplaceable. Viral infections that attack neurons fundamentally endanger their existence, but only the most acute virus infections result in a devastating destruction of the majority of neurons. During evolution, the brain has acquired the capacity to defend itself against such devastating attacks. Some viral agents have been forced into latency, others into persistent infection. Since the immune system, in many instances, is unable to eliminate these infections, a compromise has been established to achieve neuronal survival. In the course of these chronic, non-lethal infections, viruses and/or their constituents may interfere with neuronal function, even if these viruses do not actually infect neurons but are produced in other cell types of the central nervous system. This is the case with the neurotoxic effect of the envelope protein gp120 of the human immune deficiency virus. Viral proteins may interfere with differentiated cell functions, e.g., with those of neurotransmitters, resulting in deafferentation through obstruction of receptors. Blockage of inhibitory mechanisms may result in excitotoxic neuronal death. Interference with binding of neurotrophic factors may cause subacute or chronic neuronal degeneration. These latter mechanisms, virus-induced excitotoxicity and interference with neurotrophic functions, would have deserved independent chapters in this *Current Topics in Microbiology and Immunology* volume; however, the presently available knowledge does not suffice to justify this. In contrast, an increasing amount of data are accumulating to document that apoptotic neuronal death is an important decay mechanism in neurotropic virus infections.

In fact, various agents, in various constellations, can induce programmed cell death, so that apoptosis, in addition to necrosis, seems to represent a *final common pathway* of neuronal destruction.

Immune reactions are aimed at diverting and eliminating infection. Unfortunately, in the delicate structure of the brain, cellular immune reactions often cause more harm than benefit. Due to the irreplaceable nature of neurons, T cell-mediated cytotoxicity would be a catastrophe to the nervous system; fortunately, in the course of evolution, neurons have become able to suppress class I expression and, by doing so, they have been preserved for higher development. Virus infections often induce host autoimmunity reactions, whose targets are myelin and oligodendrocytes; however, by consecutive axonal damage, severe interference with neuronal function ensues.

Not infrequently, the pathogenetic mechanisms discussed in this volume act in a combined way; nevertheless, in many instances one or the other mechanism prevails. Thus, overlap between the subject matter of various chapters could not be avoided. In addition, regarding some subjects, complete agreement does not exist between different authors, and views regarding certain questions may even diverge. These circumstances reflect the fast and dynamic expansion of knowledge in this field of neurovirology. The reviews of this volume attempt to provide insight into the complexity of this process. Thanks are due to Hilary Koprowski for supporting this project and to Doris Walker and Anne Clauss for their most valuable help in assembling this volume.

Berlin, February 2001   G. GOSZTONYI

List of Contents

G. Gosztonyi and H. Koprowski
The Concept of Neurotropism and Selective
Vulnerability ("Pathoclisis") in Virus Infections
of the Nervous System – A Historical Overview   1

J.R. Anderson
The Mechanisms of Direct, Virus-Induced Destruction
of Neurons   15

U.G. Liebert
Slow and Persistent Virus Infections of Neurones –
A Compromise for Neuronal Survival   35

K. Borchers and H.J. Field
Neuronal Latency in Human
and Animal Herpesvirus Infections   61

J. Fazakerley and T.E. Allsopp
Programmed Cell Death in Virus Infections
of the Nervous System   95

G. Gosztonyi and H. Ludwig
Interactions of Viral Proteins
with Neurotransmitter Receptors May Protect
or Destroy Neurons   121

B. Dietzschold, K. Morimoto, and D.C. Hooper
Mechanisms of Virus-Induced Neuronal Damage
and the Clearance of Viruses from the CNS   145

M. Hornig, M. Solbrig, N. Horscroft,
H. Weissenböck, and W.I. Lipkin
Borna Disease Virus Infection of Adult and
Neonatal Rats: Models for Neuropsychiatric Disease . . .   157

V.J. Sanders, C.A. Wiley, and R.L. Hamilton
The Mechanisms of Neuronal Damage
in Retroviral Infections of the Nervous System   179

A. Giese and H.A. Kretzschmar
Prion-Induced Neuronal Damage –
The Mechanisms of Neuronal Destruction
in the Subacute Spongiform Encephalopathies 203

R. Dörries
The Role of T-Cell-Mediated Mechanisms
in Virus Infections of the Nervous System 219

P.J. Talbot, D. Arnold, and J.P. Antel
Virus-Induced Autoimmune Reactions in the CNS 247

Subject Index. 273

List of Contributors

(Their addresses can be found at the beginning of their respective chapters.)

ALLSOPP, T.E. 95
ANDERSON, J.R. 15
ANTEL, J.P. 247
ARNOLD, D. 247
BORCHERS, K. 61
DIETZSCHOLD, B. 145
DÖRRIES, R. 219
FAZAKERLEY, J.K. 95
FIELD, H.J. 61
GIESE, A. 203
GOSZTONYI, G. 1, 121
HAMILTON R.L. 179
HOOPER, D.C. 145

HORNIG, M. 157
HORSCROFT, N. 157
KOPROWSKI, H. 1
KRETZSCHMAR, H.A. 203
LIEBERT, U.G. 35
LIPKIN, W.I. 157
LUDWIG, H. 121
MORIMOTO, K. 145
SANDERS, V.J. 179
SOLBRIG, M. 157
TALBOT, P.J. 247
WEISSENBÖCK, H. 157
WILEY, C.A. 179

The Concept of Neurotropism and Selective Vulnerability ("Pathoclisis") in Virus Infections of the Nervous System – A Historical Overview

G. GOSZTONYI* and H. KOPROWSKI**

1 Emergence of the Concept of Neurotropism 1
2 Neurotropism as the Basis of the Classification of the Encephalitides 3
3 The Panencephalitides 6
4 Impact of Electron Microscopy and Immunohistochemistry 7
5 Virus Receptors 9
6 Outlook 11
References 11

1 Emergence of the Concept of Neurotropism

Nervous diseases that we regard now to be of viral origin were known and described already in antiquity and in medieval times. Rabies, poliomyelitis and yellow fever belong to this category. The contagious nature of these diseases was suspected for centuries, and their symptomatology suggested that the brain is preferentially targeted by these infections. Not long after the advent of virology, the viral etiology of these diseases was proven. That such agents may have a special affinity to neural structures was first documented by experimental studies on rabies. CANTANI (1888), Professor of Internal Medicine in Naples, laid down the concept of the neural spread of rabies: transection of limb nerves after peripheral inoculation prevented the evolution of the disease. One year later, two of his pupils, DI VESTEA and ZAGARI (1889), published a more elaborate study on this subject in the *Annales de l'Institut Pasteur*; the recognition of the neural spread of rabies is attributed to these researchers in the literature. SCHAFFER (1890), in Budapest, provided evidence for the neural spread of rabies in humans based on histological

---
*Department of Neuropathology, University Clinics Benjamin Franklin, Freie Universität Berlin, Hindenburgdamm 30, 12165 Berlin, Germany
e-mail: gegos@zedat.fu-berlin.de
**Center for Neurovirology, Department of Microbiology and Immunology, Jefferson Alumni Hall, 1020 Locust Street, Philadelphia, PA 19107-6799, USA

studies: the most severe changes developed in spinal cord segments corresponding to the site of the animal bite. These early studies clearly established that the agent of rabies has such an elementary affinity to neural structures that it spreads exclusively along these pathways to the central nervous system (CNS).

The relation of these infectious diseases to the nervous system was further documented by early histopathological studies around the turn of the nineteenth/ twentieth century. Adventitial and perivascular lymphomonocytic infiltrates, glial stars and neuronophagic nodules were the hallmarks of the inflammatory character of these diseases. In a few of these encephalitides, the appearance of inclusion bodies, e.g. the neuronal cytoplasmic inclusions (Negri bodies) in rabies (NEGRI 1903) and the nuclear inclusions in Borna disease of horses (JOEST and DEGEN 1909), were interpreted as indications of the viral nature of the infectious process. It became clear that for viral infections nonpyogenic inflammation was distinctive. The nature of the histological signs, however, apart from inclusion bodies in a few types of infections, was not characteristic for individual infectious diseases. What was actually peculiar for the disease types that was the *distribution* of the inflammatory signs.

The study of the distribution pattern of the lesions led CONSTANTIN LEVADITI (1874–1953, Fig. 1) to the first formulation of the specific affinity of viruses to neural structures. Levaditi was a prominent microbiologist of Rumanian origin, a pupil of Elie Metschnikoff, who spent most of his career at the Pasteur Institute in Paris and became its *Chef de Service* in 1926. He denoted the diseases with this specific affinity *ectodermoses neurotropes* on the basis of the observation that their agents have a variably expressed, dual affinity to ectodermal structures, i.e. to the epidermis and cornea, on the one hand, and to the invaginated part of the ectoderm, the brain and spinal cord, on the other (LEVADITI 1921, 1922).

Fig. 1. Constantin Levaditi (1874–1953)

Levaditi set up a scale of neurotropic agents as follows: vaccinia, herpes virus, agent of the epidemic encephalitis (von Economo), furthermore, of rabies and poliomyelitis (Heine-Medin). While with the first two agents the epidermal and corneal affinities prevailed over that to the brain, in epidemic encephalitis there seemed to be an equal affinity to both ectodermal germinal layers. Finally, in rabies and poliomyelitis the epidermal and corneal affinities were suppressed compared to the affinity to the brain and spinal cord. A few years later, Levaditi complemented his scale with diseases with ectodermal affinities discovered in the meantime: stomatitis vesicularis, Japanese, American and equine encephalomyelitides and Borna disease (Fig. 2; LEVADITI and VOET 1935; LEVADITI 1938).

2 Neurotropism as the Basis of the Classification of the Encephalitides

The quality and the distribution of the histological changes in the encephalitides has been the basis for the study of neurotropism for decades. At the same time, nevertheless, these features have also been the basis for the classification of the inflammatory processes of the nervous system. At the end of the 1920s, HEINRICH PETTE (1887–1964, Fig. 3) devoted much attention to these phenomena. Pette was a neurologist and neuropathologist as well as an outstanding researcher of inflammatory diseases of the nervous system. From 1934 on, he was Professor and Director of the University Clinic for Neurology in Hamburg-Eppendorf. In 1948 he founded an Institute for the Study of Poliomyelitis and Multiple Sclerosis at

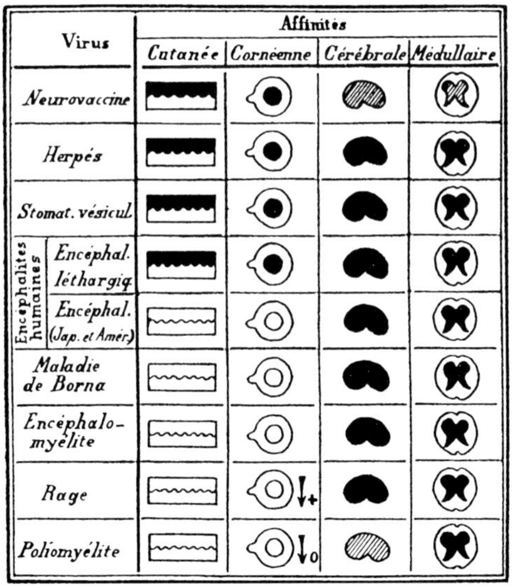

Fig. 2. Classification of the "*ectodermoses neurotropes*" from LEVADITI and VOET 1935

Fig. 3. Heinrich Pette (1887–1964)

the University of Hamburg. After his death, this research institute adopted his name (BAUER 1998).

In 1929, Pette established his concept of classification of the inflammatory diseases of the nervous system (TETTE 1929). He recognized that these diseases can be divided into two groups: (1) acute inflammatory diseases predominantly of the gray matter, and (2) acute inflammatory diseases predominantly of the white matter. The first group incorporated three neurotropic diseases, which Levaditi also included into his *ectodermoses neurotropes*: poliomyelitis, rabies, and epidemic encephalitis. Qualitatively, the histological picture of these diseases was very similar, characterized by adventitial/perivascular lymphomonocytic infiltrates, glial nodules, neuronal degeneration and neuronophagias. There are, however, essential differences in the distribution of the lesions. Poliomyelitis predominates in the anterior horns of spinal cord segments, rabies in the brain stem and spinal cord, and epidemic encephalitis in the periaqueductal gray matter, substantia nigra and the wall of the third ventricle. The histology of the second group was qualitatively quite different: perivenular or more extensive demyelinating foci with moderate inflammation and intense, mainly focal microglial proliferation. The diseases belonging to this second group were recognized in the 1920s; they were acute inflammatory CNS diseases presenting some time after vaccination (most frequently with vaccinia against smallpox) and following exanthematous diseases (measles, chickenpox, rubella). The same histological picture was described also in an acute, sometimes relapsing CNS disease: acute disseminated encephalomyelitis. Pette clearly characterized the distinctive features of both groups: in the first one the neurons were damaged primarily, the myelin remained intact and the brunt of the changes was in the gray matter; in the second, neurons remained intact, myelin was destroyed and the pathological changes were restricted mainly to the white matter. He gave also indications as to the etiology: poliomyelitis, rabies and epidemic encephalitis were clearly primary viral diseases; behind the demyelinating group he suspected constitutional and immunological factors as uniform etiological agents, despite the fact that these diseases were precipitated by contact with different types of viruses. A few years later Rivers and coworkers (1933, 1935) reported that by injection of brain extracts to monkeys a demyelinating disease can be induced, *experimental allergic encephalitis* (EAE). In his comprehensive monograph on encephalitides, PETTE (1942) also applied the adjective *allergic* to characterize the human demyelinating diseases. This way, these diseases, due to their specific features and unique etiology, have been unequivocally separated from the primary viral encephalitides.

In 1930, Hugo Spatz, in his comprehensive chapter on the morphology of encephalitides in the *Handbuch der Geisteskrankheiten* (Handbook of Mental Diseases), adopted the classification proposed by Pette, separating the acute inflammatory diseases affecting predominantly the gray matter, viz. the white matter of the CNS. Spatz, however, applied the much shorter term *polioencephalitis*[1] to the first group (SPATZ 1930, 1931). For the second group, that of inflammatory diseases affecting predominantly the white matter, progressively the term *leukoencephalitis* has been adopted. The restriction of the inflammatory process to the gray matter in the polioencephalitis group was explained by the presumption that viruses are present mainly in the cell bodies of neurons, as they have an affinity for the nerve cells themselves: *gangliocytotropism* or *neurocytotropism* (KÖRNYEY 1933). SPATZ (1930, 1931) particularly emphasized the similarities between members of the polioencephalitis group, i.e. poliomyelitis, rabies and epidemic encephalitis, their affinities for various levels of the spinal cord and brain stem, and the discontinuous, patchy distribution of the inflammatory lesions. SPATZ (1930, 1931) also included the Borna disease of horses in this group and, with Seifried, performed a comprehensive comparative study of these polioencephalomyelitides (SEIFRIED and SPATZ 1930). In this study, special emphasis was placed on analogous features of epidemic encephalitis (von Economo) and Borna disease of horses. The extensive involvement of the mesencephalon, in particular, was most impressive. The authors suggested that the agents of these two diseases might be closely related. This assumption, however, could never be proven, since, in spite of many attempts, the agent of epidemic encephalitis could not be identified. No closer relations could be assessed between other members of the polioencephalitis group either; their agents belong to quite different taxonomic groups. Apparently, there are other factors that determine the specific affinity of viruses for definite neuronal formations. In the 1930s it was generally accepted that this affinity can be quite strict. PETTE (1938, 1942) characterized this feature with the term *special neurotropism* (spezielle Neurotropie), in contrast to *general neurotropism* (allgemeine Neurotropie), an overall affinity of viruses for neural tissue.

In the meantime, experimental studies on viral encephalitides made great progress. The distribution patterns of inflammatory lesions in various neurotropic

---

[1] The term "polioencephalitis" was already widely used to characterize a group of nonpurulent encephalitides at the end of the nineteenth century (VOGTH 1912); it was, however, applied mainly to denote "pseudoencephalitides", as e.g. the polioencephalitis haemorrhagica superior WERNICKE.

virus infections and in various phases of these infections were monitored by histological techniques. It was realized that the distribution patterns in the early phases of infection depended greatly on the portal of entry of the virus into the nervous system (SABIN and OLITSKY 1938). KÖRNYEY (1939) called attention to the importance of the time factor in the formation of the distribution pattern in the course of the evolution of the encephalitic process: the localization of the histological changes in the fully developed phase of the disease becomes independent of the portal of entry, and progressively the neurotropic features of the agent become decisive. In certain types of encephalitides, however, the portal of entry remains the decisive factor throughout the entire course. Herpes simplex virus may cause two characteristic types of encephalitides in humans. The more frequent manifestation is an acute, necrotizing inflammatory process in the frontobasal and temporal regions bilaterally, but with a unilateral preponderance. It was assumed that this distribution pattern is the consequence of penetration of the virus through the olfactory nerves and its intracerebral spread along neuronal chains of the limbic system (JOHNSON and MIMS 1968). The less frequent form manifests itself as a brain stem encephalitis, which might be the consequence of a centripetal spread of activated herpesvirus from the latently infected trigeminal ganglion. According to another view, however, fronto- and temporobasal infection results from spread of herpesvirus from the gasserian ganglia along trigeminal nerve fibers innervating the meninges of the anterior and middle cranial fossae (DAVIS and JOHNSON 1979).

3 The Panencephalitides

By the 1930s, various types of encephalitides, transmitted predominantly by arthropodes as vectors, were known worldwide. In the histological picture inflammatory infiltrations were found mainly in the gray matter, but the white matter was not spared either. This group included *Japanese encephalitis* (also called encephalitis B, in contradistinction to epidemic encephalitis, also known as encephalitis A, at that time still occurring sporadically), *St. Louis encephalitis* and *American equine encephalitides*. In contrast to these encephalitides, which occurred in epidemics, in Europe a new but sporadically presenting encephalitis appeared. Initially, three types of this disease were described: *inclusion body encephalitis* (DAWSON 1933), *indigenous panencephalomyelitis* (einheimische Panencephalomyelitis, PETTE and DÖRING 1939), and *subacute sclerosing leukoencephalitis* (leucoencéphalite sclérosante subaiguë, VAN BOGAERT 1945). The findings of these types were correlated and, mainly on the recommendation of GREENFIELD (1950), the name *subacute sclerosing panencephalitis* (SSPE) was accepted for the entire complex. This is an invariably fatal disease predominantly of children, with insidious onset of behaviorial changes and cognitive deterioration followed by myoclonic jerks and ataxia. Pathologically, inflammatory infiltrates are found both in the gray and white matter; furthermore, diffuse demyelination, astroglial sclerosis and inclusion bodies in neurons and oligodendrocytes are prominent features. Later morphological and virological studies documented that SSPE is a slow infection by measles virus with reactivation years after the initial attack and production of incomplete viral particles in CNS cells (BOUTEILLE et al. 1965; CONOLLY et al. 1967; BARBANTI-BRODANO et al. 1970; TER MEULEN et al. 1972; KATZ and KOPROWSKI 1973). PETTE (1942) recognized early the similarities between the Japanese and American encephalitides and the sporadic European encephalitides. Based on the histological feature that both gray and white matter are involved in these diseases, he coined the term *panencephalitis* for these diseases. Pette also included the encephalitis associated with the rickettsiosis, exanthematic typhus, to this group. The term *panencephalitis* was quickly and generally accepted.

In a comprehensive monograph, RADERMECKER (1956), neurologist and electroencephalographist of the Bunge Institute in Antwerpen, summarized and extended the systematics of encephalitides. For this purpose, he adopted the principles laid down by PETTE (1929) and SPATZ (1930). His schematic figures clearly and elegantly demonstrate the characteristic distribution patterns of lesions in various neurotropic virus infections. Furthermore, the findings of at that time a rather new diagnostic procedure, electroencephalography, were also laid down and correlated with the clinical and anatomical features in the various encephalitides.

4 Impact of Electron Microscopy and Immunohistochemistry

During the middle of the twentieth century, novel histological techniques emerged that complemented conventional methods and opened new horizons in the study of the phenomenon of neurotropism. Electron microscopy and immunohistochemistry offered new data on viral infections of the nervous system, so that our views about neurotropism were fundamentally altered and extended. Electron microscopy allowed visualization of virus particles, the various phases of their assembly, their localization in various compartments of the cell, and, most importantly, determination of the cell type that harbors the virus. The assessment of this *cytotropism* within the CNS has led to a better understanding of both the distribution patterns and the pathogenesis of various virus infections. Classic examples are SSPE and progressive multifocal leukoencephalopathy (PML). In SSPE, electron microscopy disclosed the lack of production of complete measles virus particles within the CNS and the presence of paramyxovirus nucleocapsids in the nuclei and cytoplasm of neurons and in the nuclei of oligodendroglial cells (BOUTEILLE et al. 1965). This double tropism of the agent explained why SSPE is a panencephalitis and why widespread demyelination is present in the white matter. In PML, ZURHEIN and CHOU (1965) discovered crystalline arrays of papovavirus particles in characteristically altered nuclei of oligodendrocytes. This finding documented that myelin breakdown can also ensue as a direct consequence of the cytopathic effect of a primary virus infection, not only as a sequel of an autoimmune process, as in the postvaccinial/parainfectious encephalomyelitides and in acute disseminated encephalomyelitis. Furthermore, papovavirus particles have occasionally been found also in astrocytes (MAZLO and HERNDON 1977; MAZLO and TARISKA 1982).

Since electron microscopic techniques are limited in space, the impact of immunohistochemistry was greater, since it allowed the survey of large brain areas and good assessment of the extension and distribution of the virus infection. When the distribution patterns of virus antigens were compared with morphological changes, as shown on conventionally stained preparations, it could be seen that the viral antigens were more widely distributed than the morphological alterations. Neurons with cytopathic changes almost always harbored viral antigens, but many neurons that were light microscopically normal in appearance were also antigen-positive. This means that viruses may exhibit tropism toward a great number of neurons; in some, the presence of the virus does not grossly interfere with the functions of the cell, but in others the virus is *neuropathogenic* and its replication leads to disintegration of the host neuron. Consequently, immunohistochemistry enabled us to differentiate between *neurotropism*, on the one hand, and *selective*[2] *vulnerability*, on the other. The latter phenomenon is a manifestation of the great diversity in the architecture of the CNS. That diverse neuronal systems react in a quite differentiated way to various noxae was first formulated as *pathoclisis* (Pathoklise) by C. and O. VOGT (1922), who defined it as a structural or constitutional propensity of certain neuronal populations to react with disease to specific pathogenetic factors. While pathoclisis predominantly referred to the "endogenous" systemic atrophies, it was also used for lesions evoked by hypoxic, vascular, viral and other exogenous factors (PETTE 1938). Later on, the term "selective vulnerability" progressively replaced the concept of pathoclisis. The neurobiological basis of selective vulnerability has been poorly elucidated. It may be that host factors regulating viral synthetic processes are expressed in different ways in various neuronal populations. On the other hand, viral products may interfere with cell functions that are specific for definite neuronal types.

As to cell tropism, immunohistochemistry enriched our knowledge even more than electron microscopy. Thus, it could be documented that in rabies viral antigens are almost exclusively harbored by neurons (GOSZTONYI et al. 1993). By contrast, in Borna disease viral antigens were found not only in neurons, but also in astrocytes, oligodendrocytes, and ependymal and plexus epithelial cells, both in naturally and experimentally infected animals (LUDWIG et al. 1985, 1988; GOSZTONYI et al. 1993). Accordingly, in Borna disease both gray and white matter are involved; thus, Borna disease is a panencephalitis, in contradiction to SPATZ (1930), SEIFRIED and SPATZ (1930) and others. Rabies, however, remains a classical polioencephalitis.

---

[2] The adjective *selective* is the standard expression to characterize this phenomenon in the Anglo-American literature. In the French and German literature rather the adjective *elective* is being used ("vulnerabilité élective", viz. "elektive Vulnerabilität")

5 Virus Receptors

An answer for the basis of viral tropism to certain types of cells and tissues had to be sought in specific cell surface receptors. While these normally mediate critical cellular functions, they are also used by viruses for their attachment and entry into the cell and thus are important determinants of tissue tropism and virus host range. The role of surface receptors in connection with a neurotropic virus was first raised in studies on poliomyelitis. The restriction of poliovirus infection to primates and the insusceptibility of nonprimates to this infection was explained by the presence of specific poliovirus receptors on the cell surface (HOLLAND et al. 1959; HOLLAND 1961). The role of receptors was underlined by the observation that enterovirus RNA extracted with phenol infected nonprimate cells and animals insusceptible to whole virus as such (HOLLAND et al. 1959). An important principle was already defined in the course of studies with enteroviruses: The host range of various tissues in vitro was much wider than in vivo, so that in the assessment of tissue tropisms the in vivo affinity was decisive (HOLLAND 1961).

For the binding of a virus to a cell receptor, certain structures on the virus surface, the *viral attachment proteins*, have an equally important role. From the early 1960s, studies on reoviruses greatly enriched our understanding of the phenomenon of neurotropism. Infection of newborn mice with reovirus type 3 results in a severe, almost invariably fatal encephalitis with diffuse neuronal degeneration and necrosis (STANLEY et al. 1964). Infection with reovirus type 1 and 2, however, induced a less severe disease, in which the CNS lesions were less prominent (WALTERS et al. 1965). Four years after this assessment, MARGOLIS and KILHAM (1969) observed that reovirus type 1 induces obstructive hydrocephalus as a consequence of a selective attack of this agent upon the ependyma in suckling mice, rats and hamsters. Thus, one reovirus type is *neuronotropic*, and the other type is *ependymotropic*. Subsequently, in a series of studies, the group of B.N. Fields at the Harvard Medical School explored the molecular basis of this disparate tropism of the two reovirus serotypes. It was established that the factor responsible for the differing cell tropisms was the $\sigma 1$ outer capsid polypeptide (which also functions as the viral hemagglutinin), coded by the S1 genome segment (WEINER et al. 1977; FIELDS and GREENE 1982; SPRIGGS et al. 1983; TYLER et al. 1985). The $\sigma 1$ polypeptide secured binding to neurons of type 3 reovirus, and binding to ependymal cells with type 1. Furthermore, the Harvard group successfully applied anti-idiotypic anti-receptor antibodies (antibodies to the F'ab segment of monoclonals raised against viral surface epitopes) mimicking viral attachment determinants to explore cellular receptors for reoviruses (NEPOM et al. 1982; Co et al. 1985a,b; GAULTON and GREENE 1989; SAUVÉ et al. 1993).

For the concept of neurotropism it was a great step forward when LENTZ et al. (1982) published their observation that nicotinic acetylcholine receptors may serve as portals of entry for rabies virus, a strict neurotropic agent. Binding of rabies virus to the chick neuromuscular junction could be prevented by α-bungarotoxin and d-tubocurarin. Soon thereafter, based on tissue culture studies, doubt was cast on the acetylcholine receptor hypothesis (REAGAN and WUNNER 1985). Nevertheless, this hypothesis has been widely accepted, since it offers a plausible explanation for the affinity of rabies virus to motor nerve endings and striated muscle and for the very wide host range of the agent.

In the early 1980s, the acetylcholine receptor hypothesis gave rise to the idea that if an agent has a very strict affinity for the nervous system, its cellular receptor has to be sought among surface structures that are highly specific for, and occur almost exclusively in neural tissue. Based on the assessments of Lentz and coworkers and on findings that Borna disease virus (BDV) apparently has an affinity for receptors of the excitatory amino acids (glutamate and aspartate) in the CNS, it has been postulated that viral neurotropism may be explained by the affinity of the infectious agents for *neurotransmitter receptors* (GOSZTONYI and LUDWIG 1984).

The late 1980s and the 1990s brought an explosive increase in our knowledge concerning virus receptors. Nonetheless, the list of agents with affinities for neurotransmitter receptors remained meager, with only three representatives: rabies virus, with affinity for nicotinic acetylcholine receptors (LENTZ et al. 1982); BDV, with affinity for excitatory amino acid receptors (GOSZTONYI and LUDWIG 1984, 1995; GOSZTONYI et al. 1993) and reovirus type 3, with affinity for β-adrenergic receptors (CO et al. 1985a,b). However, it has to be kept in mind that the list of very strictly neurotropic agents is also scant.

Although great progress was achieved in defining the poliovirus receptor, the strict tropism of this agent to central and peripheral motor neurons remained without sufficient explanation. It was already established, in 1974, that chromosome 19 bears the gene for the poliovirus receptor (GREEN 1974). In 1989 it was shown that the poliovirus receptor is a member of the immunoglobulin superfamily. While its physiological function remains unknown, its gene has been cloned and its nucleotide sequence determined (MENDELSOHN et al. 1989). The restriction of poliovirus replication to a few sites (intestine, pharyngeal lymphoid tissue, and motor neurons of the CNS, HOLLAND 1961) is in sharp contrast to the wide expression of the members of the immunoglobulin superfamily in human tissues. This and similar controversies support the emerging view that, although receptors play an important role in viral tropism, they are, with a few exceptions, not the sole determinants. In the early 1980s, it was already suspected that possession of the appropriate receptors is no guarantee that a cell can be infected (DIMMOCK 1982). MARSH and HELENIUS (1989) also emphasized that receptor expression is not the only factor that determines cell tropism. Additional factors or receptor modifications are needed to permit virus attachment (MENDELSOHN et al. 1989).

In the 1990s, an intense search was performed to identify these additional factors. For some viruses secondary receptors, or coreceptors, are needed for adhesion and entry (CALLEBAUT et al. 1993; WEISS and TAILOR 1995). A few viruses can bind to more than one receptor (HAYWOOD 1994), and different viruses may use the same receptor for binding (MARSH and HELENIUS 1989). It has become clear that virus attachment and penetration is a very complex, multistep process and not a simple ligand/receptor relationship.

Perhaps the most thoroughly studied virus/receptor relationship is that of the retrovirus HIV-1. This agent invades the CNS in about 20% of patients with AIDS and is harbored in macrophages/microglial cells. HIV-1 antigens have never been convincingly demonstrated in neurons (CIARDI et al. 1990; GOSZTONYI et al. 1994); instead, neuronal damage was shown to ensue by an indirect, in many details not fully elucidated mechanism. In spite of the fact that a non-neuroectodermal cell, the macrophage, is the target of infection in the CNS, the concept of neurotropism is being used in this context. This usage is justified by the fact that certain variants of HIV-1, designated neurotropic or monocytotropic (see PETITO 1996), have a stronger, while others a weaker or absent, liability to invade the brain.

6 Outlook

While great progress has been made in revealing the mechanisms of attachment and penetration of viruses into host cells in general, these mechanisms are still incompletely elucidated for the neurotropic viruses. To make up for these shortages, a more intense application of modern molecular biological techniques and better knowledge of neuroanatomy and neurotransmitter physiology is required. In particular, thorough correlative studies on distribution patterns of neurotransmitter receptors and viral antigens on neurons and glial cells have to be instituted. Ultimately, a better understanding of the phenomenon of neurotropism depends greatly upon progress in molecular neurobiology.

References

Barbanti-Brodano G, Oyanagi S, Katz M, Koprowski H (1970) Presence of two different viral agents in brain cells of patients with subacute sclerosing panencephalitis. Proc Soc Exp Biol (NY) 134: 230–236
Bauer HJ (1998) Heinrich Pette (1887–1964) In: Schliack H, Hippius H (eds) Nervenärzte – Biographien. Georg Thieme Verlag, Stuttgart, New York, pp 129–137
Bouteille M, Fontaine C, Vedrenne CL, Delarue J (1965) Sur un cas d'encéphalite subaigue à inclusions. Étude anatomoclinique et ultrastructurelle. Rev Neurol 118:454–458
Callebaut C, Krust B, Jacotot E, Hovanessian AG (1993) T cell activation antigen, CD26, as a cofactor for entry of HIV in CD4+ cells. Science 262:2045–2050
Cantani A (1888) Ueber die Verbreitung des Wuthgiftes längs der Nerven und Pasteur's Schutzimpfungen. Wiener Med Wschr 38:1061–1062
Ciardi A, Sinclair E, Scaravilli F, Harcourt-Webster NJ, Lucas S (1990) The involvement of the cerebral cortex in human immunodeficiency virus encephalopathy: a morphological and immunohistochemical study. Acta Neuropathol 81:51–59
Co MS, Gaulton GN, Fields BN, Greene MI (1985a) Isolation and biochemical characterization of the mammalian reovirus type 3 cell-surface receptor. PNAS USA 82:1494–1498
Co MS, Gaulton GL, Tominaga A, Homcy CJ, Fields BN, Greene MI (1985b) Structural similarities between the mammalian β-adrenergic and reovirus type 3 receptors. Proc Natl Acad Sci USA 82:5315–5318

Conolly JH, Allen IV, Hurwitz LJ, Millar JHD (1967) Measles virus antibody and antigen in subacute sclerosing panencephalitis. Lancet 1:542–544

Davis LE, Johnson RT (1979) An explanation for the localization of herpes simplex encephalitis? Ann Neurol 5:2–5

Dawson JR Jr (1933) Cellular inclusions in cerebral lesions of lethargic encephalitis. Am J Pathol 9:7–15

Dimmock NJ (1982) Initial stages of infection with animal viruses. J Gen Virol 59:1–22

Di Vestea A, Zagari G (1889) La transmission de la rage par voie nerveuse. Ann Inst Pasteur 3:237–248

Fields BN, Greene MI (1982) Genetic and molecular mechanism of viral pathogenesis: implications for prevention and treatment. Nature 300:19–23

Gaulton GN, Greene MI (1989) Inhibition of cellular DNA synthesis by reovirus occurs through a receptor-linked signaling pathway that is mimicked by antiidiotypic, antireceptor antibody. J Exp Med 169:197–211

Gosztonyi G, Ludwig H (1984) Neurotransmitter receptors and viral neurotropism. Neuropsychiatr Clin 3:107–114

Gosztonyi G, Ludwig H (1995) Borna disease – neuropathology and pathogenesis. Curr Top Microbiol Immunol 190:39–73

Gosztonyi G, Artigas J, Lamperth L, Webster H deF (1994) Human immunodeficiency virus (HIV) encephalitis: study of 19 cases with combined use of in situ hybridization and immunocytochemistry. J Neuropathol Exp Neurol 53:521–534

Gosztonyi G, Dietzschold B, Kao M, Rupprecht CE, Koprowski H (1993) Rabies and Borna disease: a comparative pathogenetic study of two neurovirulent agents. Lab Invest 68:285–295

Green H (1974) The gene for the poliovirus receptor. New Eng J Med 290:1018–1019

Greenfield JG (1950) Encephalitis and encephalomyelitis in England and Wales during the last decade. Brain 73:141–166

Haywood AM (1994) Virus receptors: binding, adhesion strengthening, and changes in viral structure. J Virol 68:1–5

Holland JJ (1961) Receptor affinities as major determinants of enterovirus tissue tropism in humans. Virology 15:312–326

Holland JJ, McLaren LC, Syverton JT (1959) The mammalian cell virus relationship. IV. Infection of naturally insusceptible cells with enterovirus ribonucleic acid. J Exp Med 110:65–79

Joest E, Degen K (1909) Über eigentümliche Kerneinschlüsse der Ganglienzellen bei der enzootischen Gehirn-Rückenmarksentzündung der Pferde. Z Inf Krkh Haustiere 6:348–356

Johnson RT, Mims CA (1968) Pathogenesis of viral infections of the nervous system. New Eng J Medicine 278:23–30, 84–92

Katz M, Koprowski H (1973) The significance of failure to isolate infectious viruses in cases of subacute sclerosing panencephalitis. Arch Ges Virusforschung 41:390–393

Környey St (1933) Die Bedeutung der mesodermalen Reaktion und der Systemelektivität in der Pathologie der Poliomyelitis. Z Neur 146:724–746

Környey St (1939) Die primär neurotropen Viruskrankheiten des Menschen. Fortschr Neurol Psychiat 11:82–100 and 146–166

Lentz TL, Burrage TG, Smith AL, Crick J, Tignor GH (1982) Is the acetylcholine receptor a rabies virus receptor? Science 215:182–184

Levaditi C (1921) Comparaison entre les divers ultra-virus neurotropes (Ectodermoses neurotropes). C R Soc Biol 85:425–429

Levaditi C (1922) Ectodermoses neurotropes – poliomyélite, encéphalite, herpès. Étude clinique, histo-pathologique et expérimentale. Masson, Paris

Levaditi C (1938) Les ultravirus des maladies humaines. Librairie Maloine, Paris

Levaditi C, Voet J (1935) Nouvelle classification des ectodermoses neurotropes. C R Acad Sci 201:743–745

Ludwig H, Kraft W, Kao M, Gosztonyi G, Dahme E, Krey H (1985) Borna-Virus-Infection (Borna-Krankheit) bei natürlich und experimentell infizierten Tieren: ihre Bedeutung für Forschung und Praxis. Tierärztl Prax 13:421–453

Ludwig H, Bode L, Gosztonyi G (1988) Borna disease: a persistent virus infection of the central nervous system. Prog Med Virol 35:107–151

Marsh M, Helenius A (1989) Virus entry into animals cells. Adv Virus Res 36:107–151

Mazlo M, Herndon M (1977) Progressive mulitifocal leukoencephalopathy: ultrastructural findings in two brain biopsies. Neuropathol Appl Neurobiol 3:323–339

Mazlo M, Tariska I (1982) Are astrocytes infected in progressive multifocal leukoencephalopathy (PML)? Acta Neuropathol 56:45–51

Mendelsohn CL, Wimmer E, Racaniello VR (1989) Cellular receptor for poliovirus: Molecular cloning, nucleotide sequence, and expression of a new member of the immunoglobulin superfamily. Cell 56:855–865

Negri A (1903) Beitrag zum Studium der Histologie der Tollwuth. Z Hyg Infektionskrankheiten 43: 507–528

Nepom JT, Weiner HL, Dichter MA, Tardieu M, Spriggs DR, Gramm CF, Powers ML, Fields BN, Greene MI (1982) Identification of a hemagglutinin-specific idiotype associated with reovirus recognition shared by lymphoid and neural cells. J Exp Med 155:155–167

Petito CK (1996) Neuropathology of human immunodeficiency virus: questions and answers. Editorial. Human Pathol 27:623–624

Pette H (1929) Akute Infektion und Nervensystem. Münch Med Wochenschr 76:225–230

Pette H (1938) Die akut entzündlichen Erkrankungen des Zentralnervensystems. Verh Deutsch Ges Inn Med, Fünfzigster Kongress, 486–540

Pette H (1942) Die akut entzündlichen Krankheiten des Nervensystems. Thieme, Leipzig

Pette H, Döring G (1939) Über einheimische Panencephalitis vom Charakter der Encephalitis japonica. D Zeitschr Nervenheilk 149:7–44

Radermecker J (1956) Systématique et électroencéphalographie des encéphalites et encéphalopathies. EEG Clin Neurophysiol Suppl No. 5. Masson, Paris

Reagan KJ, Wunner WH (1985) Rabies virus interaction with various cell lines is independent of the acetylcholine receptor. Arch Virol 84:277–282

Rivers TM, Sprunt DH, Berry GP (1933) Observations on attempts to produce acute disseminated encephalomyelitis in monkeys. J Exp Medicine 58:39–53

Rivers TM, Schwentker FF (1935) Encephalomyelitis accompanied by myelin destruction experimentally produced in monkeys. J Exp Med 61:689–702

Sabin AB, Olitsky PK (1938) Influence of host factors on neuroinvasiveness of vesicular stomatitis virus: III. Effect of age and pathway of infection on the character and localization of lesions in the central nervous system. J Exp Med 67:201–228

Sauvé GJ, Saragovi HU, Greeve MI (1993) Reovirus receptors. Adv Virus Res 42:325–341

Schaffer K (1890) Pathologie und pathologische Anatomie der Lyssa. Beitr Pathol Anat 7:189–144

Seifried O, Spatz H (1930) Die Ausbreitung der encephalitischen Reaktion bei der Bornaschen Krankheit der Pferde und deren Beziehungen zu der Encephalitis epidemica, der Heine-Medinschen Krankheit und der Lyssa des Menschen. Z Neurol Psychiat 124:317–382

Spatz H (1930) Einteilung der echten Encephalitiden vom morphologischen Stand punkt aus. In: Handbuch der Geisteskrankheiten, Bd. 11, Spezieller Teil VII, Die Anatomie der Psychosen. Springer, Berlin, pp. 196–224

Spatz H (1931) Über Encephalitis und Encephalitiden. Nervenarzt 4:466–472, 531–542

Spriggs DR, Bronson RT, Fields BN (1983) Hemagglutinin variants of reovirus type 3 have altered central nervous system tropism. Science 220:505–507

ter Meulen V, Katz M, Müller D (1972) Subacute sclerosing panencephalitis. A review. Curr Top Microbiol Immunol 57:1–38

Tyler KL, Bronson RT, Byers KB, Fields B (1985) Molecular basis of viral neurotropism: experimental reovirus infection. Neurology 35:88–92

van Bogaert L (1945) Une leucoencéphalite sclérosante subaiguë? J Neurol Neurosurg Psychiat 8:101–120

Vogt C, Vogt O (1922) Erkrankungen der Grosshirnrinde im Lichte der Topistik, Pathoklise und Pathoarchitektonik. J Psychol Neurol 28:1–171

Vogt H (1912) Encephalitis nonpurulenta. In: Lewandowsky M (ed) Handbuch der Neurologie. Dritter Band, Spezielle Neurologie II. Berlin, Springer, pp 229–276

Walters MNI, Leak PJ, Joske RA, Stanley NF, Perret DH (1965) Murine infection with reovirus. III. Pathology of infection with types 1 and 2. Br J Exp Pathol 46:200–212

Weiner HL, Drayna D, Averill DL Jr, Fields BN (1977) Molecular basis of reovirus virulence: role of the S1 gene. PNAS USA 74:5744–5748

Weiss RA, Tailor CS (1995) Retrovirus receptors. Cell 82:531–533

ZuRhein GM, Chou SM (1965) Particles resembling papovaviruses in human cerebral demyelinating disease. Science 148:1477–1479

The Mechanisms of Direct, Virus-Induced Destruction of Neurons

J.R. ANDERSON

1	Introduction	16
2	General Histopathological Features of Acute Viral Encephalitis	16
3	Herpes Simplex Encephalitis	17
3.1	Pathology	17
3.1.1	Macroscopic Appearance	17
3.1.2	Histopathogical Changes	17
3.2	Limbic Location	18
3.2.1	Olfactory Pathway	19
3.3	Intracerebral Latency	19
3.4	Neonatal Infection	20
3.5	Virus Entry and Replication	20
3.5.1	Virus Entry into Neurons	20
3.5.2	Transcription	21
3.5.3	Information from Animal Models	22
4	Mechanisms of Cell Death in Acute Viral Encephalitis	22
4.1	Apoptosis	22
4.1.1	Control of Apoptosis	23
4.1.2	Apoptosis in Sindbis Virus Infection	23
4.1.3	Apoptosis in Herpes Simplex Infection	24
4.2	Host Immune Responses and Cell Death	24
4.2.1	Microglial Activation	24
4.2.1.1	Cytokines	25
4.2.2	Nitric Oxide	25
4.2.3	T Cell Responses	26
5	Characterisation of the Immune Responses in Human HSV Encephalitis	27
5.1	In CNS Parenchyma	27
5.2	In Cerebrospinal Fluid	27
6	Vasculitis and Ischaemic Necrosis	28
7	Rabies	28
8	Viral Encephalitis in the Immunosuppressed	29
9	Conclusion	29
References		30

Department of Histopathology, Addenbrooke's Hospital, Cambridge, CB2 2QQ, UK
e-mail: jra20@cam.ac.uk

1 Introduction

The mature neuron is a highly specialised, irreplaceable, post-mitotic cell and neuronal death leaves a permanent deficit. Therefore, irrespective of the cause, any extensive neuronal loss will have profound, if not fatal, consequences for the organism as a whole. The central nervous system (CNS) in humans and all higher orders of the animal kingdom is normally protected by both anatomical barriers and systemic physiological barriers of innate and specific immunity. A limited number of viruses have evolved strategies to penetrate these defences and to gain access to neurons that may result in direct neuronal destruction. However, the outcome of any infection is determined by the interplay between the tactics of the invader and the host response and this balance is equally relevant to viral infection of the CNS. This chapter will concentrate upon herpes simplex virus (HSV) encephalitis, which is still the commonest cause of sporadic acute encephalitis in immunocompetent individuals living in temperate parts of the world. In naturally occurring disease it is rarely possible to dissociate completely the direct cytopathic effects of the virus from cellular injury inflicted by the host immune system, but animal experiments that allow manipulation of certain facets have helped to elucidate the complex interactions and the mechanisms of neuronal destruction. Other chapters will deal more extensively with neurotropism but it will also be considered here in as much as a portal of entry into the CNS, together with susceptibility of CNS cell populations to permissive infection, are the essential prerequisites of virally mediated neuronal destruction.

2 General Histopathological Features of Acute Viral Encephalitis

In the peripheral nervous system several neurotropic viruses achieve latency, but in the immunocompetent host acute permissive viral infection of the CNS invariably results in neuronal lysis and elicits a stereotyped response, which involves microglial activation, neuronophagia and perivascular cuffing with lymphocytes and some plasma cells. Whilst different forms of acute viral encephalomyelitis are not readily distinguished by the character of the inflammatory changes, the geographical distribution within the CNS may provide diagnostic clues. It has been frequently hypothesised that distribution reflects selective vulnerability of certain neuronal populations, but more often it would appear simply to highlight a portal of entry into the CNS. Viruses that reach the CNS via a haematogenous pathway and penetrate the blood–brain barrier are likely to be widely distributed, whereas viruses which travel along nerves follow neuroanatomical pathways, at least initially. The early spinal neuronal involvement in most cases of rabies is attributable to axonal transport and entry via the spinal motor nerves supplying peripheral muscles. Likewise, the apparent susceptibility of anterior horn cells to poliovirus is not due to any unique receptor status but to peripheral axonal delivery of the virus. In contrast, the widespread involvement of grey matter in arbovirus encephalitis invokes a blood-borne route. Viral inclusion bodies provide additional diagnostic clues. By light microscopy, only the Negri body of rabies is virtually pathognomonic by dint of its cytoplasmic location, but electron microscopy will aid in the distinction between different intranuclear inclusions.

3 Herpes Simplex Encephalitis

3.1 Pathology

3.1.1 Macroscopic Appearance

Post-mortem examination of the brains of children or adults dying in the acute phase of HSV encephalitis usually reveals characteristic temporal lobe pathology. The lobes are soft, swollen and may appear haemorrhagic. The changes are generally bilateral but often asymmetrical. Other regions of the cerebral cortex are frequently involved, particularly the frontal lobes and cingulate gyri, but almost invariably, the temporal lobes are most severely affected. The brain stem is largely spared, apart from rare cases with a clinical brain stem syndrome, in which this region is severely and sometimes exclusively affected (AYUSO BLANCO et al. 1994; ROSE et al. 1992). Occasionally, there is a concomitant necrotising retinitis due to spread of virus from the brain along the optic nerves to the eyes. Examination of the brains of long-term survivors of acute encephalitis may reveal severely shrunken temporal lobes with gross thinning and yellow discolouration of the cortex and cystic changes in the underlying white matter. In neonatal HSV encephalitis temporal lobe predominance is unusual and cerebral pathology is widespread. Extensive white matter destruction in the cerebral hemispheres may result in encephalomalacia and parenchymal calcification (BENATOR et al. 1985).

3.1.2 Histopathogical Changes

The evolution of symptoms and clinical deterioration corresponds with an expanding focus of virus replication in the temporal lobe that leads to progressive destruction of both neurons and glial cells and elicits an intense inflammatory reaction. The cellular events have been examined in human biopsy and post-mortem material (ESIRI 1982) and in animal models of HSV encephalitis (ANDERSON and FIELD 1983). Intracellular HSV antigen can be identified with the immunoperoxidase technique, in both nucleus and cytoplasm of infected neurons and glial cells, several hours prior to inflammatory cell infiltration. The initial focus of virus replication rapidly expands as HSV spreads from cell to cell, creating large, confluent foci of infected cortical neurons and glial cells in adjacent white matter.

Neurons, astrocytes, oligodendrocytes and microglia all show cytolytic infection with HSV, but the dentate fascia of the hippocampus appears to be particularly susceptible. The apparent vulnerability of this layer is probably a reflection of high cell density as the closely packed cell bodies readily permit cell to cell transfer. The earliest inflammatory response is a transient influx of polymorphoneutrophil leucocytes at a time when neurons are beginning to appear abnormally eosinophilic and slightly shrunken. An immunoperoxidase for CD68 or other macrophage markers highlights early hypertrophy and hyperplasia of microglia (ESIRI et al. 1995). These cells gradually accumulate, and small eosinophilic intranuclear inclusions can be found. Infected cells without distinct inclusions often show margination of the chromatin and an 'empty' nucleus. Nucleocapsids can be demonstrated in both nuclei and cytoplasm by electron microscopy. The initial neutrophilic response is rapidly followed by the more conspicuous, perivascular infiltrate of lymphocytes, macrophages and small numbers of plasma cells, which coincides with obvious cell death. By now immunoperoxidase staining for virus antigen reveals confluent infected cells in large areas of cortex and white matter. The infected neurons and glia disintegrate and are phagocytosed by the infiltrating macrophages. The neuropil appears oedematous and petechial haemorrhages appear as red cells leak from damaged capillaries with swollen endothelium. By the time cell death and inflammation reach a peak, often about 7–10 days after onset in humans, detectable virus antigen is waning. At this stage the widespread cell death in large areas of temporal cortex, combined with breakdown of subcortical white matter and massive influx of macrophages, resembles a recent infarct.

3.2 Limbic Location

There is no universally accepted explanation for the limbic topography of human HSV encephalitis. HSV utilises multiple receptors to enter host cells and replicates with equal facility in different neuronal populations. Thus, viral tropism cannot readily be explained by specific receptor distribution. The propensity of the HSV to spread retrogradely along axons and the initial temporal lobe selectivity are undisputed and together suggest that a virus travels along a neuroanatomical pathway to the temporal lobe. The medial aspect of the temporal lobe, also referred to as the limbic lobe, has profuse connections with the olfactory system, the hypothalamus, thalamus and various areas of the neocortex, which together constitute the limbic system (PARENT 1996). The olfactory connections have received the greatest attention in HSV encephalitis as these alone provide an almost direct link between the external environment and the temporal lobe. Despite the acknowledged frequency of latency and reactivation from the trigeminal ganglion, an alternative hypothesis that invokes passage of HSV along meningeal branches of the trigeminal nerve to the surface of the temporal lobe is largely discounted because meningitis is neither an early nor significant feature of the human disease. Equally, spread along the central portion of the fifth nerve cannot explain the intracerebral distribution except, possibly, in the very rare cases of herpetic brain stem encephalitis.

3.2.1 Olfactory Pathway

Both human post-mortem studies and animal models lend credence to the olfactory route. In brains of persons dying with acute HSV encephalitis, the immunoperoxidase method revealed HSV antigen in the olfactory pathway and limbic system, but little elsewhere (ESIRI 1982; OJEDA et al. 1983). The distribution of HSV antigen following different routes of inoculation in mice suggests that the portal of entry is a key determinant and that intranasal inoculation can initiate temporal lobe infection (ANDERSON and FIELD 1983). Bipolar primary olfactory neurons lying within the nasal mucosa extend short peripheral processes almost to the mucosal surface and long central axons which pass through the cribriform plate of the ethmoid bone to synapse with mitral cells in the olfactory bulbs on the inferior surface of the frontal lobes. The axons of the mitral cells form the lateral olfactory tract, which passes backwards and divides into lateral and medial olfactory striae. The lateral stria terminates in the piriform cortex and amygdaloid complex of the limbic lobe. The medial stria is continuous with the septal area of the cortex and septal nuclei from which both afferents and efferent fibres connect to the hippocampus via the fornix. The vulnerability of the olfactory pathway may be related to certain biological peculiarities. Primary olfactory neurons are almost unique amongst mammalian nerve cells in that they have the capacity to regenerate and are continuously replaced from stem cells throughout adult life. The extracerebral axons of the olfactory nerve are not sheathed by Schwann cells but by astrocyte-like cells that contain glial fibrillary acidic protein (BARBER and LINDSAY 1982) whilst other cranial nerves, except the optic nerve, are encased by Schwann cells. Astrocytes are susceptible to HSV infection, whereas infection is usually abortive in Schwann cells. The primary olfactory neuron is in some respects a peripheral extension of the CNS.

3.3 Intracerebral Latency

Identification of this portal of entry still fails to explain why this ubiquitous virus only establishes permissive CNS infection in a tiny minority of persons exposed to it. There is no evidence that the viruses isolated from cases of encephalitis have had unusual neurovirulent properties, and epidemics of HSV encephalitis do not occur. Therefore it is likely that host factors determine susceptibility. Immune responsiveness may play a role, but the relatively infrequent occurrence of HSV encephalitis compared to cutaneous lesions in the immunosuppressed implies another explanation. There is some evidence that HSV can establish latency in intracerebral neurons as well as in trigeminal ganglion cells. Stable HSV-specific DNA sequences have been detected in the CNS of mice after recovery from acute infection (EFSTATHIOU et al. 1986). Nevertheless, with few possible exceptions, which have generally entailed heavy immunosuppression, CNS reactivation has been unsuccessful (KASTRUKOFF et al. 1981). A few studies claim to have identified latent HSV in the brains of a minority of elderly persons, although only by the highly sensitive PCR method (JAMIESON et al. 1992; BARINGER and PISANI 1994). A link with Alzheimer's disease is postulated (ITZHAKI et al. 1997). However, even if HSV DNA persists at low levels there is no firm evidence that the virus ever genuinely reactivates in the CNS in vivo.

3.4 Neonatal Infection

When herpetic encephalitis occurs in neonates it is usually in the context of disseminated infection with HSV 2 acquired from the maternal genital tract. Widespread cell death and inflammation in the CNS suggest haematogenous spread, and electron microscopy has identified viral replication in arterial endothelial cells (PHINNEY et al. 1982). A comparison of sera from neonates with HSV infection and sera from older children with herpes simplex encephalitis lent support to viraemic spread in neonates. PCR assays to detect HSV DNA in the sera revealed positive sera in the majority of neonates but in none of the children with HSE (KIMURA et al. 1991).

3.5 Virus Entry and Replication

The complex interaction of host and HSV viral proteins which control binding and penetration and initiate replication are described as an essential prelude to the final stages of replication which culminate in cell death.

3.5.1 Virus Entry into Neurons

Herpes simplex virus encephalitis is characterised by permissive infection in both neurons and glia. Immunoperoxidase staining readily reveals virus antigen, and the almost universal staining of all parenchymal cells in the affected temporal lobe serves to emphasise the rapid cell to cell spread of the virus. Entry is a multistep process effected by attachment and interaction between components of the virion envelope and cell surface molecules prior to penetration. Study of embryonic chick neuronal cells in culture suggests requirements for entry may differ between neurons and other cells types. Heparin sulfate glycosaminoglycans has been shown to serve as both a neuronal and epithelial cell surface receptor. However, there are differences. Competitive inhibition of binding by the heparin sulfate analogue heparin is more efficient in epithelial than neuronal cells (IMMERGLUCK et al. 1998). The initial step in epithelial cell binding is mediated by HSV glycoprotein C (gC). Neuronal binding is not impaired in HSV gC mutants, nevertheless a marked reduction in immediate early (IE) gene expression suggests this glycoprotein plays a different role in facilitating neuronal infection. Neuron to neuron transmission, but not entry from the extracellular space, appears to require expression of the HSV glycoprotein hetero-oligomer gE-gI (DINGWELL 1995). HSV mutants unable to express either gE or gI have been shown to spread inefficiently within rat retina, from retina to brain and from cell to cell in neuronal cultures. The mechanism whereby cell to cell transmission is enhanced is unknown. Speculatively, the gE-gI complex may promote interactions between HSV and cell receptors concentrated at intercellular junctions or synapses or sorting of virus particles to these locations. Penetration occurs through fusion of the virion envelope with the plasma membrane and also requires HSV glycoproteins. Uncoating of virus particles occurs in the cytoplasm with release of the transcriptional activator VP16. The capsid is transported to the nuclear envelope and docks at fibres emanating from the nuclear pore complex. When viral DNA enters the nucleus it is preferentially deposited close to discrete, nuclear domains (MEREDITH et al. 1994). These bodies, named ND10 to reflect the average number in a nucleus, are approximately 0.3–0.5µm in diameter and visible by electron microscopy. They are present in uninfected cells but disappear during lytic infection. The nuclear replication cycle appears to begin at these pre-existing nuclear domains.

3.5.2 Transcription

During lytic infection HSV-1 gene expression occurs in a regulated fashion beginning with synthesis of IE gene products followed by synthesis of early and late gene products (SAMANIEGO et al. 1995). Three of the five HSV-1 IE proteins synthesised, namely infected cell peptides ICP0, 4, and 27, affect the subsequent expression of viral genes. Regulation of these HSV IE genes is thought to be a critical feature in determining the outcome of HSV infection. Transcription of the IE genes is activated by VP16, a component of the virion tegument, which is complexed with Oct-1 and other cellular proteins and binds at specific sites in the IE promoters. Vmw110 is a protein encoded by gene IE-1 that appears to serve an important role at the onset of both lytic infection and reactivation from latency. Vmw110 isolated from infected cells, including neurons, is complexed with a host cell protein of approximately 135kDa that is localised to discrete nuclear domains (MEREDITH et al. 1994). Mutations of the C-terminal portion of Vmw110 disrupt the ability of the protein to localise to ND10 and to bring about their apparent disappearance. A major component of the nuclear domains is the PML body. The PML protein (named as such because the chromosomal translocation which results in PML-RAR α-fusion protein causes promyelocytic leukaemia) probably normally functions as a tumour suppressor. Vmw110 accumulates throughout infection and is responsible for the dispersal of the PML-body-associated antigens. The initial replication sites associated with ND10 expand and fuse to fill the nucleus eventually corresponding with the intranuclear inclusions evident by light microscopy. The localisation of Vmw110 to ND10 and its specific binding is an essential prerequisite of lytic infection. It may also be the point of no return beyond which neuronal death is inevitable. Ubiquitination is a cytoprotective mechanism whereby damaged or abnormal intracellular protein is tagged for degradation. The discovery that the 135kDa protein bound to Vmw110 is a novel member of the ubiquitin-specific protease family has stimulated the hypothesis that Vmw110 may serve to direct the protease activity and prevent ubiquitination of early viral proteins (EVERETT et al. 1997).

3.5.3 Information from Animal Models

As the expression of the viral genes is controlled by host-viral protein interaction, potentially differing availability of specific host cell DNA-binding proteins may influence the outcome of infection in different cell types. An investigation of factors required for transcription from HSV-1 IE promoter ICP4 in transgenic mice, in which the promoter was linked to bacterial β-galactosidase, revealed that in the absence of viral proteins, including VP16, baseline levels of promoter activity varied considerably between different neuron populations and varied with age in the same group (MITCHELL 1995). Nevertheless, there is no clear evidence that such ab initio up-regulation of the promoter enhances HSV replication and promoter negative neurons are also readily infected. Animal models have failed to identify any viral function uniquely required for neuronal infection. Attenuated mutants with reduced neurovirulence generally show similar diminished replicative capacity in other tissues. Mutants lacking the $\gamma 34.5$ gene are totally avirulent after intracerebral inoculation in mice although able to replicate in some cell culture systems. In the absence of this gene, premature shut-off of host cell protein synthesis not only causes host cell death but also prevents viral replication and further spread (VALYI-NAGY et al. 1994). In experimental models HSV-1 neuroinvasiveness is influenced by amino acid changes in HSV glycoproteins D and B, but a comparison of human CSF-derived virus and peripheral isolates showed no striking differences between these glycoproteins (SIVADON et al. 1998). The functions of the different HSV glycoproteins and other HSV gene products can be examined in vitro but these results are not necessarily directly applicable to humans.

4 Mechanisms of Cell Death in Acute Viral Encephalitis

4.1 Apoptosis

The factors controlling neuroinvasion by HSV in humans remain elusive but when the virus gains access to the CNS, as with all cytopathic viruses, irrespective of host and cell type, cellular destruction is attributable to apoptosis or programmed cell death. This intrinsically controlled process of self-destruction can occur in all mammalian cell types. It can be triggered in a variety of ways by both intracellular and extracellular factors and occurs as both a normal physiological process and as a result of disease (WYLLIE 1997). Apoptosis plays an important role in modelling developing organs during embryogenesis and it occurs during hormone-dependent involution. It is the mechanism whereby auto-reactive lymphocytes are normally eliminated in the developing immune system. The removal of damaged cycling cells by this mechanism prevents the propagation of harmful mutations, whereas the failure of apoptosis permits survival and replication of mutated cells and may play an important role in oncogenesis. Apoptotic cell death results in stereotyped cytological changes that begin within the nucleus and are morphologically distinct from cell necrosis. The nuclear chromatin shows peripheral condensation and the nucleolus and nuclear lamina disintegrate. The cytoplasm shrinks and the cell membrane becomes convoluted. The cell breaks up into membrane-bound almost spherical bodies containing compacted intact organelles. Apoptotic bodies do not stimulate any acute inflammatory response and usually are quickly phagocytosed. The whole process occurs with great rapidity. In contrast, necrosis, exemplified by cell death due to hypoxia, results from autodegradation. Catalytic lysosomal enzymes that are released into the cytoplasm digest organelles and nuclear contents, causing cell swelling, plasma membrane rupture and discharge of contents which stimulate the acute inflammatory cascade.

4.1.1 Control of Apoptosis

Apoptosis is controlled by a set of cellular genes (WYLLIE 1997). The effector pathway is a proteolytic cascade mediated by interdependent cysteine proteases, which progressively cleave cytoskeletal proteins. Mammalian cells possess multiple factors that both oppose and augment the pathway, e.g. *Bcl*-2 is a proto-oncogene that inhibits apoptosis, whereas *p53* is an oncosuppressor gene which can induce it. Many different events, both physiological and pathological, can disturb the balance between the controlling factors and so trigger the same final common pathway. Viruses may exhibit both apoptosis-inducing and apoptosis-inhibitory effects. Induction of apoptotic host cell death may be advantageous and facilitate the exit of viral progeny, and in some instances this property correlates with virulence. Conversely, premature cell death would abort viral replication and thus many viruses have evolved molecular mechanisms to circumvent this process until mature virions have been produced. For example, Adenovirus E1B 55K and 19K proteins bind, respectively, to *p53* or functionally initiate *bcl*-2 through binding its natural inhibitors (DEBBAS and WHITE 1993).

4.1.2 Apoptosis in Sindbis Virus Infection

Sindbis virus, originally isolated from Culex mosquitoes and closely related to human pathogens including equine encephalitis viruses, is a small single-stranded RNA virus that causes encephalitis in mice. Following intracerebral inoculation of mice, in situ hybridisation for viral RNA, combined with the TUNEL (terminal deoxynucleotidyltransferase-mediated dUTP nick end-labelling) technique for detection of DNA cleaved by endonucleases, reveals that virus RNA colocalises with neuronal apoptosis (LEWIS et al. 1996). Different strains of Sindbis virus show variable age-dependent neurovirulence, which correlates with the level of cerebral neuronal apoptosis. In cultured cells transfection with the human *bcl*-2 oncogene suppresses growth of the avirulent strain but it offers no protection against apoptosis induced by the neurovirulent strain. This property of neurovirulence is determined by a single amino acid substitution in the Sindbis virus E2 glycoprotein coat. Analysis of cultured cells indicates that progeny virus production precedes the apoptotic morphological changes by several hours (LEWIS et al. 1996).

4.1.3 Apoptosis in Herpes Simplex Infection

Herpes simplex virus 1 is able both to induce and to block host cell apoptosis. HSV-1 encodes a protein, $\gamma_1 34,5$, that blocks phosphorylation of eukaryotic initiation factor 2α. HSV-1 regulatory protein ICP4 is also shown to block apoptosis. Although HSV prevents apoptosis at the outset of infection, host cell death is only delayed and it is the inevitable outcome of permissive infection and successful replication. The disintegration of neurons and glia that is very evident histologically is almost entirely attributable to apoptosis. Examination of the brains of immunocompetent patients dying of acute HSV combined with immunocytochemical demonstration of HSV antigen clearly reveal the characteristic nuclear apoptotic changes in large numbers of infected cells prior to the influx of host inflammatory cells. The occurrence of atypical HSV limbic encephalitis in patients with depressed cellular immunity due to HIV infection, organ transplantation or neoplasia (CHRETIEN et al. 1996; SCHIFF and ROSENBLUM 1998) also helps to clarify the relative contributions of virus and host to neuronal damage. Widespread apoptotic neuronal destruction occurs in the limbic system of these patients despite the paucity of lymphocytic infiltration. These observations suggest that significant neuronal and glial cell apoptosis is mediated by the virus prior to advent of the host cellular immune response.

4.2 Host Immune Responses and Cell Death

4.2.1 Microglial Activation

The host immune responses may exacerbate and even prolong tissue injury, and effective viral elimination may only be achieved after extensive host cell death. Viral infections initially stimulate innate immunity and in the immunocompetent individual this is followed by specific immune responses. In the CNS, microglia, the resident macrophages, form a network throughout the tissue and are the first cell type to respond swiftly to all types of CNS injury, including viral infection. Resting and activated microglia can be readily identified by immunocytochemical labelling with monoclonal antibodies such as CD68-directed cells of macrophage lineage. When CNS injury is severe and involves breakdown of the blood–brain barrier, as in herpes encephalitis, the resident population is boosted by blood-borne monocytes. Within the tissue the haematogenous recruits are indistinguishable. Microglial activation occurs within hours of CNS injury and involves a stereotypic graded pattern of responses beginning with proliferation and phenotypic transformation to mobile migratory cells that show strong expression of various receptor molecules including the CR3 complement receptor, cell adhesion molecules and amyloid precursor protein. Up-regulation of MHC class II antigens in rodents

(HLA-DR in humans) converts them into antigen-presenting cells. At the peak of HSV-1 encephalitis, MHC class II expression is detectable on microglia throughout the rat brain, including areas apparently remote from active infection (WEINSTEIN et al. 1990). Activated microglia release a myriad of potential neurotoxins, including nitric oxide, superoxide, platelet activating factor (PAF), pro-inflammatory cytokines and cysteine (LIPTON 1996). These substances are barely detectable in the normal healthy brain (MORRIS and ESIRI 1998). The complete transformation to potentially cytotoxic phagocytic cells occurs in response to cell death, notably neuronal (KREUTZBERG 1996). Endogenous microglia and circulating leucocytes are recruited by certain chemokines, small potent chemoattractant cytokines (HESSELGESSER and HORUK 1999). Microglia both express receptors and, when activated, secrete chemokines, thereby amplifying their effect. In mice infected with LCMV the pattern of chemokine gene expression in the brain appears to have great influence on the cerebral leucocytic inflammatory response and outcome of the CNS infection. (ASENSIO and CAMPBELL 1997). Microglial activation is further amplified by the autocrine effect of TNF-α and at a later stage it is induced by interferon (IFN)-γ released from activated lymphocytes participating in the specific immune response. Astrocytes can also be induced to secrete monocyte chemoattractants that may be important in an early phase (GLABINSKI et al. 1996). The precise mechanisms responsible for initiating the microglial response within hours of any CNS insult are not fully elucidated but these cells express a wide variety of ion channels (EDER 1998) and are very sensitive to depolarisation. Microglial ion channels may be activated by ATP released from damaged cells (KETTENMANN 1993).

4.2.1.1 Cytokines

Cytokines secreted by activated microglia not only amplify the inflammatory response but some, e.g. TNF-α and PAF, have demonstrable neurotoxicity in vitro (WESTMORELAND et al. 1996). The speculative role of microglial products has received greatest attention in HIV encephalopathy where T cell infiltration is minimal, and although productive HIV infection in the CNS is almost entirely confined to cells of the microglial/macrophage lineage, significant apoptotic neuronal loss also ensues (BELL 1998). Likewise, typical inflammatory HSV limbic encephalitis rarely occurs in the profoundly immunosuppressed although a more diffuse process characterised by myelin breakdown may ensue. The paucity of inflammation correlates with an abundance of neuronal HSV inclusions, but one study showed no definite relationship between neuronal apoptosis and intranuclear inclusions. This finding suggested that direct virally induced neuronal apoptosis may be augmented by macrophages-/microglia-derived cytokines (CHRETIEN et al. 1996).

4.2.2 Nitric Oxide

Many studies, both in vitro and in vivo, have shown a link between nitric oxide (NO) and inhibition of viral replication (REISS et al. 1998). The enzyme nitric oxide synthase is constitutively present in neurons (cNOS) and induced in activated microglia (iNOS) by cytokines, notably IFN-γ, interleukin (IL)-1 and TNF. Following induction of iNOS, NO synthesised in the cytosol of microglia is released and acts in a paracrine manner on neighbouring cells. The effects of diverse interaction with different molecules at different sites in the target cells range from physiological messenger to cytotoxic agent. In different circumstances NO can cause necrotic cell death through interaction with cellular enzymes, or apoptosis by direct interaction with DNA. The toxicity of NO is related to formation of peroxynitrite from interaction between NO oxide and superoxide anions (BECKMAN et al. 1994). Peroxynitrite causes oxidation and nitration of amino acid residues of proteins and guanine of DNA, lipid peroxidation, and DNA cleavage. NO-mediated cytotoxicity has been widely demonstrated in viral infections. Experimental study has shown that mice deficient in iNOS are more susceptible to HSV-1 infection of dorsal root ganglia (MACLEAN et al. 1998). A role for constitutive NOS in clearance of virus from the CNS has also been demonstrated in NOS knockout mice infected with vesicular stomatitis virus (KOMATSU et al. 1999). However, as with other chemical messengers, NO may be a two-edged sword in the CNS. In animal models of viral encephalitis NO has been shown variously to inhibit viral replication and to exacerbate cell injury in the CNS. After intranasal inoculation of HSV-1 in Lewis rats, iNOS was detected in microglia within a few days and its distribution coincided with viral propagation (FUJII et al. 1999). Treatment of infected animals with the NOS inhibitor nomega-monomethyl-l-arginine (l-NMMA) ameliorated clinical symptoms and reduced mortality. In a similar murine model, high levels of iNOS were shown to persist for several months after viral titres in the CNS had declined (MEYDING-LAMADE et al. 1998); thus adverse effects may be prolonged after the initiating stimulus has been eliminated. Prolonged high-level production of NO may contribute to neurologic damage in HIV disease. NO is amongst the multiple neurotoxins secreted by infected microglia (LIPTON 1996) but astrocytes are also considered an important source of intracerebral NO in AIDS-related dementia (HORI et al. 1999).

4.2.3 T Cell Responses

The specific cellular immune response effectively terminates the majority of acute viral infections by elimination of virally infected cells. This role of T cell cytotoxicity in controlling the cell to cell spread of HSV is amply illustrated by the occurrence of disseminated mucocutaneous herpetic infection in the immunosuppressed. The cellular immune response is relatively late event in acute viral encephalitis but, as in the periphery, it halts further intracerebral spread of virus. However, in acute HSV and other human viral encephalitides, augmentation of CNS injury, including neuronal death, may prove too high a price. The old concept of the CNS as a privileged site shielded from the immune system was overturned by the discovery of continuous surveillance by immunocompetent cells and the ability of activated T lymphocytes to cross the endothelial blood–brain barrier and migrate into the parenchyma. Activated microglia strongly expressing MHC II antigens also produce the co-stimulatory signal B7. T cell interaction with microglia thus drives clonal expansion of activated T_H cells, which secrete an array of cytokines including IL-2 and IFN-γ, which activate both cytotoxic T cells and recruit further microglia and macrophages. The inflammatory response is thus massively amplified by both the autocrine and paracrine effects of its cellular components. Cells expressing viral antigen and MHC I can be eliminated by virus-specific cytotoxic T lymphocytes. Although neurons may evade this attack by dint of low expression of MHC class I (RALL et al. 1995), glial and microglial cells show up-regulation of MHC class and become susceptible. Animal models of viral encephalitis lend credence to this mechanism (DRESCHER et al. 1999). Apoptosis is also the effector mechanism of cytotoxic T lymphocytes mediated either by their lytic granules or via the attachment of Fas ligand to the Fas receptor on the target cell.

5 Characterisation of the Immune Responses in Human HSV Encephalitis

5.1 In CNS Parenchyma

Characterisation of the cellular immune components of human HSV encephalitis in situ using a panel of monoclonal antibodies (SOBEL 1986) revealed a predominance of T lymphocytes over B cells and natural killer cells, together with neutrophil granulocytes and abundant macrophages in the neuropil. CD4+ and CD8+ T cells were found in approximately equal proportions and many cells expressed the activation marker IL-2 receptor. Cellular expression of both class I and class II MHC molecules was also demonstrable in the CNS parenchyma. In human acute retinal necrosis, mediated by HSV-1, cytolytic CD4+ and CD8+ T cells with specific HSV reactivity were isolated from intra-ocular fluid (VERJANS et al. 1998). Although, under certain circumstances, glial cells can be induced to express MHC class I or II molecules and endothelial cells to express MHC class II, several studies of human viral encephalitides suggest the predominant parenchymal cells showing MHC expression are the microglia and infiltrating macrophages (KENNEDY and GAIRNS 1992; SASAKI and NAKAZATO 1992; AN et al. 1996).

5.2 In Cerebrospinal Fluid

More recently, CSF analyses have recognised three phases of the intrathecal immune response. Interleukin-6 and IFN-γ are elevated during the first week of human encephalitic illness and then rapidly decline. Levels of IL-6 may exceed serum levels 100-fold (AURELIUS et al. 1994). Some 2–6 weeks after onset of symptoms, TNF-α, which is cytotoxic to neurons and oligodendrocytes, is elevated, as are late markers of the adaptive immune response, soluble IL-2 receptor and soluble CD8 antigen. In the late recovery phase the latter persists for many months.

6 Vasculitis and Ischaemic Necrosis

When the inflammatory response in HSV encephalitis is well-established, there is invariably histological evidence of small vessel vasculitis and red cell extravasation to such an extent that the limbic lesion is often macroscopically haemorrhagic. At this stage, immunocytochemical staining of tissue sections reveals HSV antigen in occasional capillary endothelial cells. Viral particles have also been identified in retinal vascular endothelium in human HSV (RUMMELT et al. 1994). Several in vitro experimental studies demonstrate up-regulation of various adhesion molecules and increased leucocyte and platelet adherence to HSV-infected endothelial cells (BOK et al. 1993; BRANKIN et al. 1995). Acute vasculitis is also evident in arbovirus encephalitis (ESIRI 1997). Thus, virally induced endothelial injury may eventually cause vascular occlusion and genuine ischaemic necrosis of neurons.

7 Rabies

In other acute viral encephalitides, e.g. poliomyelitis and rabies, neurons are the only permissive cell type in the CNS. Apart from this restriction, which is generally attributed to receptor availability, the common pathogenic effect of apoptosis and the stereotyped host response equally applies. In contrast to the ubiquitous HSV, rabies virus regularly penetrates the CNS of its many mammalian hosts, effecting increased aggression, which leads to a bite and further transmission. Unless post-exposure vaccination is instituted promptly, infection in humans is invariably fatal. Rabies virus enters cells by a process of receptor-mediated endocytosis. The viral envelope fuses with the endosomal membrane to allow release of the viral nucleocapsid into the cytoplasm. Fusion is mediated by the single viral membrane glycoprotein G (GAUDIN et al. 1999). Acetylcholine receptors of the motor end plate provide a receptor for this neurotropic virus (HANHAM et al. 1993; TUFFEREAU et al. 1998) which is then transported along axons as an unenveloped nucleocapsid. Replication begins in nerve cell bodies in the spinal cord from whence virus is transported intra-axonally into the CNS. The key histological feature of rabies is the neuronal cytoplasmic location of the eosinophilic viral inclusion body, the Negri body, formed by aggregates of nucleocapsids (KRISTENSSON et al. 1996). Negri bodies are most obvious in large neurons, particularly anterior horn cells, pyramidal cells of the hippocampus, in cerebral cortex and cerebellar Purkinje cells (ESIRI 1997). A murine model shows that brain cell apoptosis coincides with rabies virus replication and antedates the host lymphocytic response. Massive apoptosis followed by paralysis occurs in infected immunosuppressed mice (THEERASURA-KARN and UBOL 1998). Availability of specific receptors restricts intracerebral infection to neurons (TUFFEREAU et al. 1998), but apoptotic cell death is not an inevitable consequence of permissive neuronal infection. Although neuronal inclusion bodies may be very widespread, in the immunocompetent the host inflammatory reaction is concentrated at sites of maximal neuronal death in the brain stem and spinal cord. This restricted distribution implies selective vulnerability other than that determined by surface receptors. The effect on the function of infected cells that are apparently surviving is unknown. In rat brain, although iNOS activity is strongly increased during rabies virus infection, a concomitant significant reduction in cNOS could not be adequately explained by neuronal loss, as choline acetyltransferase levels remained unchanged (AKAIKE et al. 1995).

8 Viral Encephalitis in the Immunosuppressed

In states of altered or deficient immunity, CNS viral proliferation may be prolonged and neurons containing readily detectable viral antigen persist for longer. Neuronal inclusions are abundant in HSV encephalitis in the immunosuppressed. Likewise, when cytomegalovirus (CMV) infection occurs in the CNS of AIDS patients inclusions are numerous and present in many cell types, including neurons. In subacute sclerosing panencephalitis (SSPE), in which the combination of defective expression of measles virus proteins and deficient host cell-mediated response leads to persistent CNS infection, neurons and glial cells containing measles virus antigen may be widespread. The extent and nature of the functional impairment of these cells is largely unknown but ultimately they are also doomed to apoptotic destruction (McQUAID 1997). The apoptotic death of neurons in HIV encephalopathy despite the absence of neuronal infection clearly involves alternate mechanisms and the possible role of cytokines has already been mentioned. A more direct toxic effect is also postulated, as the virus coat protein gp120 has been shown to induce neuronal apoptosis in cultured human neurons (LANNUZEL et al. 1997).

9 Conclusion

This chapter has concentrated on HSV encephalitis to examine the mechanisms of neuronal damage in virus infections of the nervous system. Traditionally, the changes of HSV encephalitis were described as necrotising and, prior to the inflammatory response, infected neurons have a shrunken, eosinophilic appearance, similar to that seen in ischaemia. However, it is now clear that the major mechanism of neuronal death in HSV and other viral encephalitides is apoptosis, initially triggered directly by the virus and subsequently intensified by complex innate and specific immune mechanisms. CNS injury may eventually be compounded by ischaemic necrosis secondary to endothelial cell infection. It must be stressed that, although the functions and interplay of the multiple viral and host factors involved can be dissected in vitro and by genetic manipulation of both viruses and animal hosts, the results are not necessarily directly applicable to humans. In HSV and other human encephalitides neuroinvasiveness and neurovirulence are far from fully elucidated. The phenomenon of virally mediated neuronal destruction is frequently used as a measure of HSV virulence in animal models, but it should not be forgotten that encephalitis is the exception in humans and chimpanzees, who together comprise the only natural hosts. In the usual HSV host-virus relationship, permissive infection is restricted to epithelial cells whilst sensory neurons harbour latent virus. This ability of peripheral virally infected neurons to evade immunological destruction can be perceived as an evolutionary advantage that permits cell survival at the price of persistence of the viral genome. Whether or not the CNS shares this capacity is a matter of great fascination that is not fully resolved. The rapid propagation of cell death that follows introduction of HSV into the brain suggests that access to CNS neurons is normally restricted. In contrast, studies that reveal low levels of HSV DNA in elderly human brains imply that cell death is not inevitable and strategies to repress reactivation may also operate. There are fundamental and unexplained differences between the CNS and peripheral nervous system, and unique mechanisms may protect the brain from a ubiquitous but potentially lethal virus.

References

Akaike T, Weihe E, Schaefer M, Fu ZF, Zheng YM, Vogel W, Schmidt H, Koprowski H, Dietzschold B (1995) Effect of neurotropic virus infection on neuronal and inducible nitric oxide synthase activity in rat brain. J Neurovirol 1:118–125

An SF, Ciardi A, Giometto B, Scaravilli T, Gray F, Scaravilli F (1996) Investigation on the expression of major histocompatibility complex class II and cytokines and detection of HIV-1 DNA within brains of asymptomatic and symptomatic HIV-1-positive patients. Acta Neuropathol 91:494–503

Anderson JR, Field HJ (1983) The distribution of herpes simplex type 1 antigen in mouse central nervous system after different routes of inoculation. J Neurol Sci 60:181–195

Asensio VC, Campbell IL (1997) Chemokine gene expression in the brains of mice with lymphocytic choriomeningitis. J Virol 71:7832–7840

Aurelius E, Andersson B, Forsgren M, Skoldenberg B, Strannegard O (1994) Cytokines and other markers of intrathecal immune response in patients with herpes simplex encephalitis. J Infect Dis 170:678–681

Ayuso Blanco T, Gimenez Mas JA, Omenaca Teres M, Campello Morer I, Marta Moreno ME, Martinez Lanao D (1994) Brain stem encephalitis due to herpes simplex. Neurologia 9:112–114

Barber PC, Lindsay RM (1982) Schwann cells of the olfactory nerves contain glial fibrillary acidic protein and resemble astrocytes. Neuroscience 7:3077–3090

Baringer JR, Pisani P (1994) Herpes simplex virus genomes in human nervous system tissue analyzed by polymerase chain reaction [see comments]. Ann Neurol 36:823–829

Beckman JS, Chen J, Crow JP, Ye YZ (1994) Reactions of nitric oxide, superoxide and peroxynitrite with superoxide dismutase in neurodegeneration. Prog Brain Res 103:371–380

Bell JE (1998) The neuropathology of adult HIV infection. Rev Neurol (Paris) 154:816–829
Benator RM, Magill HL, Gerald B, Igarashi M, Fitch SJ (1985) Herpes simplex encephalitis: CT findings in the neonate and young infant. Am J Neuroradiol 6:539–543
Bok RA, Jacob HS, Balla J, Juckett M, Stella T, Shatos MA, Vercellotti GM (1993) Herpes simplex virus decreases endothelial cell plasminogen activator inhibitor. Thromb Haemost 69:253–258
Brankin B, Hart MN, Cosby SL, Fabry Z, Allen IV (1995) Adhesion molecule expression and lymphocyte adhesion to cerebral endothelium: effects of measles virus and herpes simplex 1 virus. J Neuroimmunol 56:1–8
Chretien F, Belec L, Hilton DA, Flament-Saillour M, Guillon F, Wingertsmann L, Baudrimont M, de Truchis P, Keohane C, Vital C, Love S, Gray F (1996) Herpes simplex virus type 1 encephalitis in acquired immunodeficiency syndrome. Neuropathol Appl Neurobiol 22:394–404
Debbas M, White E (1993) Wild-type p53 mediates apoptosis by E1 A, which is inhibited by E1B. Genes Dev 7:546–554
Drescher KM, Murray PD, David CS, Pease LR, Rodriguez M (1999) CNS cell populations are protected from virus-induced pathology by distinct arms of the immune system. Brain Pathol 9:21–31
Eder C (1998) Ion channels in microglia (brain macrophages). Am J Physiol 275:C327–342
Efstathiou S, Minson AC, Field HJ, Anderson JR, Wildy P (1986) Detection of herpes simplex virus-specific DNA sequences in latently infected mice and in humans. J Virol 57:446–455
Esiri MM (1982) Herpes simplex encephalitis. An immunohistological study of the distribution of viral antigen within the brain. J Neurol Sci 54:209–226
Esiri MM (1997) Viruses and rickettsiae. Brain Pathol 7:695–709
Esiri MM, Drummond CW, Morris CS (1995) Macrophages and microglia in HSV-1 infected mouse brain. J Neuroimmunol 62:201–205
Everett RD, Meredith M, Orr A, Cross A, Kathoria M, Parkinson J (1997) A novel ubiquitin-specific protease is dynamically associated with the PML nuclear domain and binds to a herpesvirus regulatory protein [corrected and republished article originally printed in Embo J (1997 Feb 3; 16:566–77)]. Embo J 16:1519–1530
Fujii S, Akaike T, Maeda H (1999) Role of nitric oxide in pathogenesis of herpes simplex virus encephalitis in rats. Virology 256:203–212
Gaudin Y, Tuffereau C, Durrer P, Brunner J, Flamand A, Ruigrok R (1999) Rabies virus-induced membrane fusion. Mol Membr Biol 16:21–31
Glabinski AR, Balasingam V, Tani M, Kunkel SL, Strieter RM, Yong VW, Ransohoff RM (1996) Chemokine monocyte chemoattractant protein-1 is expressed by astrocytes after mechanical injury to the brain. J Immunol 156:4363–4368
Hanham CA, Zhao F, Tignor GH (1993) Evidence from the anti-idiotypic network that the acetylcholine receptor is a rabies virus receptor. J Virol 67:530–542
Hesselgesser J, Horuk R (1999) Chemokine and chemokine receptor expression in the central nervous system. J Neurovirol 5:13–26
Hori K, Burd PR, Furuke K, Kutza J, Weih KA, Clouse KA (1999) Human immunodeficiency virus-1-infected macrophages induce inducible nitric oxide synthase and nitric oxide (NO) production in astrocytes: astrocytic NO as a possible mediator of neural damage in acquired immunodeficiency syndrome. Blood 93:1843–1850
Immergluck LC, Domowicz MS, Schwartz NB, Herold BC (1998) Viral and cellular requirements for entry of herpes simplex virus type 1 into primary neuronal cells. J Gen Virol 79:549–559
Itzhaki RF, Lin WR, Shang D, Wilcock GK, Faragher B, Jamieson GA (1997) Herpes simplex virus type 1 in brain and risk of Alzheimer's disease [see comments]. Lance 349:241–244
Jamieson GA, Maitland NJ, Wilcock GK, Yates CM, Itzhaki RF (1992) Herpes simplex virus type 1 DNA is present in specific regions of brain from aged people with and without senile dementia of the Alzheimer type. J Pathol 167:365–368
Kastrukoff L, Long C, Doherty PC, Wroblewska Z, Koprowski H (1981) Isolation of virs from brain after immunosuppression of mice with latent herpes simplex. Nature 291:432–433
Kennedy PG, Gairns J (1992) Major histocompatibility complex (MHC) antigen expression in HIV encephalitis [published erratum appears in Neuropathol Appl Neurobiol (1992 Dec; 18:627)]. Neuropathol Appl Neurobiol 18:515–522
Kettenmann H, Banati R, Walz W (1993) Electrophysiological behavior of microglia. Glia 7:93–101
Kimura H, Futamura M, Kito H, Ando T, Goto M, Kuzushima K, Shibata M, Morishima T (1991) Detection of viral DNA in neonatal herpes simplex virus infections: frequent and prolonged presence in serum and cerebrospinal fluid. J Infect Dis 164:289–293

Komatsu T, Ireland DD, Chen N, Reiss CS (1999) Neuronal expression of NOS-1 is required for host recovery from viral encephalitis. Virology 258:389–395

Kreutzberg GW (1996) Microglia: a sensor for pathological events in the CNS. Trends Neurosci 19:312–318

Kristensson K, Dastur DK, Manghani DK, Tsiang H, Bentivoglio M (1996) Rabies: interactions between neurons and viruses. A review of the history of Negri inclusion bodies. Neuropathol Appl Neurobiol 22:179–187

Lannuzel A, Barnier JV, Hery C, Huynh VT, Guibert B, Gray F, Vincent JD, Tardieu M (1997) Human immunodeficiency virus type 1 and its coat protein gp120 induce apoptosis and activate JNK and ERK mitogen-activated protein kinases in human neurons. Ann Neurol 42:847–856

Lewis J, Wesselingh SL, Griffin DE, Hardwick JM (1996) Alphavirus-induced apoptosis in mouse brains correlates with neurovirulence. J Virol 70:1828–1835

Lipton SA (1996) Similarity of neuronal cell injury and death in AIDS dementia and focal cerebral ischemia: potential treatment with NMDA open-channel blockers and nitric oxide-related species. Brain Pathol 6:507–517

MacLean A, Wei XQ, Huang FP, Al-Alem UA, Chan WL, Liew FY (1998) Mice lacking inducible nitric-oxide synthase are more susceptible to herpes simplex virus infection despite enhanced Th1 cell responses. J Gen Virol 79:825–830

McQuaid S, McMahon J, Herron B, Cosby SL (1997) Apoptosis in measles virus-infected human central nervous system tissues. Neuropathol Appl Neurobiol 23:218–224

Meredith M, Orr A, Everett R (1994) Herpes simplex virus type 1 immediate-early protein Vmw110 binds strongly and specifically to a 135-kDa cellular protein. Virology 200:457–469

Meyding-Lamade U, Haas J, Lamade W, Stingele K, Kehm R, Fath A, Heinrich K, Storch Hagenlocher B, Wildemann B (1998) Herpes simplex virus encephalitis: long-term comparative study of viral load and the expression of immunologic nitric oxide synthase in mouse brain tissue. Neurosci Lett 244:9–12

Mitchell WJ (1995) Neurons differentially control expression of a herpes simplex virus type 1 immediate-early promoter in transgenic mice. J Virol 69:7942–7950

Morris CS, Esiri MM (1998) The expression of cytokines and their receptors in normal and mildly reactive human brain. J Neuroimmunol 92:85–97

Ojeda VJ, Archer M, Robertson TA, Bucens MR (1983) Necropsy study of the olfactory portal of entry in herpes simplex encephalitis. Med J Aust 1:79–81

Phinney PR, Fligiel S, Bryson YJ, Porter DD (1982) Necrotizing vasculitis in a case of disseminated neonatal herpes simplex infection. Arch Pathol Lab Med 106:64–67

Rall GF, Mucke L, Oldstone MB (1995) Consequences of cytotoxic T lymphocyte interaction with major histocompatibility complex class I-expressing neurons in vivo. J Exp Med 182:1201–1212

Reiss CS, Plakhov IV, Komatsu T (1998) Viral replication in olfactory receptor neurons and entry into the olfactory bulb and brain. Ann NY Acad Sci 855:751–761

Rose JW, Stroop WG, Matsuo F, Henkel J (1992) A typical herpes simplex encephalitis: clinical, virologic, and neuropathologic evaluation. Neurology 42:1809–1812

Rummelt V, Rummelt C, Jahn G, Wenkel H, Sinzger C, Mayer UM, Naumann GO (1994) Triple retinal infection with human immunodeficiency virus type 1, cytomegalovirus, and herpes simplex virus type 1. Light and electron microscopy, immunohistochemistry, and in situ hybridization. Ophthalmology 101:270–279

Samaniego LA, Webb AL, DeLuca NA (1995) Functional interactions between herpes simplex virus immediate-early proteins during infection: gene expression as a consequence of ICP27 and different domains of ICP4. J Virol 69:5705–5715

Sasaki A, Nakazato Y (1992) The identity of cells expressing MHC class II antigens in normal and pathological human brain. Neuropathol Appl Neurobiol 18:13–26

Schiff D, Rosenblum MK (1998) Herpes simplex encephalitis (HSE) and the immunocompromised: a clinical and autopsy study of HSE in the settings of cancer and human immunodeficiency virus-type 1 infection [see comments]. Hum Pathol 29:215–222

Sivadon V, Lebon P, Rozenberg F (1998) Variations of HSV-1 glycoprotein B in human herpes simplex encephalitis. J Neurovirol 4:106–114

Theerasurakarn S, Ubol S (1998) Apoptosis induction in brain during the fixed strain of rabies virus infection correlates with onset and severity of illness. J Neurovirol 4:407–414

Tuffereau C, Benejean J, Alfonso AM, Flamand A, Fishman MC (1998) Neuronal cell surface molecules mediate specific binding to rabies virus glycoprotein expressed by a recombinant baculovirus on the surfaces of lepidopteran cells. J Virol 72:1085–1091

Valyi-Nagy T, Fareed MU, O'Keefe JS, Gesser RM, MacLean AR, Brown SM, Spivack JG, Fraser NW (1994) The herpes simplex virus type 1 strain 17+ gamma 34.5 deletion mutant 1716 is avirulent in SCID mice. J Gen Virol 75:2059–2063

Verjans GM, Feron EJ, Dings ME, Cornelissen JG, Van der Lelij A, Baarsma GS, Osterhaus AD (1998) T cells specific for the triggering virus infiltrate the eye in patients with herpes simplex virus-mediated acute retinal necrosis. J Infect Dis 178:27–34

Weinstein DL, Walker DG, Akiyama H, McGeer PL (1990) Herpes simplex virus type I infection of the CNS induces major histocompatibility complex antigen expression on rat microglia. J Neurosci Res 26:55–65

Westmoreland SV, Kolson D, Gonzalez-Scarano F (1996) Toxicity of TNF alpha and platelet activating factor for human NT2N neurons: a tissue culture model for human immunodeficiency virus dementia. J Neurovirol 2:118–126

Wyllie AH (1997) Apoptosis: an overview. Br Med Bull 53:451–465

Slow and Persistent Virus Infections of Neurones – A Compromise for Neuronal Survival

U.G. LIEBERT

1	Introduction	35
2	Virus–Cell Interactions in the CNS	36
2.1	Acute Infections	37
2.2	Persistent Infections of the Nervous System	38
3	Impact of Viral Infection on Specific Cell Functions	42
4	Immune-Mediated Antiviral Mechanisms	44
4.1	The Cell-Mediated Immune Response	48
4.2	Virus-Induced Cell-Mediated Autoimmune Reactions Against Brain Antigens	51
5	Consequences of Viral Persistence in Neurones	52
References		53

1 Introduction

Infections of the central nervous system (CNS) with intracellular pathogens are different in many respects from infections in other parts of the body due to both the anatomical and functional properties of the brain and the biological basis of immune surveillance in the CNS. Damage to brain cells might have severe consequences for the entire body and, in many instances, would conceivably interfere with vital functions. The CNS is particularly vulnerable to pathological stimuli since it consists of highly differentiated cell populations with complex functionally integrated cell-to-cell connections and specialised cytoplasmic membranes. Furthermore, CNS tissue is unique in its high metabolic rate and relative lack of capacity to regenerate. While persistent infection by a non-cytopathogenic virus in cells of an organ with a low-energy requirement and a high rate of regeneration may be tolerated, in CNS tissue such infections may interfere with normal function, especially when neurones are affected (JOHNSON 1982). From this point of view, the paucity of lymphatic drainage and the lack of constitutive expression of immune-regulatory molecules, e.g. MHC class II and even class I, make sense. Fortunately, the participation of the CNS in a viral infection is relatively uncommon, but it may

Institute of Virology, University of Leipzig, Johannisallee 30, 04103 Leipzig, Germany develop as a complication of many systemic viral infections. The special situation in CNS with its quasi-syncytium favours persistent infection without immediate destruction of the infected target cells. In the long run, however, functional deficits, progressive disease and eventually death of the individual ensue. The invasion of the brain by viruses does not result per se from a pathogen's specific tropism for neural tissue, since neurotropic viruses usually cause infections without involvement of this organ. Several viruses, such as herpes simplex (HSV) or mumps virus, usually result in rather mild illnesses that may become severe only when the CNS is infected, while other viruses, e.g. certain enteroviruses, lead to symptoms of disease only when the brain or the spinal cord is invaded. Many neurotropic viruses more readily invade the CNS of the young. The reasons for this are obvious: (a) immaturity of the immune response, (b) reduced capacity to produce interferon, (c) dependence of susceptibility to viral infection on the level of cell differentiation, and (d) age-specific distribution of receptor proteins. Thus the viral biology plays an important part in the establishment of CNS infections. However, it is not only the direct virus–host relationship that may affect brain function and integrity; in many instances the resulting damage is caused by the immune system rather than by the virus itself. Considering these general points, viral infections of the CNS represent a competitive process in which properties of the infecting virus and those of the host, i.e. cellular and immunological factors, interact simultaneously.

In cases of acute lytic virus infections of the CNS, a rapid and efficient elimination of the pathogenic agent is necessary because nerve cells lack the capacity for regeneration, and loss of the highly specialised neurones may rapidly reach the threshold beyond which survival of the individual is endangered. On the other hand, the chances for viruses to persist and at least temporarily escape immune surveillance are facilitated. In the long run, however, persistent infections of the CNS lead to cell death, disturbed brain cell function, immunopathological disease and, if vital structures of the brain are involved, to the death of the individual.

2 Virus–Cell Interactions in the CNS

Although the CNS complications are usually self-limiting, in rare instances they may be severe and even life threatening. Viral infections of the CNS are not the result of a direct viral tropism for neural tissue, since in most cases the CNS is not the primary site of viral replication. After multiplication in the periphery and the ensuing viraemia, viral particles reach the CNS by various pathways. The major route is via the bloodstream (JOHNSON 1982). The blood–brain barrier formed by endothelial cells of the cerebral capillaries, astrocyte foot-processes and the basal lamina usually prevents viral particles from entering the CNS directly. In contrast, the brain is relatively easily accessible to activated lymphocytes and macrophages, which, if infected, may serve as vehicles carrying viruses across the barrier. This transport mechanism for viruses, also referred to as the 'Trojan horse' mechanism, is responsible for CNS infections with several viruses such as paramyxoviruses (APPEL et al. 1978; FOURNIER et al. 1985; WOLINSKY et al. 1976) and lentiviruses (HAASE 1986; NARAYAN and CLEMENTS 1989). Other viruses invade the CNS by direct infection of endothelial cells, a mechanism proposed for poliovirus (BLINZINGER et al. 1969), Sindbis virus (JOHNSON 1965), Semliki Forest virus (PATHAK and WEBB 1974) and murine retroviruses (SWARZ et al. 1981). Other viruses, e.g. mumps or eastern and western equine encephalitis virus, invade the CNS by direct infection of the stroma of the choroid plexus, or by passive transport into the cerebrospinal fluid, where ependymal cells of the ventricle walls provide the basis for further spread in the CNS (HERNDON et al. 1974; LIU et al. 1970).

The second major pathway of CNS infection involves infection of neurones at nerve endings in the periphery and subsequent retrograde axonal transport into the CNS. Particularly rabies virus (MURPHY and BAUER 1974; IWASAKI et al. 1985) has been shown to use the neural route of CNS invasion. In the case of herpes simplex virus (HSV) infection of human beings, the primary site of replication is the epithelial cells of the skin or mucosa before the virus enters the peripheral ganglia (MARTIN and DOLIVO 1983). Viruses remain hidden from the immune system during the entry process into the CNS via direct cellular connections until the CNS infection is established. It has been demonstrated that axonal transport of herpes and rabies viruses occurs also in tissue culture of primary neurones and can be specifically inhibited in vitro and in vivo (LYCKE et al. 1984; LYCKE and TSIANG 1987; KRISTENSSON et al. 1986).

The interaction of infectious viruses and susceptible cells leads in general to cell destruction, persistent infection or cell transformation. This is largely determined by the genetic constitution of the host and by the type of viral agent. After an infectious virus has reached the CNS, symptoms of disease develop only if sufficient numbers of susceptible cells are infected to cause brain dysfunction and virus spread within the CNS is accomplished.

2.1 Acute Infections

Acute cytolytic infections inevitably lead to the death of individual cells and the release of progeny virus. Cell destruction is induced by the products of the viral genome or their effect on the regulatory mechanism of the cell. For several viruses the molecular events taking place during a lytic infection have been thoroughly described in tissue culture systems, and it is assumed that similar processes occur in CNS infections. A classical example of destructive infection of CNS tissue is poliomyelitis, an Enterovirus that exists in three serotypes (MELNICK 1996). The portal of entry for poliovirus is the alimentary tract, as experimental studies in monkeys have shown. The virus replicates initially in tonsils, neck lymph nodes, and Peyer's patches of the small intestine. Viraemia subsequently develops, which can lead to infection of the spinal cord or brain. Alternatively, poliovirus may spread along the axons of peripheral nerves to the CNS, as has been demonstrated in experimental infection (LA MONICA et al. 1987; MELNICK 1996). Once poliovirus has entered the CNS, neuronal damage and destruction ensues as the consequence of intracellular replication, particularly of anterior horn cells of the spinal cord. Poliovirus contains a genome of single-stranded, positive-sense RNA that closely resembles a cellular messenger RNA in structure and function (KITAMURA et al. 1981; SABIN and BOULGER 1973; TOYODA et al. 1984; HOGLE et al. 1985). The initial step of poliovirus replication is its binding to a specific cellular receptor, a typical transmembrane protein of the immunoglobulin superfamily (MENDELSOHN et al. 1989). Receptor mRNA can be detected in many tissues that do not bind poliovirus or support its replication. Since several alternatively spliced mRNAs of the receptor exist, it is conceivable that not all of them direct the translation of fully functional poliovirus receptors (KOIKE et al. 1990; MENDELSOHN et al. 1989). In addition, another surface molecule of human cells, CD44, is functionally associated with the susceptibility of cells for poliovirus, although poliovirus does not bind directly to CD44 (SHEPLEY et al. 1994).

In order to understand the neurovirulence of poliovirus, nucleotide sequencing and gene cloning were carried out to identify the genetic differences between wild-type virulent strains of poliovirus and the life-attenuated strains used as vaccines (ALMOND 1991). Significant changes were observed in the $5'$ nontranslated region (NTR) at positions 480, 481 and 472 in poliovirus types 1, 2 and 3, respectively. These mutations were linked to the ability of the virus to replicate in mouse brain or in neuroblastoma cells in culture (LA MONICA et al. 1987). Detailed investigation of the structure and function of the mutated NTR region revealed that attenuating mutations in the vaccine strains act through the disruption of a stem structure in the region of positions 470–540. Computer-generated analysis suggested that the greater the disruption of the stem, the greater the temperature sensitivity and the extent of attenuation. The intracellular functions of these mutations have not yet been identified completely, but the data available suggest that the $5'$ NTR interact with cellular factors influencing the tropism of the virus. Understanding the intracellular events may eventually provide an explanation as to why motor neurones are the specific target. The tissue tropism, neurovirulence and species specificity of the poliovirus infection are being studied in a transgenic mouse model (HORIE et al. 1994; REN et al. 1990). The control of polio infection depends on the humoral immune system. Although many details of the virus–cell interaction in this disease are well understood, the host factors that in the majority of cases prevent poliovirus from entering the CNS and infecting the neural tissue are still unknown.

2.2 Persistent Infections of the Nervous System

A variety of virus–cell and virus–host interactions may lead to chronic or persistent virus infections of the CNS. Firstly, latent viral infections are characterised by intermittent episodes of viral replication and formation of infectious virus (see the chapter by Borchers and Field, this volume). This process may either remain clinically silent or result periodically in clinical disease. In between such episodes, the virus remains in a quiescent form. Secondly, in chronic viral infections, virus can be continuously recovered from the host. Overt clinical disease may or may not develop, and the ensuing symptoms are caused either by viral replication or by reaction of the immune system. Thirdly, in slow virus infections, after a long incubation period of months to years, a slowly progressive disease course develops that is usually fatal.

An example illustrating the fact that closely related viruses may induce acute as well as persistent CNS infections is Theiler's murine encephalomyelitis virus (TMEV) infection. This infection serves as a model to understand poliomyelitis and multiple sclerosis. TMEV belongs to the picornaviridae and has an RNA genome structure very similar to that of the polioviruses. There are two groups of isolates, which cause either acute, rapidly fatal encephalitis (virulent strains) or biphasic chronic persistent disease with demyelination (avirulent strains). Despite their distinct biological properties, both strains are highly homologous with 95% identity at the genome and 90% at the protein level (PEVEAR et al. 1987, 1988). Regions of the genome associated with neurovirulence and persistence are similar to those of poliovirus and have been mapped to the 5' NTR and to the region coding for the VP1 capsid protein (CALENHOFF et al. 1990; MCALLISTER et al. 1990). Isolation of monoclonal antibody escape mutants with a single amino-acid change in the VP1 underline the importance of the protein for the viral phenotype (ZURBRIGGEN et al. 1989).

Acute infection with the virulent TMEV strains affects predominantly neurones in the cerebral cortex and the ventral horns of the spinal cord (as in poliovirus infections). The virus reaches the CNS primarily by axonal transport, but endothelial cell infection has also been shown. In contrast, a persistent infection of primarily oligodendrocytes develops in the majority of mice surviving the infection with avirulent strains of TMEV, where productive virus infection is associated with demyelinating lesions. The avirulent virus finds its access into the CNS via macrophages as a 'Trojan horse'. Macrophage-like cells constitute about 10% of infected cells in the brain. The mechanism of persistence depends on interferon-induced blockage of viral RNA replication at the level of negative-strand RNA synthesis, and the generation of antigenic variants by antibody escape. The susceptibility to TMEV infection is associated with MHC class I and two non-MHC genes encoded on chromosomes 3 and 6 (RODRIGUEZ et al. 1986; MELVOLD et al. 1990). $CD8^+$ T cells are involved in antiviral immunity during acute infection. They also play an important role in immune surveillance of persistently infected CNS cells, but they are not vital for recovery from acute infection (BORROW et al. 1992). Essential for the control of TMEV infection are $CD4^+$ T cells, as has been shown by depleting this cell type or by inhibiting its function in vivo using anti-CD4 or anti-Ia blocking MAbs, respectively (WELSH et al. 1987). Simultaneously, however, TMEV-specific $CD4^+$ T cells might lead to immunopathology and demyelination. There are essentially three possible explanations for the immunopathological response. Firstly, the $CD4^+$ cells directly damage MHC class II-expressing glial cells. Secondly, they induce mononuclear cell infiltration and activate macrophages, which damage myelinated nerve fibers in the so-called bystander effect. Thirdly, during the infection, T cells with autoimmune properties against nerve cell antigens could be induced, which might exacerbate ongoing pathology or induce new lesions.

In the human CNS, chronic viral infections rarely develop but latent infections are common. The classical example of this type is the infection with herpes simplex virus (HSV). The pathology inflicted on the brain tissue is the consequence of HSV reactivation from a latent state leading to complete viral replication cycles after certain stimuli. The pathology depends on the effectiveness of host defence mechanisms. Extensive molecular biological studies have defined several regions on the viral genome as specific contributors to HSV neuroinvasiveness and neurovirulence (STEVENS 1993). HSV consists of a double-stranded DNA molecule that codes for 74 genes. Half of them have been deleted without interference of the viral capacity to replicate in cell cultures. These genes, referred to as "supplemental essential genes", are associated with virus entry, sorting and augmenting the precursor DNA pool, repair of DNA, and shut-off of host macromolecular metabolism. There is evidence that several of them are linked directly to neuroinvasiveness and neurovirulence. Expression of these genes allows the virus to invade the CNS, to replicate in different brain cells and to spread efficiently from cell to cell. For its survival in a human population, HSV requires only a restricted set of genes. Interestingly, those genes dispensable for HSV replication in dividing cells, such as the thymidine kinase gene, are required for infection of non-dividing cells. Marker rescue experiments also suggest a role for the viral DNA polymerase in neuroinvasiveness and neurovirulence (DAY et al. 1988). The majority of viral deletion mutants revealed a reduced capacity to invade the CNS and to replicate in brain cells. Further changes in genes coding for structural glycoproteins as well as in immediate early genes and the long terminal repeat region are involved (THOMPSON et al. 1989; CHOU et al. 1990). The infection of neurones results in two mutually exclusive processes. Only if immediate early genes that regulate the viral lytic cascade are repressed will destruction and death of neurones be prevented. The price that the infected neurone has to pay is the establishment of viral latency with the potential reactivation of viral multiplication. The establishment of the latent phase is a function controlled and executed by the neurone rather than by the virus itself. This is concluded from observations showing that viruses selected for their ability to replicate in vitro still establish latent infection; vice versa, mutants with a deletion of the immediate-early transcription regulator gene persist in neurones (SEDARATI et al. 1993). Although no infectious virus can be isolated during latency, viral genomes are detected as multiple extrachromosomal DNA copies characteristically represented as covalently closed circles associated with nucleosomes (ROCK and FRASER 1985; DESHMANE and FRASER 1989). The only viral RNA transcripts detected in neurones during latency are the so-called latency-associate transcripts (LATs) (FRASER et al. 1992). Open reading frames are present within LAT sequences and proteins may be expressed. Further analysis revealed that competence to replicate in tissue culture and to establish latent infections in vivo is not affected in viruses with mutated LAT genes. A functional role during reactivation has been suggested by showing that LAT minus mutants were reactivated normally from sacral ganglia, but only slowly from trigeminal ganglia (SAWTELL and THOMPSON

1992). The molecular events that trigger viral reactivation are largely unknown. Clinical observations have shown that recurrence is associated with physical or emotional stress, immune suppression, UV light, or nerve damage. The role of the immune system in controlling HSV infection is not entirely clear, because recurrences can be observed even in the presence of normal cell-mediated and humoral immune responses. While there is apparently no gross interference with nerve function during latency, reactivation of HSV leads to the destruction of infected neurones. Since reactivation of HSV is successful in only a very small proportion of neurones, functional defects are usually not observed.

Among the best-studied slow virus diseases of humans is subacute sclerosing panencephalitis (SSPE) (TER MEULEN et al. 1983; LIEBERT 1997). The disease develops on the basis of a persistent measles virus (MV) infection in brain cells months to years after acute measles. How and when the virus reaches the CNS remains unknown, as do the mechanisms that trigger the disease. The tropism of MV for human tissue including brain cells is determined by its receptor, the ubiquitously expressed complement receptor CD46 (DÖRIG et al. 1993; NANICHE et al. 1993). The persistent CNS infection is characterised by a restricted measles virus gene expression at several stages (BILLETER et al. 1991; SCHNEIDER-SCHAULIES et al. 1995; LIEBERT 1997). Electron microscopical studies revealed the presence of viral nucleocapsids in neurones and other cells of the CNS in the absence of viral budding from these cells. It was found that the viral envelope proteins are markedly underexpressed or absent in infected brain cells. Transcriptional efficiency of the corresponding mRNAs is reduced, leading to a steep expression gradient for the virus-specific monocistronic transcripts and an increase of bicistronic transcripts has been described. Mutations and hypermutations in various genes of cloned SSPE viruses were detected that prevent a complete replicative cycle of MV and hence might contribute to the establishment of persistence (HIRANO 1992; CECCALDI et al. 1993).

Additional mechanisms may contribute to the restriction of MV gene expression, supporting the establishment of persistence (SCHNEIDER-SCHAULIES and LIEBERT 1991). Firstly, high levels of intrathecal antibodies causing antibody-induced antigenic modulation; secondly, a brain cell-specific restriction of MV mRNA expression; and thirdly, the presence of IFN-α/β and interferon-inducible gene products in SSPE brains. In spite of the pathognomonic hyperimmune response against measles virus in SSPE, clearance of virus from the CNS is not observed. In tissue culture experiments and in the rat model, antibody-induced antigenic modulation with monoclonal antibodies to MV haemagglutinin induced the downregulation of viral RNA and protein expression. Complete clearance of virus was not achieved, and removal of antibodies led to reactivation of the viral infection from low copy numbers of persisting viral genomes (BARRETT et al. 1985; LIEBERT et al. 1990b; SCHNEIDER-SCHAULIES et al. 1992). Similar results were generated in Sindbis virus-infected immunodeficient SCID mice, where infectious viral particles are cleared from the CNS by antiviral antibodies, but viral RNA can persist for months in mouse brains (LEVINE and GRIFFIN 1992; LEVINE et al. 1991). Intrinsic brain cell-specific mechanisms leading to a restriction of viral gene expression were observed in the experimental rat model and in infected brain cell cultures (SCHNEIDER-SCHAULIES J et al. 1993; SCHNEIDER-SCHAULIES S et al. 1989, 1990). The induction of type-I interferon by MV infection of brain cells results in further antiviral activity (HOFMAN et al. 1991; FUJII et al. 1988) and might cause the selection of interferon-resistant MV strains that are able to persist in the brain (CARRIGAN and KNOX 1990). The interferon-inducible Mx protein inhibits the expression of MV in human monocytes and neural cells (SCHNEIDER-SCHAULIES et al. 1994). The viral restriction was shown to occur at the transcriptional or post-translational level.

3 Impact of Viral Infection on Specific Cell Functions

In response to infection, cytokines are secreted within the CNS either from infected brain cells or from the infiltrating lymphomononuclear cells. Cytokines serve as important factors in the stimulation of the humoral and cell-mediated immune response by acting on cells of the immune system as well as on surrounding brain cells, thereby inducing antiviral proteins such as Mx, or cell surface molecules such as MHC antigens (CAMPBELL 1991; PLATA-SALAMAN 1991). Several studies report on measuring cytokine levels in human diseased brain. For example, in SSPE elevated levels of IFN-α/β-, IFN-γ- and TNF-α-positive cells have been detected (COSBY et al. 1989; HOFMAN et al. 1991; SEDGWICK et al. 1993). In the brains of HIV-1-infected patients, where essentially CD4-positive microglial cells take up and propagate the virus, these cells have been found to be the source of intrathecal synthesis of IL-1, IL-6, GM-CSF, and TNF-α (JORDAN et al. 1991; MERRILL and CHEN 1991). TNF-α and IL-1 have pleiotropic effects including regulation of body temperature, sleep, stimulation of surface molecules, chronic inflammatory effects, and the stimulation of proliferation and differentiation of glial cells (GIULIAN and LACHMAN 1985; MARTINEY et al. 1992; MERRILL 1991; PLATA-SALAMAN 1991). Since these two cytokines and MHC class-II antigens were increased in gliotic areas of HIV brains, a potential role for the immune system in the pathogenesis of HIV encephalopathy has been suggested (TYOR et al. 1992). Furthermore, TNF-α, IFN-γ and IFN-β can contribute to selective virus elimination by interacting directly with infected brain cells (KARUPIAH et al. 1991; LUCCHIARI et al. 1993; SCHIJNS et al. 1991).

The data obtained from human brains reflect late or final stages of disease processes, and little is known about the expression of cytokines and their cellular sources during early phases of an infection, when virus is spreading in the brain. To study the role of cytokines in the development of CNS virus infections, animal models, particularly with RNA viruses, have been investigated with respect to cytokine expression. Induction of IL-1a, IL-2, IL-6, TNF-α, and IFN-γ mRNA synthesis has been found in Borna disease virus (BDV)-infected rat brain within 2 weeks after intranasal infection. IL-2 and IFN-γ mRNA expression correlated with the appearance of $CD4^+$ and $CD8^+$ T lymphocytes during the early stages of BDV infection (SHANKAR et al. 1992). Cytokine expression is different in acute and persistent infections of the CNS. In the CSF of mice persistently infected with lymphocytic choriomeningitis virus (LCMV), significant levels of IL-6 were detected in high-responder (NMRI) but not in low-responder (CBA/J) mouse strains. In contrast, after acute intracerebral infection both strains contained high levels of IL-6 in CSF and serum (MOSKOPHIDIS et al. 1991). The source of IL-6 in mouse brains was identified to be astrocytes and microglial cells infected with LCMV or vesicular stomatitis virus (FREI et al. 1989). In tissue cultures of rat astrocytes, Newcastle disease virus has been shown to induce TNF-α and -β, IL-6, and IFN-α and -β shortly after primary infections (LIEBERMAN et al. 1989). Acute and persistent infection of human astrocytoma cells with measles virus results in transient expression of a similar set of cytokines, namely IL-1, IFN-β, IL-6 and TNF-α (SCHNEIDER-SCHAULIES S et al. 1993). Although TNF-α and IL-1β were hardly detectable in persistently infected cells their induction was not suppressed, and such additional stimuli as diacylglycerol and calcium ionophore induced overexpression of these genes (SCHNEIDER-SCHAULIES S et al. 1993). Additionally, chemokine expression in experimentally infected rodents strongly suggests that the induction of certain cytokines and chemokines plays an important role in the activation of the host antiviral immune responses and in the pathogenesis of viral CNS infection (SAUDER et al. 2000).

Virus infections can directly lead to, or increase the amount of, MHC class II expression on the surface of astrocytes and microglia in the absence of IFN-γ, as shown for murine JHM coronavirus (JHMV) or MV-infected astrocytes (MASSA et al. 1986, 1987). Interestingly, in the TMEV model MHC induction was IFN-γ dependent and did not result from mere TMEV infection. The susceptibility to immunopathology depends on the mouse strains. MHC class II was readily inducible in the susceptible SJL and CBA strains but not in the resistant Balb/c mice (NASH 1991). Similarly, BN rats live well with experimental persistent JHMV or MV infections. A constitutive high MHC class II expression was detected in BN brain, while immunopathological (autoimmune) processes were observed in a significant proportion of Lewis rats, in which MHC class II was expressed only as a consequence of viral infection. This illustrates that early presence of immunoregulatory molecules in the brain may serve a protective purpose, while the same process could lead to pathology and disease when MHC molecules are expressed several days or weeks later (SEDGWICK and DÖRRIES 1991). This is also largely true for MHC class I expression that can be induced only under certain conditions (NEUMANN et al. 1997). Generally MHC molecules are usually absent or expressed at low levels on neural cells. This is one of the prerequisites for efficient cell-mediated defence against intracellular pathogens in the brain. That this is not sufficient, however, to result in viral elimination is well illustrated by the observations in brain tissue from patients with progressive multifocal leukoencephalopathy, a slow papovavirus infection predominantly of oligodendrocytes with extensive demyelination and bizarre glial cell changes. Here, both MHC class I and II are expressed in the lesions, yet viral clearance does not occur, probably because there are no reactive T cells generated and the patients are generally immunosuppressed. This may allow for the establishment of persistent and ultimately chronic viral infection with progressive disease (ACHIM and WILEY 1992).

Direct consequences of virus–cell interaction on the neural cell function proper may be severe, even if only restricted areas of the brain are involved and particularly when the functional integrity of the affected brain cells is vital for the host. The disturbance of neurone function has been extensively studied with rabies virus. The virus causes a nonlytic infection of brain cells that rapidly leads to the death of the infected individual. In contrast to the limited cythopathology, the death of the infected individual is apparently the result of interference with neuronal cell function(s) in vital centres of the brain regulating sleep, body temperature, and respiration (TSIANG 1993). In experimental rabies virus infections it was observed that uptake and release of gamma-amino-n-butyric acid (GABA) decreased in infected rat embryonic cortical neurons (LADIGANA et al. 1994) and that binding of 5-hydroxytryptamine to serotonin receptor subtypes is reduced in infected rat brains (CECCALDI et al. 1993). It is conceivable that disturbances of specialised receptor systems for neurotransmitter and neurohormone turnover, as well as for the generation of chemical signals and electrical potentials, is the major cause of death of the infected individual. In rat astrocytoma cells persistently infected with MV, the infection strongly reduces cAMP response following the addition of catecholamines. Furthermore, the density of β-adrenergic receptors is decreased by 50% and coupling of the receptor to G-protein is affected (HALBACH and KOSCHEL 1979; KOSCHEL and MÜNZEL 1980). The endothelin-1-induced Ca^{2+} signal was absent in cells persistently infected with MV, and 95% of the binding sites for endothelin-1 were lost (TAS and KOSCHEL 1991). Anti-MV antibodies, present in high concentrations in SSPE brain, influenced the inositol-phosphate signal transduction pathway in the cells (WEINMANN-DORSCH and KOSCHEL 1990).

4 Immune-Mediated Antiviral Mechanisms

The control of virus infection depends on the generation of both humoral and cell-mediated immune responses (see the chapter by Dörries, this volume). Antibodies, which attack predominantly extracellular virus particles released from infected cells, are essential to limit the spread of virus in the host (SISSONS and OLDSTONE 1985). However, the failure to clear viruses such as varicella zoster virus, cytomegalovirus, or MV infections in cell-mediated immunodeficiency states suggests that T-cell responses may be more important than antibodies in overcoming several virus infections (SMITH et al. 1992). The basis for any immune response to viral infections of the CNS is that it is probably initiated in peripheral lymphoid tissue and followed by the invasion of activated T cells into the cerebrospinal fluid, meninges, and brain parenchyma (SEDGWICK et al. 1991b). The investigation of these aspects of viral CNS infections requires appropriate animal model systems.

Since virtually all viral CNS infections are preceded by primary peripheral infection, it is clear that virus neutralisation and opsonisation during viraemia is one of the most efficient defence reactions that prevents viral entry into the CNS. Experimentally, the lethal encephalomyelitis induced by the JHMV in newborn or suckling rats is prevented by nursing the babies from JHMV-immunised mothers (WEGE et al. 1993; PERLMAN et al. 1987). Identical findings were published in a mouse model of murine retrovirus-induced neurological disease and in protection from LCMV-induced teratogenic effects in newborn mice (SAHA et al. 1994; BALDRIDGE et al. 1993). Moreover, experimental virus infections of the CNS in immunocompetent hosts remain regularly subclinical if a strong virus-neutralising antibody response is mounted (TYLER et al. 1989; JUBELT et al. 1991; RIMA et al. 1991; SCHWENDER et al. 1991).

Nevertheless, viruses sometimes escape neutralisation in the periphery and succeed in entering the CNS. During retrograde axonal transport, the virus is inaccessible to the immune system and the immune system is no longer aware of the invading agent. Neither virus-specific antibodies nor cytotoxic T lymphocytes (CTL) can interfere with the axonal transport of viruses to the CNS, especially because nerve cells most likely are unable to up-regulate MHC class I molecules upon viral infection (MOMBURG et al. 1986). In addition, by using monocytes as 'Trojan horses', viruses may escape neutralisation by antibodies and enhance their probability of reaching the perivascular space in the CNS, because perivascular microglial cells are frequently exchanged by peripheral monocytes (SEDGWICK et al. 1991a, 1993). After invading the brain, viruses replicate and spread within the CNS, as long as the infected host does not succeed in recruiting immune effector cells into the brain parenchyma. Among the effector cells, B and plasma cells home toward virus-infected areas, where they secrete virus-specific antibodies (DÖRRIES et al. 1991; SCHWENDER et al. 1991). Also virus-specific $CD4^+$ helper T cells enter the CNS and are detectable in infected brain regions (LIEBERT and KOLOKYTHAS 2000). To prevent the formation of secondary virus-infected foci following extracellular spread of virus, intracerebral secretion of virus-specific antibodies in close spatial arrangement to virus-infected cells is needed. In several animal models of virus-induced encephalitis, specific antibodies will significantly restrict extracellular viral spread in the CNS (PERLMAN et al. 1989; PATICK et al. 1990; SCHWENDER et al. 1991; YOKOMORI et al. 1992; FAZAKERLY et al. 1992; ATHERTON 1992). In addition, dissemination from cell to cell, e.g. by fusion, can be interrupted by antibodies, as shown for mice infected with rabies virus (DIETZSCHOLD et al. 1992). The underlying molecular mechanism is not fully understood, but it is suggested that upon uptake of antibody-complexed virus into the infected nerve cell, viral RNA transcription is severely disturbed. In line with this concept are data obtained in vitro and in vivo with antibody-induced antigenic modulation (BARRETT et al. 1985; LIEBERT et al. 1990; SCHNEIDER-SCHAULIES et al. 1992) which demonstrate abrogation of MV transcription by antiviral antibodies. Morphological studies by immune electron microscopy supported the view that viral replcation is disturbed on the transcriptional or translational level, because the typical arrangement of virus-specific proteins in perinuclear cytopathic vacuoles and the amount of rough endoplasmatic reticulum was drastically reduced in antibody-treated neurones after virus infection.

The successful combat of viral CNS infections by the humoral immune response requires rapid recruitment of pre-existing virus-specific antibody-secreting cells into the brain parenchyma. This is usually achieved if viral CNS infection occurs concomitantly with the acute peripheral infection. In contrast, when viral CNS infection occurs late after primary infection or remains unrecognised by the immune system, the immune response has to be initiated in peripheral lymphatics such as the cervical lymph nodes. This gives the virus considerably more time to travel through the tissue, until humoral effector systems reach the brain parenchyma. Recruitment of antibody-secreting cells is determined by the genetic background of the host, as is clearly evident from studies of coronavirus JHM (JHMV)-induced encephalomyelitis in rats. While clinically resistant Brown Norway rats rapidly differentiate virus-specific $CD4^+$ T cells and recruit virus-specific antibody-secreting cells into virus-infected areas of the brain, this process is significantly delayed in highly susceptible Lewis rats (SCHWENDER et al. 1991; IMRICH et al. 1994). Furthermore, specificity and effectiveness of the recruited humoral response is important, as seen in severe acute Sindbis virus encephalitis in mice (TYOR and GRIFFIN 1993). In the JHMV rat model, an individual plasma cell from the brain of infected but clinically healthy BN rats synthesises approximately five times more effective specific virus-neutralising antibodies, compared with a plasma cell from the CNS of infected and severely diseased Lewis rats (SCHWENDER et al. 1991).

The effect of a late or a rather unspecific recruitment of humoral immunity to the CNS is followed by viral spread within the CNS and destruction of important neural cells. Moreover, large virus-infected areas will cause a more vigorous infiltration of virus-specific T cells, as is usually seen in cases of limited viral spread. This intimate relationship between virus-specific antibody response and T-cell-mediated immunopathology was observed in acute LCMV-induced encephalitis of mice. In this model, passive administration of MAbs between 1 day before and 2 days after viral infection resulted in protection of mice from lethal encephalitis. This was accompanied by a diminished CTL response and clearance of the virus from the brain with less tissue damage than usually seen in unprotected mice (WRIGHT and BUCHMEIER 1991). Similar results were obtained in the measles encephalitis model in rats and mice (LIEBERT and FINKE 1995). Here, the induction of MV-neutralising antibodies may completely suppress the development of disease in weanling animals (MALVOISIN and WILD 1990; BRINCKMANN et al. 1991) and maternal antibodies transferred during gestation are protective in newborn animals. These findings are consistent with the resistance to encephalitis observed in BN rats that mount an early high level of MV-specific humoral immune response (LIEBERT and TER MEULEN 1987). The inevitably fatal acute disease can also be prevented by passive immunisation of newborn animals with neutralising monoclonal antibodies, however, at the price of converting the infection into one of a persistent nature (RAMMOHAN et al. 1983; LIEBERT et al. 1990b). Thus, the presence of mere antibody, even if neutralising the infectivity of a virus, may be generally insufficient to eliminate measles virus from the infected brain when there are not also MV-reactive T cells available. On the contrary, even without available virus-neutralising antibodies animals can be protected. This was demonstrated by immunisation with recombinant vaccinia viruses expressing either internal nucleocapsid protein, or haemagglutinin, or fusion glycoprotein that prevented the disease upon subsequent challenge with MV in nonimmunised rats (BANKAMP et al. 1991; BRINCKMANN et al. 1991). This was further shown in μMT mice that have an inherent defect for antibody production (KITAMURA et al. 1991; Liebert and Geißendörfer, unpublished data).

Although, in general, humoral immunity does not contribute to the pathology of viral CNS infection, indirect tissue destruction might occur during viral encephalitis. Activation of macrophages or microglia may be triggered by engagement of the FcR, which is expressed in high densities on these cells. Binding of immune complexes to the FcR of macrophages can stimulate these cells to release toxic substances that will cause severe bystander destruction of 'innocent' healthy cells in the surroundings of virus-infected areas. This assumption is supported by the observation in vitro of macrophage-dependent oligodendroglia cell degeneration in mixed glial cell cultures that were treated with immune complexes formed by canine distemper virus and CDV-specific antibodies (BOTTERON et al. 1992).

Incomplete elimination of virus from the CNS will result in chronic persistent infection. Usually, a long-lasting intrathecal antibody synthesis with specificity for viral proteins accompanies viral persistence (TER MEULEN et al. 1983; SONNERBORG et al. 1989; TYOR et al. 1992) and efficiently prevents reactivation of the infection (LEVINE and GRIFFIN 1992). Over time there is selection of the best-fitting antibody clones to the virus. Thus high-avidity antibodies prevail eventually, and the respective clones are preferentially recruited to the CNS. In this case, isoelectric focusing of cerebrospinal fluid specimens will show a restricted 'oligoclonal pattern' of antibody clones compared with the polyclonal distribution detectable in paired serum specimens. Presence of these oligoclonal bands can continue over decades after primary infection of the CNS and thus is used as a diagnostic marker of viral CNS infections (FELGENHAUER and REIBER 1992).

Besides the fact that intrathecal virus-specific antibody synthesis is a relevant indicator of viral CNS infection, long-lasting presence of these antibodies in high titres is supposed to interfere with viral replication, thereby probably contributing to selection of virus variants. Direct evidence for selection of neurotropic variants by antibodies has been provided by the demonstration of changes in the cell tropism of neurotropic JHMV when grown in the presence of virus-neutralising monoclonal antibodies (BUCHMEIER et al. 1984). When the neurotropic virus and the MAb were inoculated simultaneously, viral target cells were primarily oligodendrocytes, in contrast to animals inoculated with virus alone, where neurones were the major target. In the rat model of MV-induced encephalitis (LIEBERT and TER MEULEN 1987), MV-neutralising monoclonal antibodies modified the disease process when administered intraperitoneally, and MV variants emerged that were no longer neutralisable by the monoclonal antibody used for treatment, but by other monoclonal antibodies (LIEBERT et al. 1994). From these data it has to be concluded that only rapid and effective elimination of virus-infected CNS cells will prevent long-lasting antibody-controlled persistence of the virus and thereby the potentially dangerous development of viral variants with altered neurotropism.

4.1 The Cell-Mediated Immune Response

In contrast to the effect of antibody-mediated antiviral mechanisms, which, at least in vivo, either act predominantly against the virus itself or interact with cell surface molecules without damaging the cell integrity, the T-cell immune protection is mediated generally by cell destruction, i.e. pathology (see also the chapter by Dörries, this volume). In experimental infection of mice with the lymphocytic choriomeningitis virus (LCMV), the number of infected brain cells has been shown to constitute an important factor directing beneficial or harmful effects mediated by effector lymphocyte activity (ALLAN and DOHERTY 1985). Since the timely T-cell immune response encounters a limited number of infected cells, the pathology inflicted should be little in most instances of acute viral encephalitis, and the beneficial effects will usually outweigh the harmful effects of the cell-mediated immune response. Paradoxically, the elaborate system of the host immune response to virus infection that may be protective outside the CNS can be destructive when operating within and may injure the host while helping to clear virus, particularly in a persistent infection. This also illustrates the importance of the balance between the kinetics of immune responses and the virus host interaction. The outcome of a viral infection is also determined in part by the type of neural cell infected. This is an important factor in the pathogenesis of disease, not only for the potential injury caused by viral infection per se but also for the potential interactions of the infected cells with the immune system. Thus the effect of even local damage can be dramatic when neurones are infected which are vital to host survival and cannot be replaced once injured or destroyed. In view of these considerations, the task of the immune response during CNS infection is to either quickly eliminate the viral agent or arrange to tolerate the infection within neural cells.

Although both $CD4^+$ and $CD8^+$ T cells can be relatively easily isolated and grown in culture from the CSF of patients with viral encephalitis and meningoencephalitis, their relative importance in, and contribution to, combating an infection is uncertain. From the observations made in several animal models it appears that the presence and function of $CD4^+$ rather than $CD8^+$ T cells is required to overcome viral CNS infections. This contrasts with the situation in other organs where, during acute and chronic viral hepatitis, cytotoxic MHC class-I-restricted $CD8^+$ T cells attack virus-infected hepatocytes and thus mediate protection as well as immunopathology or cell destruction.

The role of T cells in overcoming measles virus infection was analysed in the rodent MV-encephalitis model. For this purpose lymphocyte subpopulations were depleted by in vivo administration of MAbs directed against the CD4 or CD8 surface molecules of lymphocytes (BANKAMP et al. 1991; FINKE and LIEBERT 1994), or MV-primed T cells were intravenously transferred into MV-infected rats (REICH et al. 1992; LIEBERT and GEISSENDÖRFER 2000). Both approaches demonstrated that $CD4^+$ cells are apparently indispensable for achieving viral clearance from the CNS, while $CD8^+$ cells were not vital for recovery from the acute infection. When $CD8^+$ T cells are depleted, rats are still completely protected by adoptive transfer of immune-primed viral antigen $CD4^+$ T cells without local production of neutralising antibody (LIEBERT et al. 1993). From these results it was concluded that neither $CD8^+$ T cells nor antibodies are necessary for efficient elimination of MV and protection from disease in the encephalitis model. The results were surprising, because the susceptibility of mice for MV encephalitis correlates with their ability to generate an MHC class-I (L^d)-restricted, $CD8^+$ T-cell-mediated cytotoxic immune response (NIEWIESK et al. 1993). Furthermore, the Balb/c^{dm2} mouse strain, that fails to express L^d but is otherwise genetically identical to MV-resistant Balb/c mice, also eliminates MV efficiently from the CNS, although no MV-reactive $CD8^+$ T cells are primed in these mice. The apparent explanation for the observation is that $CD8^+$ cells are unable to interact with neurones that lack MHC class-I expression after viral infection (MOMBURG et al. 1986; Müller, Löffler and Liebert, unpublished data). At this point it has to be remembered that the generation of virus-specific immune effector cells depends on the species infected, and even within a single species there is no unique T-cell subset used in the antiviral defence. In several virus infections of rodents, the importance of $CD8^+$ CTL for combatting infection has been consistently shown (AHMED et al. 1988; BENDER et al. 1992; MOSKOPHIDIS et al. 1987; OLDSTONE et al. 1986; NASH et al. 1987). In contrast, the $CD8^+$ T-cell-mediated clearance of JHMV from the CNS requires CD4 help (WILLIAMSON and STOHLMANN 1990; KÖRNER et al. 1991; FLORY et al. 1993; WEGE et al. 1993). In the protective immunity to retroviruses both $CD8^+$ and $CD4^+$ T cells were partially effective, but only the combination of both led to full protection (HOM et al. 1991). Recovery from acute murine cytomegalovirus infection can proceed in the absence of the $CD8^+$ subset and is mediated by $CD4^+$ T cells, which develop a compensatory protective activity that is absent in normal mice (JONJIC et al. 1990). $CD4^+$ T cells appear to be required for maintenance of the spontaneous recovery from Friend virus-induced leukaemia (ROBERTSON et al. 1992). These examples illustrate that there is no general assignment of a determinative role in vivo to either T-lymphocyte subset in the recovery from viral infections. Instead, a detailed examination is required for every virus infection in relation to its susceptible host, and hardly any prediction can be made even for related viruses. It appears, however, that in the CNS antiviral cell-mediated activity is dependent largely on $CD4^+$ T cells.

A detailed characterisation in vitro revealed that protective T cells in the experimental MV-encephalitis model produce high amounts of IL2, IFN-γ, and TNF-α, but not IL4 or IL6, defining them as TH_1 cells (REICH et al. 1992; FINKE et al. 1995). Interestingly, the predominant generation of TH_2 cells seen after both natural MV infection and vaccination against measles in human beings has led to the hypothesis that the lack of virus-specific TH_1 cells may contribute to the immunosuppression seen after infection or vaccination (WARD and GRIFFIN 1993). Data obtained in the mouse model are consistent with this concept, as from the highly susceptible C3H mouse strain preferentially TH_2 $CD4^+$ T-cell lines can be isolated but not TH_1 cells. If cytokines secreted by TH_1 cells were important, two requirements would have to be fulfilled. Firstly, virus-primed T cells would have to invade the CNS and home toward sites of infection, and secondly, blocking cytokine function should abolish protection. By exploiting a genetic marker, it was shown in adoptive transfer experiments into MV-infected rats and mice that MV-specific $CD4^+$ T cells from a donor animal enter the brain of the host. These cells accumulated in infected areas, whereas in an immunocompetent animal they never comprised more than 5% of the infiltrating T cells (LIEBERT and KOLOKYTHAS 2000). It is unlikely that MHC class-II-restricted cytolysis plays a major role in eliminating MV from the brain, because the major cell population in the infected rodent brain is neurones. Rather, virus-specific MHC class-II-restricted lymphocytes – as shown before in the LCMV-infection model – induce immunopathological foci in the brain (MULLER et al. 1992). The observed paucity of virus-specific donor T cells in the brains of MV-infected mice and rats leads to the conclusion that recruitment of further cells is essential for combat of virus in an infected brain. Accordingly, the neutralisation of IFN-γ by administration of anti-IFN antibody rendered all mice susceptible to MV-induced acute encephalitis. Irrespective of the mouse strain, anti-IFN-treated animals died of infection, suggesting that cytokines may play an important role in the immune surveillance of the CNS. It was also shown that the infection of brain cells led to a differential induction of cytokines in primary and persistent MV-infection in human glial cell cultures and in rodent neurones (SCHNEIDER-SCHAULIES S et al. 1993; Mosch, Löffler and Liebert, unpublished data). The mechanism of cytokine action is conceivably to assist in the recruitment of effector cells into the CNS. For example, IFN-γ and TNF enhance the expression of VCAM-1 on brain endothelial cells to which stimulated T cells bind before they enter the brain and encounter viral antigen, which leads to further events of activation and the secretion of cytokines (BARON et al. 1993). The candidate prime source of IFN-γ and TNF-α in MV infection of the murine CNS are $CD4^+$ T cells and microglia. The potential contribution of other cytokines and the role of mononuclear phagocytes that are consistently present in infected foci have not yet been clarified.

In the murine TMEV model, the susceptibility to infection is MHC associated and maps to the class-I locus H-2D (NASH 1991). In susceptible strains of mice, $CD8^+$ T cells apparently fail to recognise viral antigens in the context of MHC class I, and so the virus persists and eventually causes disease. In in vivo depletion experiments using anti-CD8-antibody it was shown that virus clearance is delayed and demyelinating disease develops. These data show that $CD8^+$ T cells are not involved in immunopathology (i.e. demyelination) and are also not vital for recovery from acute infection. They may, however, contribute to antiviral immunity in acute infection and immune surveillance of persistently infected cells. Observations made in beta 2-microglobulin-deficient transgenic mice suggest that $CD8^+$ T cells may play a role in clearing viral persistence from glial cells (PULLEN et al. 1993). Similar to the measles model, $CD4^+$ cells are essential for controlling the early stages of infection. Depletion studies of $CD4^+$ cells in the TMEV model suggest that the major role of CD4$^+$ T cells in Picornavirus infections is probably to provide help for B lymphocytes and thus enable the production of neutralising antibody (WELSH et al. 1987). Little is known about the possible antiviral activity of CD4$^+$ T cells in TMEV infection. However, experiments in which the MHC-II-restricted CD4$^+$ T cell function was suppressed have resulted in a reduction in the incidence of demyelinating disease. Following TMEV infection and initial T-cell infiltration into the CNS, MHC class-II induction on astrocytes is a key step allowing local antigen presentation and amplification of immunopathological responses within the CNS, and hence development of demyelinating disease (BORROW and NASH 1992). A bystander effect caused by the mononuclear cell infiltration and activation of macrophages, which in turn can lead to damage on myelin sheaths, is probably responsible for the observed immunopathology.

The price for evading persistent virus infection may be development of autoantibodies and autoreactive T cells. In rats infected with JHMV or MV, myelin basic protein (MBP)-reactive CD4$^+$ T cells have been detected that could transfer experimental allergic encephalomyelitis (EAE) to naive uninfected animals (WATANABE et al. 1983; LIEBERT et al. 1988). In rats rendered tolerant to MBP, not only EAE but also the precipitation of the autoimmune subacute measles encephalitis was suppressed (LIEBERT and TER MEULEN 1993).

4.2 Virus-Induced Cell-Mediated Autoimmune Reactions Against Brain Antigens

A co-factor role for MV in the development of EAE was suggested by early observations that showed that the course of EAE and its severity were potentiated in MV-infected hamsters (MASSANARI et al. 1979). Interestingly, after infection with measles virus some Lewis rats develop a disease process that is characterised by an inflammation in the CNS in the absence of MV antigen or viral nucleic acid. The lesions are very similar to those of naive rats receiving MBP specific CD4$^+$ T-lymphocytes. Oligoclonal immunoglobulins with restricted heterogeneity were detected in the cerebrospinal fluid of the animals, which probably react to brain antigens (DÖRRIES et al. 1988). Lymphocytes isolated from these animals were found to proliferate in vitro in the presence of MBP or PLP (LIEBERT et al. 1988). The intravenous transfer of MBP-reactive MHC class II-restricted CD4$^+$ T cell lines isolated from bulk cell populations induced a disease in naive syngeneic recipients with clinical and histopathological signs identical to T cell mediated EAE. The analysis of the antigenic fine specificity revealed that MBP-specific T cells from MV-infected as well as from MBP-challenged rats displayed an identical pattern of reactivity to a panel of synthetic peptides (LIEBERT et al. 1990a). The high degree of antigenic specificity was further supported by the failure of the T cell lines to proliferate in the presence of disrupted measles virions, isolated MV proteins, or other control antigens or peptide sequences. Vice versa, MV-specific T cell lines did not proliferate when MBP or synthetic MBP peptides were added to the cultures. The disease induced was clearly not due to activation of MV in the brain of immunised Lewis rats, because virus could not be isolated from brain material and measles antigen was not detectable (LIEBERT and TER MEULEN 1993). The interaction between MBP-peptide and MV-infection was not observed when rats were infected intraperitoneally or when inactivated MV was used. Obviously, at least initially, some viral replication in the brain is required to enhance the vulnerability of the brain to autoimmune aggression. If autoimmune mechanisms participate in the pathogenesis of virus-induced encephalomyelitis, susceptibility to measles encephalitis and EAE should parallel in different rat strains depending on the genetic background. This is indeed the case, as Brown Norway rats that are resistant to EAE did not develop a subacute clinical disease, although they are generally able to replicate MV (LIEBERT and TER MEULEN 1987). The susceptibility of rats to the development of MV-induced CNS changes and disease is multifactorial, with the development of an MBP-specific CMI response representing a major factor.

5 Consequences of Viral Persistence in Neurones

In summary, establishment and maintenance or eventual elimination of persistent viral infections in the CNS is not the result of a single factor. Neither viral, nor host, nor immune defence functions alone are sufficient; it is rather the highly complex interrelation between those components that results in the temporary symbiosis of viruses and CNS cells. In many instances this leads to latent or slow viral infections with little or no specific immune response. The ongoing presence of viral proteins and/or nucleic acid, however, threatens to reactivate the viral replication process with the emergence of new symptoms and severe damage to brain substance and function. These symptoms may be the result of destructive virus–cell interaction or of the antiviral immune response in form of immunopathology. Since immune responses are generated in the periphery the combat of CNS virus infections is hindered. When neurones are infected, direct interaction of T cells with the infected host cell is not possible due to the lack of MHC expression. In any case, this would not be desirable, because immune responses, while being beneficial and effective in the periphery, inflict enormous pathology when attacking cells that lack the capacity to regenerate. Hence, a rapid elimination of virus (infected cells) is necessary, whereas a delayed immune response may allow the virus to spread in the CNS and, even if ultimately effective against the virus, may be destructive to the host. Because of this, precautions exist to prevent potentially damaging immune responses during persistent infections when, at least temporarily, the virus does not destroy its host cell. Obviously to the advantage of the host, a delicate balance is normally maintained between the requirements for the morphological and functional integrity of the CNS and the pretension of the immune system to combat virus and eliminate virus-infected cells.

From extensive studies in human and animal CNS infections the following facts have become clear: (a) The immune response to viral infections is initiated in peripheral lymphoid tissues, followed by entry of activated end-differentiated T and B cells into the cerebrospinal fluid, meninges, and brain parenchyma. (b) During viral infections, the cytokines and chemokines induced vary between different strains of mice and in different cell types of the same mouse strain or human individual. (c) Interferon-induced proteins such as MA may contribute to the establishment of persistent infections, which are accompanied in certain cases by down-regulation of viral replication and/or restriction of viral gene expression. (d) During viral infections, MHC class antigens are expressed on astrocytes and oligodendrocytes and extensively on microglia that present viral antigen produced by infected cells. (e) In many viral infections T cells are required for viral elimination; sometimes clearance of virus also depends on the timely presence of virus-specific antibodies. (f) The synergistic interaction of all components of the adaptive immune system is required for both limitation of virus spread within the CNS and ultimate elimination of virus from brain cells. (g) However, in established persistent infection, immunopathology and/or autoimmunity may develop as a result of immune-mediated damage from inappropriate T-cell responses generated during attempted viral clearance.

References

Achim CL, Wiley CA (1992) Expression of major histocompatibility antigens in the brains of patients with progressive multifocal leukoencephalopathy. J Neuropathol Exp Neurol 51:257–263

Ahmed R, Butler LD, Bhattli L (1988) T4+ T helper cell function in vivo: differential requirement for induction of antiviral cytotoxic T-cell and antibody response. J Virol 62:2102–2106

Allan JE, Doherty PC (1985) Immune T cells can protect or induce fatal neurological disease in murine lymphocytic choriomeningitis. Cell Immunol 90:401–407

Almond JW (1991) Poliovirus neurovirulence. Semin Neurosci 3:101–108

Almond PS, Bumgardner GL, Chen S, Platt J, Payne DW, Matas AJ (1991) Immunogenicity of class I$^+$, class II hepatocytes. Transplant Proc 23:108–109

Atherton SS (1992) Protection from retinal necrosis by passive transfer of monoclonal antibody specific for herpes simplex virus glycoprotein D. Curr Eye Res 11:45–52

Baczko K, Liebert UG, Billeter MA, Cattaneo R, Budka H, ter Meulen V (1986) Expression of defective measles virus genes in brain tissue of patients with subacute sclerosing panencephalitis. J Virol 59:472–478

Baer GM, Bellini WJ, Fishbein DB (1990) Rhabdoviruses. In: Fields BN, Knipe DM (eds) Virology, 2nd edn. Raven, New York, pp 883–930

Baldridge JR, Pearce BD, Parekh BS, Buchmeier MJ (1993) Teratogenic effects of neonatal arenavirus infection on the developing rat cerebellum are abrogated by passive immunotherapy. Virology 197:669–677

Bankamp B, Brinckmann UG, Reich A, Niewiesk S, ter Meulen V, Liebert UG (1991) Measles virus nucleocapsid protein protects rats from encephalitis. J Virol 65:1695–1700

Baron JL, Madri JA, Ruddli NH, Hashim G, Janeway CA (1993) Surface expression of a4 integrin by CD4T cells is required for their entry into brain parenchyma. J Exp Med 177:57–68

Barrett PN, Koschel K, Carter M, ter Meulen V (1985) Effect of measles virus antibodies on a measles SSPE virus persistently infected C6 rat glioma cell line. J Gen Virol 66:1411–1421

Bender B, Croghan T, Zhang L, Small P (1992) Transgenic mice lacking class I major histocompatibility complex-restricted T cells have delayed viral clearance and increased mortality after influenza virus challenge. J Exp Med 175:1143–1145

Billeter MA, Cattaneo R (1991) Molecular biology of defective measles viruses persisting in the human central nervous system. In: Kingsbury D (ed) The paramyxoviruses. Plenum, New York, pp 323–345

Blinzinger K, Simon J, Magrath D, Boulger L (1969) Poliovirus crystals within the endoplasmic reticulum of endothelial and mononuclear cells in the monkey spinal cord. Science 163:1336–1337

Borrow P, Nash AA (1992) Susceptibility to Theiler's virus-induced demyelinating disease correlates with astrocyte class II induction and antigen presentation. Immunology 76:133–139

Borrow P, Tonks P, Welsh CJR, Nash AA (1992) The role of $CD8^+$ T cells in the acute and chronic phases of Theiler's virus-induced disease in mice. J Gen Virol 73:1861–1865

Botteron C, Zurbriggen A, Griot C, Vandevelde M (1992) Canine distemper virus-immune complexes induce bystander degeneration of oligodendrocytes. Acta Neuropathol Berl 83:402–407

Brinckmann UG, Bankamp B, Reich A, ter Meulen V, Liebert UG (1991) Efficacy of individual measles virus structural proteins in the protection of rats from measles virus encephalitis. J Gen Virol 72: 2491–2500

Buchmeier M, Lewicki H, Talbot P, Knobler R (1984) Murine hepatitis virus-4 (strain JHM) induced neurologic disease is modulated in vivo by monoclonal antibody. Virology 132:261–270

Calenhoff MA, Faaberg KS, Lipton HS (1990) Genomic regions of neurovirulence and attenuation in Theiler's murine encephalomyelitis virus. Proc Natl Acad Sci USA 87:978–982

Campbell IL (1991) Cytokines in viral diseases. Curr Opin Immunol 3:486–491

Carrigan D, Knox KK (1990) Identification of interferon-resistant subpopulations in several strains of measles virus: positive selection by growth of the virus in brain tissue. J Virol 64:1606–1615

Cattaneo R, Rebmann G, Schmid A, Baczko K, ter Meulen V, Billeter M (1987) Altered transcription of a defective measles virus genome derived from a diseased human brain. EMBO J 6:681–687

Ceccaldi PE, Fillion MP, Ermine A, Tsiang H, Fillion G (1993) Rabies virus selectively alters 5-HT1 receptor subtypes in rat brain. Eur J Pharmacol 245:129–138

Chou J, Kern ER, Whitley RJ, Roizman B (1990) Mapping of herpes simplex virus-1 neurovirulence to $\gamma_1 34.5$, a gene nonessential for growth in culture. Science 250:1262–1266

Cosby SL, Macquaid S, Taylor MJ, Bailey M, Rima BK, Martin SJ, Allen IV (1989) Examination of eight cases of multiple sclerosis and 56 neurological and non-neurological controls for genomic sequences of measles virus. J Gen Virol 70:2027–2036

Day SP, Lausch RN, Oakes JE (1988) Evidence that the gene for herpes simplex virus type 1 DNA polymerase accounts for the capacity of an intertypic recombinant to spread from eye to central nervous system. Virology 163:166–173

Deshmane SJ, Fraser JW (1989) During latency, herpes simplex virus type I DNA is associated with nucleosomes in a chromatin structure. J Virol 63:943–947

Dietzschold B, Kao M, Zheng YM, Chen ZY, Maul G, Fu ZF, Rupprecht CE, Koprowski H (1992) Delineation of putative mechanisms involved in antibody-mediated clearance of rabies virus from the central nervous system. Proc Natl Acad Sci USA 89:7252–7256

Dörig RE, Marcil A, Chopra A, Richardson CD (1993) The human CD46 molecule is a receptor for measles virus (Edmonston strain). Cell 75:295–305

Dörries R, Liebert UG, ter Meulen V (1988) Comparative analysis of virus-specific antibodies and immunglobulins in serum and cerebrospinal fluid of subacute measles virus-induced encephalomyelitis (SAME) in rats and subacute sclerosing panencephalitis (SSPE). J Neuroimmunol 19:339–352

Dörries R, Schwender S, Imrich H, Harms H (1991) Population dynamics of lymphocyte subsets in the central nervous system of rats with different susceptibility to coronavirus-induced demyelinating encephalitis. Immunology 74:539–545

Fazakerley JK, Parker SE, Bloom F, Buchmeier MJ (1992) The V5A13.1 envelope glycoprotein deletion mutant of mouse hepatitis virus type-4 is neuroattenuated by its reduced rate of spread in the central nervous system. Virology 187:178–188

Felgenhauer K, Reiber H (1992) The diagnostic significance of antibody specificity indices in multiple sclerosis and herpes virus induced diseases of the nervous system. Clin Invest 70:28–37

Finke D, Brinckmann UG, ter Meulen V, Liebert UG (1995) Gamma interferon is a mediator of antiviral defense in experimental measles virus-induced encephalitis. J Virol 69:5469–5474

Finke D, Liebert UG (1994) $CD4^+$ T cells are essential in overcoming experimental murine measles encephalitis. Immunology 83:184–189

Flory E, Pfleiderer M, Stühler A, Wege H (1993) Induction of protective immunity against coronavirus-induced encephalomyelitis: evidence for an important role of $CD8^+$ T cells in vivo. Eur J Immunol 23:1757–1761

Fournier J-G, Tardieu M, Lebon P, Robain O, Rousot G, Rozenblatt S, Bouteille M (1985) Detection of measles virus RNA in lymphocytes from peripheral blood and brain perivascular infiltrates of patients with subacute sclerosing panencephalitis. N Engl J Med 313:910–915

Fraser NW, Block TM, Spivack JG (1992) The latency-associated transcripts of herpes simplex virus: RNA in search of function. Virology 191:1–8
Frei K, Malipiero UV, Leist TP, Zinkernagel RM, Schwab ME, Fontana A (1989) On the cellular source and function of interleukin-6 produced in the central nervous system in viral diseases. Eur J Immunol 19:689–694
Fujii N, Oguma K, Kimura K, Yamashita T, Ishida S, Fujinaga K, Yashiki T (1988) Oligo-2',5'-adenylate synthetase activity in K562 cell lines persistently infected with measles or mumps virus. J Gen Virol 69:2085–2091
Giulian D, Lachman LB (1985) Interleukin-1 stimulation of astroglial proliferation after brain injury. Science 228:497–499
Goodpasture E (1925) The axis cylinders of peripheral nerves as portals of entry to the central nervous sytem for the virus of herpes simplex in experimentally infected rabbits. Am J Pathol 1:11–28
Haase AT (1986) Pathogenesis of lentivirus infections. Nature 322:130–136
Halbach M, Koschel K (1979) Impairment of hormone dependent signal transfer by chronic SSPE virus infection. J Gen Virol 42:615–619
Herndon RM, Johnson RT, Davis LE, Descalzi LR (1974) Ependymitis in mumps virus meningitis: electron microscopic studies of cerebrospinal fluid. Arch Neurol 30:475–479
Hirano A (1992) Subacute sclerosing panencephalitis virus dominantly interferes with replication of wild-type measles virus in a mixed infection: implication for viral persistence. J Virol 66:1891–1898
Hofman FM, Hinton DR, Baemayr J, Weil M, Merrill JE (1991) Lymphokines and immunoregulatory molecules in subacute sclerosing panencephalitis. Clin Immunol Immunopathol 58:331–342
Hogle JM, Chow M, Filman DJ (1985) Three-dimensional structure of Poliovirus at 2.9 A resolution. Science 229:1358–1367
Hom RC, Finberg RW, Mullaney S, Ruprecht RM (1991) Protective cellular retroviral immunitiy requires both $CD4^+$ and $CD8^+$ T cells. J Virol 65:220–224
Horie H, Koike S, Kurata T, Sato-Yoshida Y, Ise I, Ota Y, Abe S, Hioki K, Kato H, Taya C, Nomura T, Hashizume S, Yonekawa H, Nomoto A (1994) Transgenic mice carrying the human poliovirus receptor: new animal model for study of poliovirus neurovirulence. J Virol 68:681–688
Imrich H, Schwender S, Hein A, Dörries R (1994) Cervical lymphoid tissue but not the central nervous system supports proliferation of virus-specific T lymphocytes during coronavirus-induced encephalitis in rats. J Neuroimmunol 53:73–81
Iwasaki Y, Liu D, Yamamoto T, Konno H (1985) On the replication and spread of rabies virus in the human central nervous system. J Neuropathol Exp Neurol 44:185–195
Johnson RT (1965) Virus invasion of the central nervous system. A study of Sindbis virus infection of the mouse using fluorescent antibody. Am J Pathol 46:929–943
Johnson RT (1982) Viral infections of the nervous system. Raven, New York
Jonjic S, Pavic I, Lucin P, Rukavina D, Koszinowski U (1990) Efficacious control of cytomegalovirus infection after long-term depletion of $CD8^+$ T lymphocytes. J Virol 64:5457–5464
Jordan CA, Watkins BA, Kufta C, Dubois-Dalque M (1991) Infection of brain microglial cells by human immunodeficiency virus type 1 is CD4 dependent. J Virol 65:736–742
Jubelt B, Ropka SL, Goldfarb S, Waltenbaugh C, Oates RP (1991) Susceptibility and resistance to poliovirus-induced paralysis of inbred mouse strains. J Virol 65:1035–1040
Karupiah R, Woodhams GCE, Blanden RV, Ramshaw IA (1991) Immunobiology of infection with recombinant vaccinia virus encoding murine IL-2. Mechanisms of rapid viral clearance in immunocompetent mice. J Immunol 147:4327–4332
Kitamura D, Roes J, Kühn R, Rajewsky K (1991) A B cell-deficient mouse by targeted disruption of the membrane exon of the immunoglobulin mu chain gene. Nature 350:423–426
Kitamura N, Semler BL, Rothberg PG, Larsen GR, Adler CJ, Dorner AJ, Emini EA, Hanecak R, Lee JJ, van der Werf S, Anderson CW, Wimmer E (1981) Primary structure, gene organization and polypeptide expression of poliovirus RNA. Nature 291:547–553
Koike S, Horie H, Ise I, Okitsu A, Yoshida N, Iizuka N, Takeuchi K, Tagegami T, Nomoto A (1990) The poliovirus receptor protein is produced both as membrane-bound and secreted forms. EMBO J 9:3217–3224
Körner H, Schliephake A, Winter J, Zimprich F, Lassmann H, Sedgwick J, Siddell S, Wege H (1991) Nucleocapsid or spike protein-specific $CD4^+$ T lymphocytes protect against coronavirus-induced encephalomyelitis in the absence of $CD8^+$ T cells. J Immunol 147:2317–2323
Koschel K, Münzel P (1980) Persistent paramyxovirus infections and behaviour of β-adrenergic receptors in C6 rat glioma cells. J Gen Virol 47:513–517

Kristensson K, Lycke E, Ryotta M, Svennerholm B, Vahlne A (1986) Neuritic transport of herpes simplex virus in rat sensory neurons in vitro. Effects of substances interacting with microtubular function and axonal flow (nocodazde, taxol and erythro-9-3(2-hydroxynonyl)adenine). J Gen Virol 67:2023–2028

Ladogana A, Bouzamondo E, Pochiari M, Tsiang H (1994) Modification of tritiated γ-amino-n-butyric acid transport in rabies virus-infected primary cortical cultures. J Gen Virol 75:623–627

La Monica N, Almond JW, Racaniello VR (1987) A mouse model for poliovirus neurovirulence identifies mutations that attenuate the virus for humans. J Virol 61:2917–2920

Levine B, Griffin DE (1992) Persistence of viral RNA in mouse brains after recovery from acute alphavirus encephalitis. J Virol 66:6429–6435

Levine B, Hardwick JM, Trapp BD, Crawford TO, Bollinger RC, Griffin DE (1991) Antibody-mediated clearance of alphavirus infection from neurons. Science 254:856–860

Levine B, Griffin DE (1993) Molecular analysis of neurovirulent strains of Sindbis virus that evolve during persistent infection of scid mice. J Virol 67:6872–6875

Lieberman AP, Pitha PM, Shin HS, Shin ML (1989) Production of tumor necrosis factor and other cytokines by astrocytes stimulated with lipopolysaccharide or a neurotropic virus. Proc Natl Acad Sci USA 86:6348–6352

Liebert UG (1997) Measles virus infections of the central nervous system. Intervirology 40:176–184

Liebert UG, Finke D (1995) Measles virus in rodents. Curr Top Microbiol Immunol 191:149–166

Liebert UG, ter Meulen V (1987) Virological aspects of measles virus-induced encephalomyelitis in Lewis and BN rats. J Gen Virol 68:1715–1722

Liebert UG, ter Meulen V (1993) Synergistic interaction between measles virus infection and MBP peptide-specific T cells in the induction of EAE in Lewis rats. J Neuroimmunol 46:217–224

Liebert UG, Linington C, ter Meulen V (1988) Induction of autoimmune reactions to myelin basic protein in measles virus encephalitis in Lewis rats. J Neuroimmunol 17:103–118

Liebert UG, Hashim GA, ter Meulen V (1990a) Characterization of measles virus-induced cellular autoimmune reactions against myelin basic protein in Lewis rats. J Neuroimmunol 29:139–147

Liebert UG, Schneider-Schaulies S, Baczko K, ter Meulen V (1990b) Antibody-induced restriction of viral gene expression in measles encephalitis in rats. J Virol 64:706–713

Liebert UG, Reich A, Bankamp B, Brinckmann UG, ter Meulen V (1993) Control of measles virus infections by virus-specific CD4$^+$ T cells. In: Thomas DB (ed) Viruses and cellular immuneresponses. Marcel Dekker, New York, pp 279–291

Liebert UG, Flanagan SG, Löffler S, Baczko K, ter Meulen V, Rima BK (1994) Antigenic determinants of measles virus hemagglutinin associated with neurovirulence. J Virol 68:1486–1493

Lindsley MD, Thiemann R, Rodriguez M (1991) Cytotoxic T cells isolated from the central nervous system of mice infected with Theiler's virus. J Virol 65:6612–6620

Liu C, Voth D, Rodina P, Shauf L, Gonzalez G (1970) A comparative study of the pathogenesis of western equine and eastern equine encephalomyelitis virus infections in mice by intracerebral and subcutaneous inoculations. J Infect Dis 122:53–63

Lucchiari MA, Modolell M, Eichmann K, Pereira CA (1993) In vivo depletion of interferon-gamma leads to susceptibility of A/J mice to mouse hepatitis virus 3 infection. Immunobiology 185:475–482

Lycke E, Tsiang H (1987) Rabies virus infection of cultured rat sensory neurons. J Virol 61:2733–2741

Lycke E, Kristensson K, Svenerholm B, Vahlne A, Ziegler R (1984) Uptake and transport of herpes simplex virus in neurites of rat dorsal root ganglia in culture. J Gen Virol 65:55–64

Malvoisin E, Wild F (1990) Contribution of measles virus fusion protein in protective immunity: anti-F monoclonal antibody neutralize virus infectivity and protect mice against challenge. J Virol 64:5160–5162

Martin X, Dolivo M (1983) Neuronal and transneuronal tracing in the trigeminal system of the rat using the herpes virus suis. Brain Res 273:253–276

Massa PT, Dörries R, ter Meulen V (1986) Viral particles induce Ia antigen expression on astrocytes. Nature 320:543–546

Massa PT, Schimpl A, Wecker E, ter Meulen V (1987) Tumor necrosis factor amplifies measles virus-mediated Ia induction on astrocytes. Proc Natl Acad Sci USA 84:7242–7245

Massanari RM, Paterson PY, Lipton HL (1979) Potentiation of experimental allergic encephalomyelitis in hamsters with persistent encephalitis due to measles virus. J Infect Dis 139:297–303

McAllister A, Tangy F, Aubert C, Brahic M (1990) Genetic mapping of the ability of Theiler's virus to persist and demyelinate. J Virol 64:4252–4257

Melnick JL (1996) Enteroviruses: polioviruses, Coxsackieviruses, echoviruses and newer enteroviruses. In: Fields BN, Knipe DM, Howley PM, et al. (eds) Virology, 3rd edn. Lippincott-Raven, Philadelphia-New York, pp 655–712

Melvold RW, Jokinen DM, Miller SD, Dal Canto MC, Lipton HL (1990) Identification of a locus on chromosome 3 involved in differential susceptibility to Theiler's murine encephalomyelitis virus-induced demyelinating disease. J Virol 64:686–690

Mendelsohn C, Wimmer E, Rancianello V (1989) Cellular receptor for poliovirus: molecular cloning, nucleotide sequence, and expression of a new member of the immunoglobulin superfamily. Cell 56:855–865

Merrill JE (1991) Effects of interleukin-1 and tumor necrosis factor-a on astrocytes, microglia, oligodendroctes, and glial precursor in vitro. Dev Neurosci 13:130–137

Merrill JE, Chen ISY (1991) HIV-1, macrophages, glial cells, and cytokines in AIDS nervous system disease. FASEB J 5:2391–2397

Momburg F, Koch N, Möller P, Moldenhauer G, Hämmerling GJ (1986) In vivo induction of H-2K/D antigens by recombinant interferon-γ. Eur J Immunol 16:551–557

Moskophidis D, Cobbold P, Waldmann H, Lehmann-Grube F (1987) Mechanism of recovery from acute virus infection: treatment of lymphocytic choriomeningitis virus-infected mice with monoclonal antibodies reveals that Lyt-2$^+$ T lymphocytes mediate clearance of virus and regulate the antiviral antibody response. J Virol 61:1867–1874

Moskophidis D, Frei K, Löhler J, Fontana A, Zinkernagel RM (1991) Production of random classes of immunoglobulins in brain tissue during persistent viral infection paralleled by secretion of interleukin-6 (IL-6) but not IL-4, IL-5, and gamma interferon. J Virol 65:1364–1369

Muller D, Koller BH, Whitton JL, Lapan KE, Brigman KK, Frelinger JA (1992) LCMV-specific, class II-restricted cytotoxic T cells in β_2-microglobulin-deficient mice. Science 255:1576–1578

Murphy FA, Bauer SP (1974) Early street rabies virus infection in striated muscle and later progression to the central nervous system. Intervirology 3:256–268

Naniche D, Varior-Krishnan G, Cervoni F, Wild F, Rossi B, Rabourdin-Combe C, Gerlier D (1993) Human membrane cofactor protein (CD46) acts as a cellular receptor for measles virus. J Virol 67:6025–6032

Narayan O, Clements JE (1989) Biology and pathogenesis of lentiviruses. J Gen Virol 70:1617–1639

Nash AA (1991) Virological and pathological processes involved in Theiler's virus infection of the central nervous system. Semin Neurosci 3:109–116

Nash AA, Jayasuriya A, Phelan J, Cobbold SP, Waldmann H, Prospero T (1987) Different roles for L3T4$^+$ and Lyt 2$^+$ T cell subsets in the control of an acute herpes simplex virus infection of the skin and nervous system. J Gen Virol 68:825–833

Nathanson N, Gonzales-Scarano F (1991) The natural history of rabies virus, 2nd edn. In: Baer GM (ed) CRC Press, Boston, pp 145–161

Neumann H, Schmidt H, Cavalié A, Jenne D, Wekerle H (1997) Major histocompatibility complex class I gene expression in single neurones of the central nervous system: Differential regulation by interferon-γ and tumor necrosis factor-α. J Exp Med 185:305–316

Niewiesk S, Brinckmann UG, Bankamp B, Sirak S, ter Meulen V, Liebert UG (1993) Susceptibility to measles-induced encephalitis in mice correlates with impaired antigen presentation to cytotoxic T lymphocytes. J Virol 67:75–81

Oldstone MBA, Blount P, Souther PJ, Lampert PW (1986) Cytoimmunotherapy for persistent virus infection reveales an unique clearance pattern from the central nervous system. Nature 321:239

Pathak S, Webb HE (1974) Possible mechanism for the transport of Semliki Forest virus into and within the mouse brain: an electron microscopic study. J Neurol Sci 23:175–184

Patick AK, Lindsley MD, Rodriguez M (1990) Differential pathogenesis between mouse strains resistant and susceptible to Theiler's virus-induced demyelination. Semin Virol 1:281–288

Perlman S, Jacobsen G, Afifi A (1989) Spread of a neurotropic murine coronavirus into the CNS via the trigeminal and olfactory nerves. Virology 170:556–560

Perlman S, Schelper R, Bolger E, Ries D (1987) Late onset, symptomatic, demyelinating encephalomyelitis in mice infected with MHV-JHM in the presence of maternal antibody. Microb Pathog 2:185–194

Pevear DC, Borkowski J, Calenhoff M, Ohn CK, Ostrowski B, Lipton HL (1988) Insights into Theiler's virus neurovirulence based on a genomic comparison of the neurovirulent GDVII and less virulent BeAn strains. Virology 165:253–259

Pevear DC, Calenhoff M, Rozhon R, Lipton HL (1987) Analysis of the complete nucleotide sequence of the picornavirus Theiler's murine encephalomyelitis virus indicates that it is closely related to cardioviruses. J Virol 61:1507–1516

Plata-Salaman CR (1991) Immunoregulators in the central nervous system. Neurosci Biobehav Rev 15:185–215

Pullen LC, Miller SD, Dal Canto MC, Kim BS (1993) Class I-deficient resistant mice intracerebrally inoculated with Theiler's virus show an increased T cell response to viral antigens and susceptibility to demyelination. Eur J Immunol 23:2287–2293

Rammohan KW, Dubois-Dalcq M, Rentier B, Paul J (1983) Experimental models to study measles virus persistence in the nervous system. Prog Neuropathol 5:343–372

Reich A, Erlwein O, Niewiesk S, ter Meulen V, Liebert UG (1992) $CD4^+$ T cells control measles virus infection of the central nervous system. Immunology 76:185–191

Ren R, Costantini F, Gorgacz EJ, Lee JJ, Rancaniello VR (1990) Transgenic mice expressing a human poliovirus receptor: a new model for poliomyelitis. Cell 63:353–362

Rima BK, Duffy N, Mitchell WJ, Summers BA, Appel MJ (1991) Correlation between humoral immune responses and presence of virus in the CNS in dogs experimentally infected with canine distemper virus. Arch Virol 121:1–8

Robertson MN, Spangrude GJ, Hasenkrug K, Perry L, Nishio J, Wehrly K, Chesbro B (1992) Role and specificity of T-cell subsets in spontaneous recovery from Friend virus-induced leukemia in mice. J Virol 66:3271–3277

Rock DL, Fraser NW (1985) Latent herpes simplex virus type 1 DNA contains two copies of the virion DNA joint region. J Virol 55:849–852

Rodriguez M, Leibowitz JL, David CS (1986) Susceptibility to Theiler's virus-induced demyelination. Mapping of the gene within the H-2D region. J Exp Med 163:620–631

Sabin AB (1956) Pathogenesis of poliomyelitis. Reappraisal in the light of new data. Science 123:1151–1157

Sabin AB, Boulger LR (1973) History of Sabin attenuated poliovirus oral live vaccine strains. J Biol Stand 1:115–118

Saha K, Hollowell D, Wong PK (1994) Mother-to-baby transfer of humoral immunity against retrovirus-induced neurologic disorders and immunodeficiency. Virology 198:129–137

Sauder C, Hallensleben W, Pagenstecher A, Schneckenburger S, Biro L, Pertlik D, Hausmann J, Suter M, Staeheli P (2000) Chemokine gene expression in astrocytes of Borna disease virus-infected rats and mice in the absence of inflammation. J Virol 74:9267–9280

Sawtell NM, Thompson RL (1992) Herpes simplex virus type 1 latency-associated transcription unit promotes anatomical site-dependent establishment and reactivation from latency. J Virol 66:2157–2169

Schijns VE, van der Neut R, Haagmans BL, Bar DR, Schellekens H, Horzinek MC (1991) Tumour necrosis factor-α, interferon-γ and interferon-β exert antiviral activity in nervous tissue cells. J Gen Virol 72:809–815

Schneider-Schaulies J, Schneider-Schaulies S, ter Meulen V (1993) Differential induction of cytokines by primary and persistent MV-infection in human glial cells. Virology 195:219–228

Schneider-Schaulies J, Dunster LM, Schwartz-Albiez R, Krohne G, ter Meulen V (1995) Physical association of moesin and CD46 as a receptor complex for measles virus. J Virol 69:2248–2256

Schneider-Schaulies S, Liebert UG (1991) Pathogenetic aspects of persistent measles virus infection in brain tissue. Semin Neurosci 3:149–156

Schneider-Schaulies S, Liebert UG, Baczko K, Cattaneo R, Billeter M, ter Meulen V (1989) Restriction of measles virus gene expression in acute and subacute encephalitis of Lewis rats. Virology 171:525–534

Schneider-Schaulies S, Liebert UG, Baczko K, ter Meulen V (1990) Restricted expression of measles virus in primary rat astroglial cells. Virology 177:802–806

Schneider-Schaulies S, Liebert UG, Segev Y, Rager-Zisman B, Wolfson M, ter Meulen V (1992) Antibody-dependent transcriptional regulation of measles virus in persistently infected neural cells. J Virol 66:5534–5541

Schneider-Schaulies S, Schneider-Schaulies J, Bayer M, Löffler S, ter Meulen V (1993) Spontaneous and differentiation-dependent regulation of measles virus expression in human glial cells. J Virol 67:3375–3383

Schneider-Schaulies S, Schneider-Schaulies J, Schuster A, Bayer M, Pavlovic J, ter Meulen V (1994) Cell type specific MxA-mediated inhibition of measles virus transcription in human brain cells. J Virol 68:6910–6917

Schneider-Schaulies S, Schneider-Schaulies J, Dunster LM, ter Meulen V (1995) Measles virus gene expression in neural cells. Curr Topics Microbiol Immunol 191:101–116

Schnorr JJ, Schneider-Schaulies S, Simon-Jödicke A, Pavlovic J, Horisberger MA, ter Meulen V (1993) MxA-dependent inhibition of measles virus glycoprotein synthesis in a stably transfected human monocytic cell line. J Virol 67:4760–4768

Schwender S, Imrich H, Dörries R (1991) The pathogenic role of virus-specific antibody-secreting cells in the central nervous system of rats with different susceptibility to coronavirus-induced demyelinating encephalitis. Immunology 74:533–538

Sedarati F, Margolis TP, Stevens JG (1993) Latent infection can be established with drastically restricted transcription and replication of the HSV-1 genome. Virology 192:687–691

Sedgwick JD, Dörries R (1991) The immune system response to viral infection of the CNS. Semin Neurosci 3:93–100

Sedgwick J, Schwender S, Imrich H, Dörries R, ter Meulen V, Butcher GW (1991a) Isolation and direct characterization of resident microglia cells from the normal and inflamed central nervous system. Proc Natl Acad Sci USA 88:7438–7442

Sedgwick JD, Mößner R, Schwender S, ter Meulen V (1991b) MHC-expressing non-hematopoietic astroglial cells prime only $CD8^+$ T lymphocytes: astroglial cells as perpetuators but not initiators of $CD4^+$ T cell responses in the central nervous system. J Exp Med 173:1235–1246

Sedgwick JD, Schwender S, Gregersen R, ter Meulen V (1993) Resident macrophages (ramified microglia) of the adult brown Norway rat central nervous system are constitutively major histocompatibility complex class II positive. J Exp Med 177:1145–1152

Shankar V, Kao M, Hamir AN, Sheng H, Koprowski H, Dietzschold B (1992) Kinetics of virus spread and changes in levels of several cytokine mRNAs in the brain after intranasal infection of rats with Borna disease virus. J Virol 66:992–998

Shepley MP, Rancaniello VR (1994) A mouse antibody that blocks poliovirus attachment recognizes the lymphocyte homing receptor CD44. J Virol 68:1301–1308

Sissons JGP, Oldstone MBA (1985) Host response to viral infections. In: Fields BN (ed) Virology. Raven, New York, pp 265–279

Smith TW, De Girolami U, Hickey WF (1992) Neuropathology of immunosuppression. Brain Pathol 2:183–194

Sonnerborg AB, von Sydow MA, Forsgren M, Strannegard OO (1989) Association between intrathecal anti-HIV-1 immunoglobulin G synthesis and occurrence of HIV-1 in cerebrospinal fluid. AIDS 3:701–705

Stevens JG (1991) Herpes simplex virus: neuroinvasiveness, neurovirulence and latency. Semin Neurosci 3:141–147

Stevens JG (1993) HSV-1 neuroinvasiveness. Intervirology 35:152–163

Swarz JR, Brooks BR, Johnson RT (1981) Spongiform poliencephalomyelopathy caused by a murine retrovirus. II. Ultrastructural localization of virus replication and spongiform changes in the central nervous system. Neuropathol Appl Neurobiol 7:365–380

Tas PW, Koschel K (1991) Loss of the endothelin signal pathway in C6 rat glioma cells persistently infected with measles virus. Proc Natl Acad Sci USA 88:6736–6739

ter Meulen V, Stephenson JR, Kreth HW (1983) Subacute sclerosing panencephalitis. In: Fraenkel-Conrat H, Wagner RR (eds) Comprehensive virology, vol 18. pp 105–159

Thompson RL, Rogers SK, Zerhusen MA (1989) Herpes simplex virus neurovirulence and productive infection of neutral cells is associated with a function which maps between 0.82 and 0.832 map units on the HSV genome. Virology 172:435–450

Toyoda H, Kohara M, Kataoka Y, Suganuma T, Omata T, Imura N, Nomoto A (1984) Complete nucleotide sequences of all three poliovirus serotype genomes. Implication for genetic relationship, gene function and antigenic determinants. J Mol Biol 174:561–585

Tsiang H (1993) Pathophysiology of rabies virus infection of the nervous system. Adv Virus Res 42:375–411

Tyler KL, Virgin HW 4th, Bassel Duby R, Fields BN (1989) Antibody inhibits defined stages in the pathogenesis of reovirus serotype 3 infection of the central nervous system. J Exp Med 170:887–900

Tyor WR, Griffin DE (1993) Virus specificity and isotype expression of intraparenchymal antibody-secreting cells during Sindbis virus encephalitis in mice. J Neuroimmunol 48:37–44

Tyor WR, Wesselingh S, Levine B, Griffin DE (1992) Long term intraparenchymal Ig secretion after acute viral encephalitis in mice. J Immunol 149:4016–4020

Ward BJ, Griffin DE (1993) Changes in cytokine production after measles virus vaccination: predominant production of IL-4 suggests induction of a Th2 response. Clin Immunol Immunopathol 67:171–177

Watanabe R, Wege H, ter Meulen V (1983) Adoptive transfer of EAE-like lesions by BMP-stimulated lymphocytes from rats with coronavirus-induced demyelinating encephalomyelitis. Nature 305:150–153

Wege H, Schliephake A, Körner H, Flory E, Wege H (1993) An immunodominant CD4+ T cell site on the nucleocapsid protein of murine coronavirus contributes to protection against encephalomyelitis. J Gen Virol 74:1287–1294

Weinmann-Dorsch C, Koschel K (1990) Coupling of viral membrane proteins to phosphatidyl-inositide signalling system. FEBS L 247:185–188

Welsh CJ, Tonks P, Nash AA, Blakemore WF (1987) The effect of L3T4 T cell depletion on the pathogenesis of Theiler's murine encephalomyelitis virus infection in CBA mice. J Gen Virol 68:1659–1667

Williamson J, Stohlmann S (1990) Effective clearance of mouse hepatitis virus from the central nervous system both requires CD4$^+$ and CD8$^+$ T cells. J Virol 64:4589–4592

Wolinsky JS, Klassen T, Baringer JR (1976) Persistence of neuroadapted mumps virus in brains of newborn hamsters after intraperitoneal inoculation. J Infect Dis 133:260–267

Wright KE, Buchmeier MJ (1991) Antiviral antibodies attenuate T-cell-mediated immunopathology following acute lymphocytic choriomeningitis virus infection. J Virol 65:3001–3006

Yokomori K, Baker SC, Stohlman SA, Lai MM (1992) Hemagglutinin-esterase-specific monoclonal antibodies alter the neuropathogenicity of mouse hepatitis virus. J Virol 66:2865–2874

Zurbriggen A, Fujinami RS (1988) Theiler's virus infection in nude mice: viral RNA in vascular endothelial cells. J Virol 62:3589–3596

Neuronal Latency in Human and Animal Herpesvirus Infections

K. BORCHERS[1] and H.J. FIELD[2]

1	Introduction	62
1.1	Definition of Terms	62
1.1.1	Latency	62
1.1.2	Recurrence and Recrudescence	62
1.1.3	Abortive Infection	62
1.1.4	Neurotropic and Neuropathogenic	63
1.2	Intention of the Review	63
2	Herpes Simplex Virus – A Paradigm for Classical Latency and Recurrence	63
2.1	Methods and Model Systems for the Study of HSV Infections	64
2.1.1	Methods for Detection of Latency in Animal Models and in Humans	64
2.1.1.1	In Vivo Reactivation	64
2.1.1.2	Reactivation by Culture or Cocultivation of Peripheral Ganglion Tissue	64
2.1.1.3	Immunohistochemistry	65
2.1.1.4	Detection of Viral Nucleic Acids	65
2.1.1.5	Defining the Genomic Conformation	65
2.1.1.6	Reporter Gene Detection	65
2.1.1.7	Detection of Latency-Associated Transcripts	66
2.1.1.8	Polymerase Chain Reaction	66
2.1.1.9	In Situ Polymerase Chain Reaction	66
2.1.2	Model Systems for the Study of HSV Latency	66
2.1.2.1	A Murine Animal Model Helps to Explain the Mechanisms of Latency	66
2.1.2.2	Further Animal Models for Latent HSV Are Developed	67
2.1.2.3	In Vitro Models for Latency	68
2.2	Herpes Simplex Virus Adaptation to the Host	68
2.3	Establishment of Latency	69
2.4	Transcriptional Activity During Latency	71
2.5	Cellular and Immunological Factors Implicated in the Establishment/Maintenance of Herpes Simplex Virus Latency	72
2.6	Reactivation from Latency	73
2.7	Latent Herpes Simplex Virus in the CNS of Animals and Humans	73
2.8	The Biological Consequences of Latent Herpes Simplex Virus in the CNS	74
2.9	The Prospects for Curing or Preventing Latency	75
3	Other Human Herpesvirus Infections	76
3.1	Varicella-Zoster Virus	76
3.2	Human Cytomegalovirus	76
3.3	Human Herpesvirus Type 6	77
4	Animal Herpesviruses	77
4.1	Herpesviruses of Nonhuman Primates: Cercopithecus Virus-1 or B Virus	78
4.2	Porcine Herpesvirus: Pseudorabies Virus	79
4.3	Equine Herpesviruses	80

[1] Institut für Virologie, Freie Universität Berlin, Königin-Luise-Strasse 49, 14195 Berlin, Germany
e-mail: borchers@zedat.fu-berlin.de
[2] Centre for Veterinary Science, Cambridge University, Madingley Road, Cambridge, CB3 0ES, UK

4.3.1	Equine Herpesvirus-1	80
4.3.2	Equine Herpesvirus-4	81
4.3.3	Equine Herpesvirus-2	83
4.4	Bovine Herpesviruses	83
4.4.1	Bovine Herpesvirus-1	83
4.4.2	Bovine Herpesvirus-2	84
4.5	Canine Herpesvirus-1 and Feline Herpesvirus-1	85
4.6	Marek's Disease Virus	85
5	The Role of Latency in Neurological Disease Caused by Members of Virus Families Other than Herpes	86
6	Conclusions	87
References		88

1 Introduction

1.1 Definition of Terms

1.1.1 Latency

"Latency is a reversibly non-productive infection of a cell by a replication-competent virus" (GARCIA-BLANCO and CULLEN 1991). Latently infected tissues contain virus that can be recovered by cocultivation, contain viral DNA and, possibly, latency-associated transcripts (LATs) but in general no viral antigen. Alternatively, latency may be defined as: "The continued presence of virus in a host in which no infectious virus can be detected but is, however, capable of reactivation to produce infectious virus" (WILDY et al. 1982).

These conditions thus distinguish 'latency' from 'persistent' infections in which there is a constant production of infectious virus, although the host may show no clinical signs for long periods. Importantly, these definitions also distinguish latency from the situation in which incomplete virus genome or other virus components persist in a form that has no potential to reactivate.

1.1.2 Recurrence and Recrudescence

We use the term recurrence to mean the shedding of infectious virus following reactivation; recurrence may be subclinical or associated with clinical signs.

Recrudescence we take to be the clinical signs associated with recurrence.
The above terms are also the definitions proposed by WILDY et al. (1982).

1.1.3 Abortive Infection

This is "an irreversibly non-productive infection" (GARCIA-BLANCO and CULLEN 1991) and should be distinguished from latency if it results in long-term retention of virus proteins and/or nucleic acids which are without the ability to reactivate infectious virus.

1.1.4 Neurotropic and Neuropathogenic

We make an important distinction between these terms. The former implies that the virus has a predilection for neural tissue wherein it may or may not cause damage. The term neuropathogenic, however, implies that infection of neural tissue results in neuropathology.

1.2 Intention of the Review

Our objective is to review the different strategies employed by viruses to achieve long-term infection of the host within neural tissues which, in some circumstances, may damage the nervous system either directly or indirectly. The most important examples are provided by members of the Alphaherpesvirinae, but we will attempt to consider other viruses in this survey. For each virus we will consider the pathogenesis of acute infection, the mechanisms by which latency is established and maintained and the possible consequences for the host.

2 Herpes Simplex Virus – A Paradigm for Classical Latency and Recurrence

Herpes simplex virus (HSV) is a paradigm for classical virus latency and is also one of the most familiar examples of a latent virus. The virus is extremely wide-spread and the majority of adults world-wide carry the infection. Two distinct types of HSV are recognised, HSV-1 and HSV-2, that may be distinguished on the basis of genotypic and phenotypic characteristics. HSV-1 and HSV-2 are usually (but not exclusively) associated with orofacial and genital herpes, respectively. HSV-1 is among the most common causes of herpes encephalitis (DENNETT et al. 1997), whereas HSV-2 encephalitis cases are relatively rare and mainly restricted to newborns (CRAIG and NAHMIAS 1973). It appears that the pattern of neurological latency with reactivation leading to recurrent virus shedding (which may or may not be accompanied by clinical signs of recurrent disease) is central to the biology of both HSV-1 and HSV-2. The same applies to varicella-zoster virus (VZV), although, in this case, the primary infection (chickenpox) is followed by shingles, which may occur as a single event several decades later. For both these human viruses and many of their animal counterparts (see sections below), periods of latency followed by reactivation and recurrence are important in perpetuating the virus in the host population. As a consequence of latency, the infection can thus be maintained in a relatively small, isolated population. The fact that a variety of stimuli including trauma, sunlight (UV light), menstruation, and other forms of stress are sometimes accompanied by the development of cold sores is common knowledge. The extent of human disease caused by cold sores and recurrent genital herpes is evidenced by the large market that exists for herpes antiviral agents, especially acyclovir, which has been reported to have been used in some 35 million patients since its introduction in the early 1980s. In some cases individuals have taken the drug prophylactically for many years in order to suppress the development of their troublesome recurrent lesions (SPRUANCE 1993).

2.1 Methods and Model Systems for the Study of HSV Infections

2.1.1 Methods for Detection of Latency in Animal Models and in Humans

2.1.1.1 In Vivo Reactivation

Following experimental inoculation of mice with HSV-1 or HSV-2 at a peripheral site, the acute infection typically lasts for 1–2 weeks during which time active virus replication is readily detected. Infectious virus is promptly cleared from the tissues of survivors, and the explantation, homogenisation and testing of tissues for infectious virus is uniformly negative thereafter. The only truly definitive test for latent HSV in such an animal model is to demonstrate infectious virus. This may be achieved by applying a suitable reactivation stimulus to the living animal and testing peripheral tissues for the presence of virions over the next few days with positive results. In practice this is difficult to achieve reproducibly in mice although ocular infections in rabbits and guinea pig genital infections readily undergo experimental reactivation (reviewed in FIELD and BROWN 1989). Alternatively, the experimental animals may be killed and the relevant tissues, i.e. the peripheral ganglia that innervate the original site of experimental inoculation, explanted.

2.1.1.2 Reactivation by Culture or Cocultivation
       of Peripheral Ganglion Tissue

The classical method to recover virus is to explant latently infected tissues (usually peripheral nerve ganglia) and maintain them in culture, with or without indicator cells, for periods ranging from a few days up to several weeks. For long-term culture, the media is replenished every few days and fresh cells added. We have found this method to be sensitive, but in some cases 70-day culture, including 14 cell passages, were required to produce positive results (THACKRAY and FIELD 1998) and the possibility of contamination or cross-infection is greater.

At the time of explantation no infectious virus or virus antigen can be detected following homogenisation of the latently infected tissue. For small animals (e.g. mice) the ganglia are small enough to be cultured whole, for other species, fragments are usually used. The homogenised explant tissue or culture fluid is tested for the presence of infectious virus. Although this test may be less sensitive than the molecular methods below (and a negative result may be questionable), a positive result is the gold standard for the detection of latency. A refinement of this method is to digest the tissue using proteolytic enzymes to produce a single cell suspension that may be cocultivated with susceptible cells, producing infectious centres (LEIB et al. 1989a).

Alternatively, the ganglia are simply maintained in culture medium at 37°C for a shorter period (e.g. 3–5 days), then homogenised and tested. In our hands (THACKRAY and FIELD 1998) this method can give positive results in 100% of mice; however it is among the least sensitive methods and poorly quantitative.

2.1.1.3 Immunohistochemistry

Latently infected tissues are snap-frozen in liquid N_2 and frozen sections prepared by means of cryostat or, alternatively, tissues may be fixed and paraffin-embedded. Classical methods for the detection of virus proteins are then applied generally using monoclonal antibodies to one or more virus products. The bound antibodies are then revealed by means of a chromophore or enzyme tag. Generally these methods have given negative results for sections of latently infected tissue, which is consistent with the paucity of HSV transcription during latency (FRASER et al. 1992; BLOCK and HILL 1997).

2.1.1.4 Detection of Viral Nucleic Acids

Herpes simplex virus DNA can readily be detected by means of Southern blot hybridisation on extracted high-molecular-weight DNA and, more recently, by means of polymerase chain reaction (PCR) techniques. In situ hybridisation is successful in tissue sections in detecting the transcriptional products, which are produced in high copy number during latency (LATs, see below). This method may, however, not be sufficiently sensitive to detect latent genome DNA.

2.1.1.5 Defining the Genomic Conformation

Whole cell DNA may be extracted from latently infected tissue and digested with an endonuclease which produces a known cleavage pattern for HSV. The products are then subjected to electrophoresis, blotted onto nitrocellulose (or similar) then hybridised to a known HSV sequence (ROCK and FRASER 1985). Using a probe derived from the junction between the unique and repeated regions of HSV DNA, this method revealed the absence of terminae in latently infected ganglia from both murine and human material (EFSTATHIOU et al. 1986). The virus DNA is not integrated into the host cell genome (MELLERICK and FRASER 1997) but organised in a structure similar to host nuclear chromatin (DESHMANE and FRASER 1989; see Sect. 2.3 below). The method is, however, poorly quantitative.

2.1.1.6 Reporter Gene Detection

An elegant method which provides a new tool for studying latency is enabled by the construction of recombinant viruses containing a suitable reporter gene. For example, the *lacZ* reporter gene has been placed under the HSV-1 LAT promoter (LACHMANN and EFSTATHIOU 1997). X-gal was then used to identify the latently infected cells in tissue sections. In this case the product of the virus expression was exported from the nucleus and blue staining was found in both nucleus and cytoplasm of many neurons. The gene expression was shown to continue for many months and forms a very useful quantitative method for direct detection of latently infected cells.

2.1.1.7 Detection of Latency-Associated Transcripts

Latency-associated transcripts may be detected directly by means of in situ hybridisation using a suitably labelled sense probe. The most sensitive probe is the major LAT product radioactively labelled with ^{3}H, although, currently, the most popular label is digoxigenin, with positive cells showing a characteristic punctate distribution of the brown stain within the neuronal nuclei (ROCK et al. 1987). More sensitive, however, is reverse transcription (RT)-PCR applied to RNA extracted from latently infected tissues (KRAMER and COEN 1995).

2.1.1.8 Polymerase Chain Reaction

This is undoubtedly the most sensitive method for the detection of HSV DNA sequences in ganglia. However, care is required in interpretation of the results. In some cases the products detected, not withstanding their validity as HSV sequences, may not be a true indication of latency as defined above, i.e. for whatever reason, they may not be capable of reactivation to produce infectious virus.

2.1.1.9 In Situ Polymerase Chain Reaction

This is a fickle method (for review see O'LEARY et al. 1996) but has great potential for the detection of HSV DNA-containing cells with much higher sensitivity than LAT (MEHTA et al. 1995). The quantitative results generated using this method are both extremely interesting and quite controversial. The general steps of this technique are summarised in Fig. 1.

2.1.2 Model Systems for the Study of HSV Latency

2.1.2.1 A Murine Animal Model Helps to Explain the Mechanisms of Latency

Our current understanding of the mechanisms of HSV latency is based on the study of animal infection models. These originate from the early part of this century (GOODPASTURE and TEAGUE 1923) and predate the isolation and characterisation of the virus itself. However, the most important single step forward occurred with the publication, in 1971, of the observations of STEVENS and COOK concerning the infection of mice with HSV-1. After inoculation of the mouse foot pad, virus

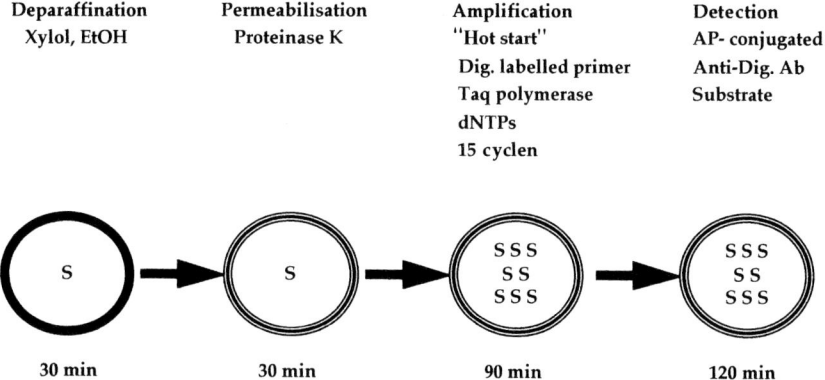

Fig. 1. Steps of the direct in situ PCR

replication was detected in the skin at the inoculation site and a few days later in the sensory ganglia innervating the skin of the foot. After a few more days the virus was cleared from the tissues of surviving mice and neither foot pad nor ganglia could be shown to contain infectious virus or virus antigens. However, when the sacral ganglia that innervated the foot were explanted and maintained in culture for several weeks, infectious virus was released into the culture medium, thus proving that the mice had been latently infected in a manner analogous to that which occurs in humans following natural exposure to HSV. The production of infectious virus from tissue provides unequivocal evidence that the intact virus had been maintained in the cells for many months or years.

Thus, latently infected murine ganglia provided a direct means to study neuronal latency. As described above, in the early 1970s a valuable in vivo murine infection model was established in which sensory ganglia containing foci of latently infected cells were available for manipulation and study. It is perhaps all the more surprising, therefore, that more than 25 years on there are still a number of important questions regarding latency and recurrence that remain to be answered.

2.1.2.2 Further Animal Models for Latent HSV Are Developed

Following the proof that the human virus, HSV-1, would establish latent infections in mice, it was shown that other routes of inoculation could be used such as the snout, neck, ear pinna, eye or genital tract and other species were also exploited. Another reproducible model of HSV-1 latency with reactivation was established in rabbits by means of corneal infection. In this case, reactivation was achieved by means of iontophoresis (KWON et al. 1981; NESBURN et al. 1983) applied to the eye. Guinea pigs are also susceptible to experimental infection and, following genital inoculation and establishment of latency in the sacral dorsal root ganglia, they were shown to undergo spontaneous reactivations (SCRIBA 1975), a model that has been widely used to study antiviral therapy (ALENEUS and OBERG 1978) and potential vaccines (STANBERRY 1991).

Among the various species, murine models are among the most convenient for the study of latency. However, murine infection models have suffered from the disadvantage that the stimulation of recurrence reproducibly has been difficult. Some success has been reported by means of UV light (BLYTH et al. 1976), mild trauma produced by application to the skin and removal of sticky tape (HILL et al. 1978), and, most recently, by means of transient hyperthermia (SATWELL and THOMPSON 1992). These techniques have enabled the study of the events in the virus-infected cells immediately before, during, and after the application of a reactivation stimulus.

2.1.2.3 In Vitro Models for Latency

Herpes simplex virus will productively infect an extremely wide range of tissue cells in culture. The result of infection is normally, however, cytolytic and the culture is quickly destroyed. Long-term survival of infected cultures was obtained using combinations of nucleoside inhibitors, high temperature or interferon (WIGDAHL et al. 1984b). More recently, in order to mimic certain aspects of herpes simplex latency, the model was successfully modified by PRESTON et al. (1994) to facilitate molecular studies of the latency phenomenon. The initial approach utilised incubation at 42°C to suppress virus replication and establish stable latency after infection of human fibroblasts at low multiplicity with HSV-2. Latent virus was reactivated efficiently by superinfection of cultures with HSV-1 or human cytomegalovirus (CMV). Improvement of the system was achieved by use of the HSV-1 mutant in1814, which has a 12-base-pair insertion in the coding sequences for the transactivating protein Vmw65 (VP16). More recent modifications include the use of mutant in1820 (a derivative of in 1814) and pretreatment of cells with interferon (IFN)-α, which enable latency to be established in approaching 100% of cells. A system was therefore developed that is claimed to be suitable for the analysis of gene expression and genome structure during latency and reactivation. The former methods have been questioned on the basis that the residual virus DNA is in a different form from that encountered in natural latency. However, the publications of Preston and colleagues are persuasive that their model is a more accurate one and has helped in the elucidation of the control of latency.

2.2 Herpes Simplex Virus Adaptation to the Host

Herpes simplex virus appears to have evolved an extremely well-balanced relationship with the human host such that it readily infects neurons and inhabits a proportion of them for life while rarely producing overt neurological damage. It is perhaps surprising, therefore, that recent clinical isolates from humans, without adaptation, appear readily to establish latency in similar anatomical sites in rodent species. The extent of neurological damage (CNS involvement characterised by weight loss, retinitis, flaccid paralysis and other neurological signs that may lead to death) tends to be greater in mice, rabbits or guinea pigs than in the natural host. Furthermore, strains of HSV-2 appear to be generally more neuropathogenic in mice than strains of HSV-1. These effects are observed following relatively high doses of inoculum, i.e. >10^4pfu, which is presumed to be much higher than the dose encountered initially by the susceptible cells in the infection site in humans during natural exposure, although the latter is extremely difficult to assess. Nonetheless, evidence to date suggests that the essential features of the establishment of, maintenance of, and reactivation from latency in the various animal models are very similar to those that occur in humans.

2.3 Establishment of Latency

The susceptible host usually encounters HSV at a mucosal site, often the mucocutaneous junction. Following adsorption to and entry into the susceptible cells, the DNA is uncoated and enters the nucleus where replication proceeds in a controlled pattern (Fig. 2a). This involves several waves of gene expression (α-, β- and γ- or immediate-early, early and late genes). The way in which each set of genes regulates the others is both complex and extremely well-documented (ROIZMAN 1996). The replication cycle is completed within 24h, with the possibility of the release of infectious virus from about 12h after infection. As shown in Fig. 2b, HSV from the inoculum itself may enter the axonal tips. Certainly, virus released from the cells in the primary infection site does so and translocates to the sensory neurons at a rate of about 1.8mm/h by means of retrograde axonal transport (reviewed by WILDY et al. 1982). There is evidence from infection of neuronal cell cultures that virus is transported in an unenveloped form to be released into the neuronal soma. It appears that replication in the neuron is restricted such that virus enters a stable state within the neuronal nucleus where it is thought to exist in an unintegrated, circular form (MELERICK and FRASER 1987; EFSTATHIOU et al. 1986). It appears that the DNA is strongly associated with histones in a chromatin-like structure which is likely to be involved in controlling genome activity during latency (DESHMANE and FRASER 1989). In animal infection models the number of latently infected neurons in the ganglion may exceed 10%. Not surprisingly, the number of neurons latently infected has been shown to correlate with the probability of reactivation in a murine model (SAWTELL 1988). More interestingly, the number of genome copies per cell has been suggested to be a factor in determining the frequency of reactivation (SAWTELL 1998; SAWTELL et al. 1998; LEKSTROM-HIMES et al. 1998). The proportion of human neurons per ganglion or the number genome copies per cell during latency in humans is unknown. However, estimations indicate that 10–1000 virus genome copies are present per latently infected neuron (HO 1992), whereas only about 1%–10% of the neurons in a ganglion seemed to be HSV-positive (NICHOLLS and BLYTH 1989; MEHTA et al. 1995). Clearly, the number of ganglionic cells that ultimately harbours latent HSV will be affected by many factors including the age and

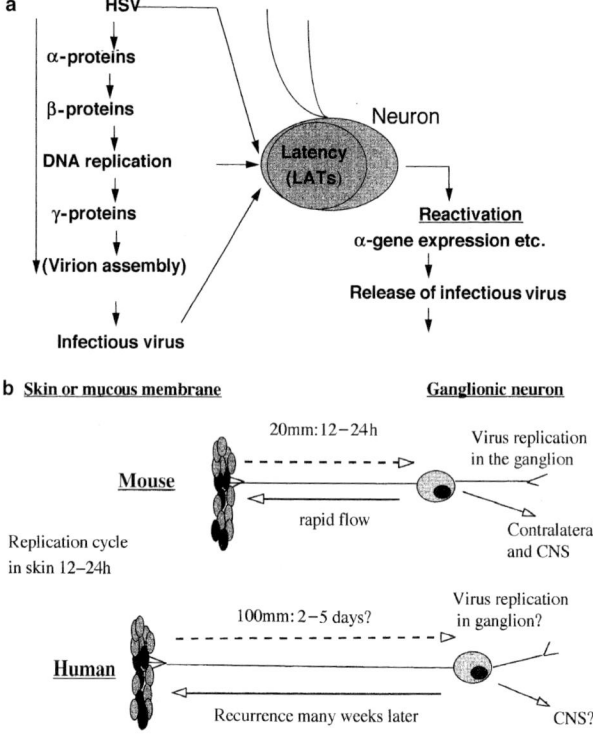

Fig. 2. a HSV infection of the host cell showing the sequence of proteins expressed during the acute HSV replication cycle in a permissive cell, e.g. epidermal cell, and indicating that entry of virus into the axon of a peripheral nerve leading to the establishment of latency in the neuronal cell body may take place at various points during the cycle. **b** HSV transport from skin to ganglion in mouse and humans, showing the temporal relationship between virus replication at the anatomical site of acute virus replication (e.g. mucous membrane) and the establishment of latency in the associated peripheral nerve ganglia which innervate the local site. (Adapted from FIELD and THACKRAY 1997)

susceptibility of the particular mouse strain and the route, dose and virulence of the virus inoculum.

Mutants that are unable to undergo DNA replication have been shown, nevertheless, to be capable of establishing and maintaining latency. For example, a temperature-sensitive DNA-negative mutant established latency and was reactivated at the permissive temperature (McLENNAN and DARBY 1980). Furthermore, inoculation into mice of mutants that are defective in immediate-early gene expression also showed that viral gene expression is not a prerequisite for the establishment of latency (ECOB-PRINCE et al. 1993). Mutants that are deficient in the HSV thymidine kinase (TK) enzyme are able to establish latency but are frequently impaired in their ability to reactivate (EFSTATHIOU et al. 1989). This failure to reactivate may be an important factor determining the continuing rarity of such mutants in clinical practice.

2.4 Transcriptional Activity During Latency

For HSV-1, transcription during latency is confined to a single region of the genome situated in the long repeat region of the linear, double-stranded DNA genome (for review see FRASER et al. 1992). The RNA products of this transcription are generally known as LATs. The sequence transcribed partially overlaps, and is antisense to, one of the immediate-early genes, ICP0 (CROEN et al. 1988). A large primary transcript (approx. 8kb) is rapidly spliced from the 5' end to produce introns and further splicing results in a stable 1.5-kb transcript (major LAT; Fig. 3). The LATs are not exported from the nucleus and very large numbers of LAT copies accumulate in neuronal nuclei during latency (ca. 50,000/cell); consequently, hybridisation techniques have proved very sensitive for the detection of HSV-infected neurons in tissue sections. The role of LATs in the mechanism of latency and reactivation remains, however, slightly uncertain. LAT^{-ve} mutants establish latency (in animal models) although their reactivation efficiency is reduced (LEIB et al. 1989b). Furthermore, there is evidence that additional cells may contain HSV DNA but in which LATs remain below the level of detection (ECOB-PRINCE 1995). It is unclear whether these cells contribute significantly to the maintenance of latency and reactivation.

For HSV-1 it does not appear that LATs code for a protein product or that such a protein is involved in regulation. Furthermore there is no persuasive evidence that the RNA acts in a simple antisense mode to regulate a coding

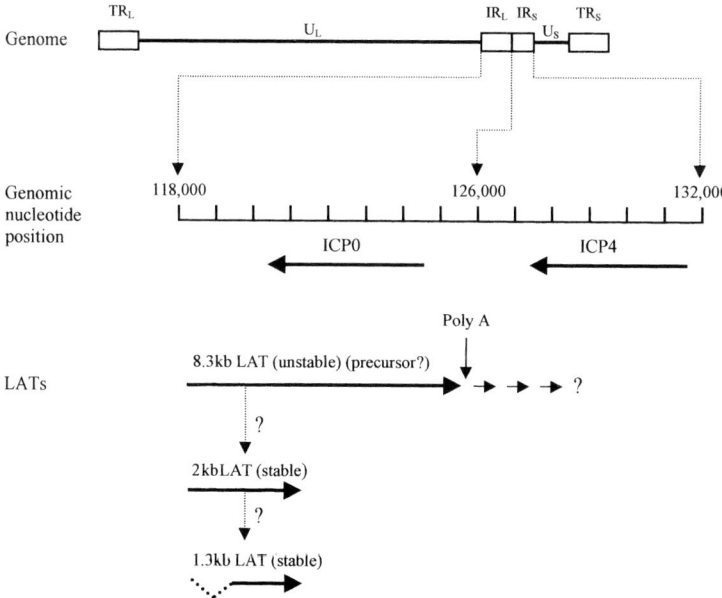

Fig. 3. HSV-1 genomic organisation and localisation of latency-associated transcripts (LATs)

sequence. The situation has been further compounded, by the growing evidence that very low levels of transcription do occur from other genes in the HSV latent DNA which are translated. It remains, however, to be seen whether this has biological significance.

2.5 Cellular and Immunological Factors Implicated in the Establishment/Maintenance of Herpes Simplex Virus Latency

Herpes simplex virus-1, like other herpesviruses, replicates in a variety of cells; however, in neurons a lytic infection does not necessarily ensue – instead latency may be established in these nonreplicating cells. Little is known about the viral or cellular factors which regulate latency, but a consensus is forming that a number of cellular gene products have an extensive role in the mechanism by which HSV establishes latency in the neuron. Gangliosides, for example, play an important role in regulation of growth and proliferation of cells possibly by protein kinase activation. In-vitro studies indicated that gangliosides, in combination with acyclovir, can exert antiviral activity (HAYASHI and NIWAYAMA 1993). Oct-1 and Oct-2 are both cellular transactivation factors binding directly to target motifs present in the HSV-1 IE promoter region. The interaction of both factors may be important in the regulation of HSV-1 replication. This question has been well-reviewed by MYERS and NASH (1994) and, more recently, by Efstathiou et al. (in press). Recently, a novel multifunctional structural domain in the HSV-1 genome has been detected. This DNA sequence might bind cellular factors and regulate viral transactivation or alter the chromatin structure to which latent viral genomes seemed to be attached (QUINN et al. 1998).

T lymphocytes are thought to be very important in the control of acute HSV-1 infection. Since many latently infected neurons survive, it appears that, if cytotoxic lymphocytes are involved, the HSV-infected neurons are not destroyed by cytotoxic activity. It is possible that T lymphocytes interfere with viral replication by nonlytic mechanisms, most likely by the production of cytokines. Indeed, T cell-associated cytokines and chemokines, i.e. interleukin (IL)-2, IL-10, and IFN-γ, have been detected in acutely and latently infected murine ganglia (HALFORD et al. 1996). However, it is unclear how the cytokine expression is triggered since, during latency, HSV-1 transcription is generally thought to be limited to LATs. Furthermore, neurons either do not express, or express only low levels of, MHC class I molecules and therefore are unlikely to present endogenous antigen to T cells. It is speculated that prolonged cytokine expression in latently infected ganglia could result from a limited HSV-1 antigen expression (KOSZ-VNENCHAK et al. 1993) and that the resulting cytokine secretion may modulate a reactivated HSV-1 infection (CANTIN et al. 1995). Alternatively, a small proportion of lytically infected neurons, or cells other than neurons, are targets for T cells. In summary, the role of T cells and the excreted cytokines such as IFN-γ in the pathogenesis of CNS infections is still very uncertain.

2.6 Reactivation from Latency

The precise mechanism leading to reactivation of HSV remains one of the most intriguing unsolved mysteries in virology. The most common stimulus for HSV-1 reactivation in humans appears to be exposure to sunlight. This can be achieved experimentally by application of UV light to the lips (BERNSTEIN et al. 1997). Experimentally, a suberythematous dose of UV applied to the skin (BLYTH et al. 1976) or corneas of mice (SHIMELD et al. 1996) was also shown to produce recurrence of virus in the ganglionic neurons followed by recurrence of infectious virus in the skin or eye, respectively. Increasing attention is now being focused on the possible role of cytokines in the mechanism of reactivation and, in the mouse, IL-6 is one of the products receiving most attention. Interestingly, this cytokine induces the same secondary pathway in neurons as nerve growth factor (NFG) following binding to its receptor (for details see MEYERS and NASH 1994). NFG is required for maintaining normal cell function. In vitro it was shown that deprivation of NFG results in reactivation of latent HSV-1 (WILCOX and JOHNSON 1987). Co-infection with other herpesviruses can also lead to reactivation, possibly by replacing viral transactivating proteins (WIGDAHL et al. 1984a; SCHECK et al. 1989; PUREWAL et al. 1992). Furthermore, it is conceivable that neurotropic viruses other than herpesviruses may lead to immune-mediated damage and consequently to a herpesvirus reactivation.

In this connection it is of interest that a previous review article dealt with the question of whether herpesvirus latency and reactivation could cause learning and behavioural deficiencies and violence in children and adults (BECKER 1995). Thus, reactivation of latent virus in the brain could lead to damage of neurons involved in memory, learning and behaviour. However, it seems also possible that latency, followed by long-term low-level stimulation of the immune system, can cause changes in the complex interaction of the immune system, the endocrine system and the CNS. This is perhaps the more likely since, in recent years, it has become clear that the immune system can be influenced by the emotions mediated, in some cases perhaps, by hormones.

2.7 Latent Herpes Simplex Virus in the CNS of Animals and Humans

The fact that a high proportion of the human population has latent HSV in their peripheral nerve ganglionic neurons begs several important questions. One important one is: What proportion of these individuals also have HSV within the cells of their CNS? Work from many laboratories has shown that, following peripheral inoculation of mice, virus rapidly migrates to the contralateral ganglia and spinal cord or brain stem and, eventually, to the mid- and forebrain (ANDERSON and FIELD 1983), including the temporal lobes. This process may lead to the clinical signs of acute CNS disease and death. However, in survivors, including mice following subclinical infection, residual virus is readily detected in the CNS by using DNA hybridisation methods (ROCK and FRASER 1983; FIELD et al. 1984). It appears to be very difficult to reactivate infectious virus from CNS tissues although this has been achieved on occasion (e.g. KNOTTS et al. 1973). There is reported to be a close parallel between the pattern of virus distribution in the brains of fatal human encephalitis cases, including the focal involvement of the temporal lobes, and that observed in mouse brains from animals following intranasal virus inoculation (ESIRI 1982). Therefore, it is of considerable interest that, recently, DNA detected by a sensitive PCR test has been reported to be present in the CNS of human patients who were seropositive for HSV but had no history of acute virus infection of the CNS (ITZHAKI and LIN 1998; ITZHAKI et al. 1998).

2.8 The Biological Consequences of Latent Herpes Simplex Virus in the CNS

The most obvious consequence of latency in the peripheral nervous system (PNS) is that of benefit to the virus. The elegant strategy of latency and recurrence (ideally subclinical) leads to ready transmission to susceptible hosts and ensures survival for the virus, even in small and isolated populations. In some infected individuals, the recurrences may be associated with diseases of hitherto uncertain origin including Bell's palsy (SCHIRM and MULKENS 1997) and severe trigeminal neuralgia (TENSER 1998).

Frank herpes encephalitis occurs in approximately $1/10^6$ adult humans (WHITLEY et al. 1981) and is an acute, life-threatening disease. In some cases, it appears to arise directly from a recently acquired HSV infection. In other cases, however, there is evidence to show that it originated from the reactivation of an existing latent infection (as evidenced by strain identification using endonuclease restriction polymorphisms to compare the virus isolated from the encephalitis patient's brain infection and his cold sores). However, it is not known whether the CNS virus originated from reactivation within the CNS itself or resulted from spread to the CNS as a consequence of PNS reactivation. As mentioned above, evidence from animal models suggests that the anatomical origin of the disease is most likely via the olfactory route.

It has been proven that both primary and recurrent HSV infections can lead to encephalitis. In patients with primary infections, an involvement of the olfactory bulb has been documented (DINN 1980; WHITLEY 1986). Neuronal transmission of reactivated virus from the trigeminal ganglia has also been proposed as a pathogenic mechanism (JOHNSON 1964; STROOP et al. 1986).

It seems probable that there exist a range of milder forms of herpes encephalitis that go unrecognised yet produce significant neurological damage. This, however, is unknown and we are unwilling to speculate further. There are a few well-documented cases of behavioural or other changes associated with recurrences of cold sores (SHEARER and FINCH 1964; RIMON and HALONEN 1969).

Animal models of HSV neural infections show considerable evidence of PNS and CNS demyelination along the neural tracts that innervate the virus-infected neurons (KRISTENSSON et al. 1979). There has been much speculation that HSV could be a factor in the demyelinating disease multiple sclerosis. In a small clinical trial of the specific antiherpes drug acyclovir, a trend to prolonged remissions was claimed, but the study has yet to be confirmed in a larger trial or by use of alternative antiherpetic drugs (LYCKE et al. 1996).

HSV has also been implicated as a causal factor in several other chronic neurological diseases. In a series of papers published during the last 2 years, Itzhaki and colleagues claimed that HSV in the CNS may be a cofactor, together with a genetic factor concerning apolipoprotein E4, in the induction of Alzheimers' disease (LIN et al. 1998). However, the data have been subject to criticism (BEFFERT et al. 1998) and the subject remains controversial.

2.9 The Prospects for Curing or Preventing Latency

Work in animal models suggests that inoculum virus may directly enter axonal tips and, in any case, virus progeny from the first rounds of virus replication in mucosal cells may be available to do so within 12–24h of exposure. Travelling at a speed estimated to be approximately 1.8mm/h, it is clear that the first neurons will have been colonised within about 24h in mice and within a few days in humans. Because of the uncertainty about the influence of latently infected cell number, their genome load, etc., it is also uncertain exactly what window of opportunity exists to influence the establishment of latency. In a series of papers by Thackray and colleagues (reviewed FIELD and THACKRAY 1997), it was shown that the nucleoside analogue prodrugs valaciclovir and famciclovir were both able to markedly influence the establishment of latency in experimentally infected mice. In the case of famciclovir, effects were observed even when the start of treatment was delayed for several days – a surprising result. This has led to a trial, currently in progress, in which human patients are being treated intensively within 72h of known exposure to HSV-2. However, to date all the known inhibitors of HSV, including acyclovir and penciclovir, have been shown to be incapable of curing latently infected cells once the latent infection is established. It has also been found that prophylaxis using a nucleoside analogue (acyclovir, its oral prodrug, valaciclovir or famciclovir) are all very effective at suppressing recurrent lesions. In some cases prophylaxis has been maintained for 10 years. However, on cessation of therapy, lesions are prone to recur. Intriguingly, evidence has been published that, in a murine infection model, very long-term acyclovir therapy is associated with specific changes in the immune responses to HSV and cytokine balance in latently infected animals (HALFORD et al. 1997). We are not aware, however, of any antiviral treatment concept to date whose rationale is to specifically eliminate latent HSV.

Several promising human HSV vaccines are currently undergoing clinical trials in humans (reviewed by SMITH 1997), including the disabled, infectious single-cycle or "DISK" vaccine which is being evaluated for a possible immunomodulatory role on recurrent herpes in recipients who are already latently infected (reviewed by ROLLINSON 1998). The growing success of antiviral therapy for the treatment of acute infections or prophylaxis to prevent recurrent lesions, may, as methodologies improve and the patient numbers increase, provide evidence for or against the role of active HSV replication in neurological diseases of hitherto unknown aetiology.

3 Other Human Herpesvirus Infections

3.1 Varicella-Zoster Virus

Regarding mechanisms of latency, by far the most information is available on HSV. In addition, VZV appears to have many parallels with HSV. The virus also establishes classical latency in the neurons of the PNS (for review see ECHEVARRIA et al. 1997). It is well known that reactivation of VZV in the form of shingles is a disease often associated with prodromal signs suggesting neural damage and further functional changes and prolonged pain as a consequence of a recurrence of virus replication in the nervous system. It is intriguing to speculate that subclinical shingles may exist and contribute to production of neurological clinical signs (GILDEN et al. 1992). VZV DNA and LATs have been detected in the cells of human nerve ganglia and is notable that satellite cells in addition to neurons have been claimed to contain evidence of latent infection (CROEN et al. 1988; MEIER and STRAUS 1992; CHORS et al. 1994, 1996; KENNEDY et al. 1998). Another difference from HSV appears to be existence of a greater variety of LATs than the small, limited range observed in HSV and several animal herpesvirus counterparts (see below). Furthermore, in latently infected human trigeminal ganglia a VZV IE protein has been detected in the cytoplasm of neurons (MAHALINGAM et al. 1996).

In contrast to HSV reactivation, which is generally not destructive to the site of latent infection, and repeated recurrences usually have no obvious effect on the function or physiology of the trigeminal ganglia, reactivation of VZV is associated with post-herpetic neuralgia reflecting inflammation and damage to neural tissues (for further details see WAGNER and BLOOM 1997).

3.2 Human Cytomegalovirus

This is usually regarded as a lymphotropic herpesvirus. It is a widespread infection in humans, with the majority of adults showing evidence of infection, but the virus very rarely produces clinical signs in otherwise healthy adults. If, however, the primary infection occurs during pregnancy, the fetus may be infected with consequential damage; virus targets include the eye, ear and CNS. Other manifestations of human cytomegalovirus (HCMV) infection are seen in immunocompromised patients, including HIV-infected individuals. In the latter, retinitis, involving the direct infection and destruction of retinal neurons, is a well-recognised complication, as well as HCMV encephalitis and polyradiculitis (for references see CINQUE et al. 1997). During these episodes HCMV may be detected in cerebrospinal fluid (CSF) using sensitive techniques such as PCR. However, it appears that these clinical manifestations do not result from the reactivation of HCMV-infected neurons but are more likely to be a consequence of infected monocytes. These cells are thought to be involved in HCMV latency and reactivation and dissemination of virus to the nervous system and other organs possibly via infected endothelial cells, whereas the bone marrow precursors of the monocyte/macrophage lineage seemed to be the primary site of latency (reviewed by SINCLAIR and SISSONS 1997; SINZGER and JAHN 1997). It has also been speculated that cytokines have an important role in mediating the CNS damage that results from the unusual pattern of HCMV infection in immunocompromised individuals. However, an aetiological role for HCMV in schizophrenia and various neuropsychiatric diseases has not been found.

3.3 Human Herpesvirus Type 6

This is a betaherpesvirus which causes the common childhood infection roseola infantum (exanthema subitum) (YAMANISHI et al. 1988). It has been reported that brains from non-neurological autopsy cases were HHV-6-DNA-positive (LUPPI et al. 1994; LIEDTKE et al. 1995). Furthermore, the virus has been detected by PCR in the CSF of children during and after primary infection as well as after reactivation (KONDO et al. 1993; CASERTA et al. 1994). These data implicate the CNS as a site of latency, however, the role of HHV-6 in the pathogenesis of neurological diseases is unknown.

4 Animal Herpesviruses

The exemplar for "classical" latency is that of HSV, described in the foregoing sections. There are, however, many animal counterparts among the Alphaherpesvirinae which exhibit forms of latency that are analogous to HSV (for review see ROCK 1990). The Alphaherpesvirinae is one of the three subfamilies of the herpesviruses, which includes, besides HSV-1 and HSV-2 and VZV, various veterinary pathogens such as B virus and simian agent 8 (SA8) of primates, equine herpesvirus types 1, 3 and 4 (EHV-1, EHV-3, EHV-4), bovine herpesvirus types 1 and 2 (BHV-1, BHV-2), pseudorabies virus (PRV) of the pig, feline herpesvirus type 1 (FHV-1), canine herpesvirus type 1 (CHV-1) and Marek's disease virus (MDV) (LUDWIG 1983; LUDWIG et al. 1983; ROCK et al. 1986, 1987; CHEUNG 1989, 1990; OHMURA et al. 1993; OKUDA et al. 1993; SLATER et al. 1994).

Although HSV is the model for HSV latency in animals, actually, based on genome sequence, EHV-1, EHV-4, BHV-1, CHV-1 and PRV are more closely related to VZV than HSV-1 and HSV-2. There are also similarities in their biological properties, e.g. the cell-associated viraemia associated with VZV is especially important in the pathogenesis of EHV-1 and to a lesser extent some of the other veterinary alphaherpesviruses. Notwithstanding, all of these viruses establish latency primarily (but not exclusively) in nervous tissues, especially in the neurons of the sensory and/or autonomic nerve ganglia. However, in addition to neuronal latency, it appears that alternative sites exist, for example lymphoid tissue (EHV-1) (CHESTERS et al. 1997) or bone marrow (PRV) (OHLINGER et al. 1987), although the biological significance of non-neuronal sites for latency is unknown. An exception to the rule is MDV, which is believed to be T cell tropic where it establishes a latent state of infection (CANTELLO et al. 1994; LI et al. 1994).

In this survey we will give most attention to PRV, EHV-1, BHV-1, and MDV of chickens, all of which are important veterinary diseases. Several other animal herpesviruses will also be given brief mention, e.g. CHV-1 and FHV-1 and the B-virus of monkeys, which provides the closest link to HSV in humans.

4.1 Herpesviruses of Nonhuman Primates: Cercopithecus Virus-1 or B Virus

Nonhuman primates are known to carry herpes infections; and at least two neurotropic alphaherpesviruses have been isolated from both New World and Old World monkeys. Two viruses from Cercopithecoids, herpesvirus simiae (B virus) and the SA8 have a close molecular and antigenic relationship to HSV (LUDWIG et al. 1983; BORCHERS and LUDWIG 1991, 1997a; BORCHERS et al. 1991; HILLIARD et al. 1989; BENNET et al. 1992; EBERLE et al. 1993). In experimental infections, these viruses have been shown to cause orofacial and genital lesions, to be shed asymptomatically in the saliva and to establish latency in sensory ganglia. B virus was found in trigeminal, lumbar and sacral ganglia of normal and seropositive animals (MALHERBE and HARWIN 1958; SLOMKA et al. 1993) In the few fatal cases examined pathologically, liver lesions, and, in the brain stem, perivascular cuffing and microglial infiltration have been described (KEEBLE 1960; LOOMIS et al. 1981). In experimentally infected monkeys virus was detected also in the spinal cord and spleen. Experimental infections with B virus have been established in rabbits and latent infection in the ganglia has been demonstrated in the rabbit model (VIZOSO 1975). The rabbit model has been used to study the efficacy of acyclovir and passive antibody administration as prophylactic therapies following human exposure to infected primates. This is important because transmission of B virus to humans by monkey bite can result in an ascending encephalomyelitis (PALMER 1987), which has proved fatal in almost 80% of reported human cases. As demonstrated in experimentally infected mice, B virus migrates intra-axonally (GOSZTONYI et al. 1992) and this seems to occur in natural infections in the monkey and in humans. Following a bite, B virus binds quickly to neural elements and thereafter it travels axonally and trans-synaptically in the nervous system. Once the virus is inside the neuronal chain, it is probably somewhat protected against the immune system. However, macaques with persistent trigeminal infection and periodical oropharyngial manifestations possess a partial immunity reducing the frequency of fatal encephalitis. Because of the antigenic relationship, it is also conceivable that HSV-specific antibodies may offer cross-protection for humans against B virus infections (UEDA et al. 1968; LUDWIG et al. 1983).

4.2 Porcine Herpesvirus: Pseudorabies Virus

This is one of the more neuropathogenic of the herpesviruses in this natural host. In adults the infection causes respiratory clinical signs and abortion and, in many cases, is subclinical. In newborn or younger animals, however, neurological signs are common. The infection translocates by intra-axonal transport as for HSV, but ganglionitis and encephalitis are common features of the pathogenesis.

Following oronasal uptake and replication, PRV invades the CNS via the trigeminal ganglia, glossopharyngeal and/or olfactory nerves (KRITAS et al. 1994, 1995). Using viral deletion mutants in experimentally infected pigs it was demonstrated that HSV-1 homologues of glycoproteins I and E play the major role in neural invasion and anterograde transport of PRV (KRITAS et al. 1994, 1995).

During latency, PRV DNA persists in the trigeminal ganglia in a nonintegrated linear and/or concatemeric form (RZIHA et al. 1986) expressing LATs antisense to HSV-1 homologue ICP4 (CHEUNG 1989, 1991).

Because of the low virus genome copy number in latently infected cells and the fact that trigeminal ganglia represent nondividing cells, studies on the physical state of latent neurotropic alphaherpesviruses are restricted. The only studies published to date were done on PRV and HSV-1 and revealed that free genomic termini of latent virus seem to be not detectable (ROCK and FRASER 1985; EFSTATHIOU et al. 1986; RZIHA et al. 1986; MELLERICK and FRASER 1987). This might result from: (a) an episomal state, (b) a concatemeric form or (c) an integration of viral genomes into chromosomes. In the case of HSV-1 it seems to be accepted that episomes are the prevalent physical state of latent genomes (MELLERICK and FRASER 1987). PRV DNA was found to exist in general in a linear and nonintegrated form in different neuronal tissues of latently infected pigs (RZIHA et al. 1986). However, the presence of circular or concatemeric molecules was also observed in a small proportion of cases (RZIHA et al. 1986).

Two general strategies have been employed to control PRV: (1) eradication of infection by the slaughter of all animals when herds are shown to contain seropositive stock; (2) a programme of vaccination using attenuated, deletion mutants. More recently, in parts of Europe, a combination of the two strategies has been employed. An interesting feature of the eradication programmes is the continued presence of a very small number of seropositive animals following "eradication". Reactors were, in some cases, shown to contain PRV DNA in their tissues following PCR. However, strenuous efforts to reactivate virus were unsuccessful, suggesting that the PRV DNA may not represent functional latency and its significance remains controversial (BASCUNANA et al. 1997).

A striking feature of PRV is that, on transmission to other species (fortunately excluding humans), the infection is both highly neurotropic and devastatingly neuropathogenic. In other species, e.g. rodents, intense pruritus is induced at the site of infection, believed to result from ganglionitis, and neurological clinical signs develop with progression to death. The neurotropism of PRV has been widely exploited as a neural tracer for the study of neural connections. One of the earliest of these was the work of McCracken, in which the rate of retrograde axonal transport was measured by means of PRV-infected calves (McCracken et al. 1973).

Recently there has been concern about the potential for PRV to be transmitted to humans from xenotransplants, for example hearts from genetically modified pigs. However, humans appear to be refractory to infection with PRV and it is considered improbable that adaptation to humans would occur under these circumstances in the unlikely event that the donor tissues were latently infected.

4.3 Equine Herpesviruses

There are at least five distinct herpesviruses common in horses, but among these the most important is EHV-1.

4.3.1 Equine Herpesvirus-1

Primary infection is associated with fever and respiratory clinical signs which, in pregnant animals, may also result in abortion. The virus establishes infection in epithelial cells in the nasal mucosa and spreads to the trachea and lower respiratory tract where the bronchiolar epithelium is infected. A prominent feature of the primary infection is cell-associated viraemia and this may result in the infection of vascular endothelial cells, which may account for the transmission of infection to the placenta and fetus (for review see Ludwig et al. 1988; Crabb and Studdert 1995).

Traditionally, latent EHV-1 was thought to be established exclusively in lymphoid tissue. More recently it has been shown that primary infection includes neurons of the trigeminal ganglia, which harbour functional latent virus (Slater et al. 1994; Baxi et al. 1996; Borchers et al. 1998a). On application of a suitable reactivation stimulus (e.g. administration of a corticosteroid) to the latently infected horse, recurrent virus may be readily detected in nasal secretions (Eddington et al. 1985). Reactivation has also been achieved in vitro by explanting the trigeminal ganglia (Slater et al. 1994). Thus, latent infections of both lymphoid tissues and trigeminal ganglia have been demonstrated by cocultivation, PCR and by detecting LATs (Baxi et al. 1995; Chesters et al. 1997; Borchers et al. 1998a). In PBL and bronchial lymph nodes of naturally and experimentally infected horses, random RT-PCR and in situ hybridisation studies indicated that gene 64 (HSV-1 ICP4 homologue) is a putative LAT region (Chesters et al. 1997). In trigeminal ganglia of experimentally infected SPF foals, using a nested PCR and the cocultivation technique latent EHV-1 was present. Furthermore, in one of the SPF trigeminal ganglia, ICP0-specific LAT equivalents could be detected by in situ hybridisation in 0.02% of the nuclei, indicating a latent infection (BAXI et al. 1995). In situ PCR and RT PCR studies on naturally infected horses corroborated that EHV-1 is neurotropic and demonstrated that EHV-1 establishes latent infection in the nerve cell nuclei of trigeminal ganglia expressing ICP0-specific antisense transcripts (BORCHERS et al. 1998a). The relative biological importance of latent EHV-1 in lymphocytes and/or trigeminal ganglia in the pathogenesis of neurological disorders or the epidemiology of the infection in the equine population are unknown.

Neurological clinical signs are a rarer, but well recognised complication of primary (and possibly reactivated) EHV-1 infection (ALLEN and BRYANS 1986; MUMFORD 1994). Paralytic disease is characterised by sudden onset of lameness and ataxia. Gross lesions are small haemorrhages in grey and white matter of the brain and spinal cord. Histological changes include foci of malacia, necrosis of arteriole walls, perivascular cuffing of monocuclear cells in brain and spinal cord, and axonal swelling in spinal cord (CHARLTON et al. 1976; LITTLE and THORSEN 1976; WHITEWELL and BLUNDEN 1992).

Since the virus also infects monocytes it is thought to spread directly from the circulating mononuclear cells to vascular endothelium without an extracellular phase, and from there to other endothelial cells (JACKSON et al. 1977). Neurological disease was explained by ischaemic injuries to nervous tissue following endothelial infection and thrombosis without direct infection of neurons (PATEL et al. 1982; EDINGTON et al. 1986). An other explanation is that vasculitis results from the deposition of circulating immune complexes that have been identified in horses with experimental EHV-1 myeloencephalopathy (KOHN and FENNER 1987).

Several workers have reported the presence of neural histopathology in equine CNS obtained from apparently normal animals. Because EHV-1 is ubiquitous it may be difficult to relate such changes to this particular infection. However, we have observed perivascular cuffing in regions of the CNS, including the choroid plexus, and demyelination in the optic nerve in specific pathogen-free ponies experimentally infected with EHV-1 (J. O'leary, J. Slater and H. Field, unpublished observations). Moreover, we have discovered that focal retinopathy associated with retinal neuronal necrosis occurs in up to 40% animals within 3 weeks of experimental EHV-1 infection (J. Slater, personal communication; SLATER et al. 1992).

4.3.2 Equine Herpesvirus-4

This Alphaherpesvirus infection is also extremely common, with the majority of horses becoming infected. The infection appears to be somewhat less pathogenic than EHV-1; again the route of infection is respiratory and the principle clinical sign is upper respiratory distress. EHV-4 has only rarely been associated with abortion or frank neurological signs.

As for EHV-1, the virus was regarded as lymphotropic and, with no neurological signs associated with acute infection, there seemed no reason to suspect a neural site for latency. However, our own studies have produced unequivocal evidence that EHV-4 does, indeed, have the classical characteristics of a neurotropic alphaherpesvirus (BORCHERS et al. 1997b). Thus in up to 60% of randomly selected, naturally infected horses, we found EHV-4 DNA by means of a PCR exclusively in the trigeminal ganglion (Fig. 4), i.e. no other tissue yielded positive results. The LATs have yet to be fully characterised; however, we have obtained evidence for transcripts complementary to genes 63 and 64. It is important to note that, by in situ RT PCR, the LATs were localised in the nerve cell nuclei. We have also obtained evidence that, similar to PRV, EHV-4 seems to establish latency in the trigeminal ganglia in a nonintegrated, concatemeric form (Borchers, in press).

Moreover, the function of latent EHV-4 in the trigeminal ganglia remains to be fully elucidated. From our current knowledge, EHV-4 most likely enters the trigeminal ganglia via the nasal epithelium, analogous to EHV-1 (BAXI et al. 1995; BARTELS et al. 1998) or PRV in experimentally infected animals (KRITAS et al. 1995). The existence of ganglionic latency would explain how reactivation of EHV-4 can result in respiratory clinical signs with no demonstrable viraemia. Furthermore, it appears that reactivation in ganglionic neurons results in anterograde axonal distribution of infectious virus to the respiratory site of recurrence, as for other alphaherpesviruses. Whether EHV-4 can result in overt neurological damage is unknown.

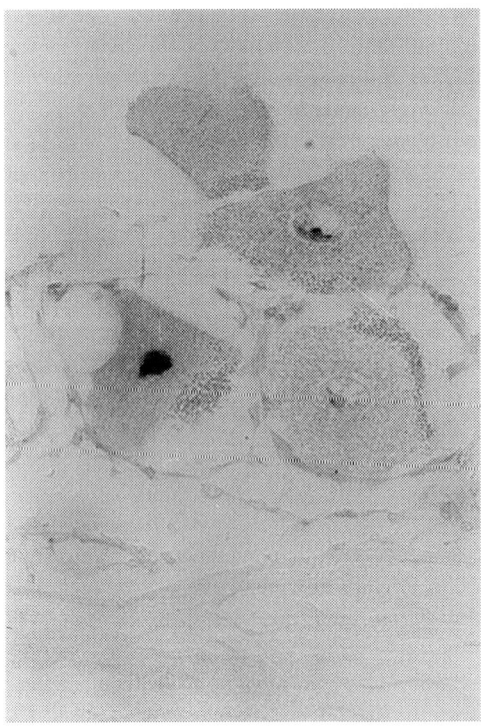

Fig. 4. Detection of EHV-4 DNA by direct in situ PCR in paraffin-embedded trigeminal ganglia

We sought an aetiological relationship for EHV-4 in equine neurological disease; however, in a small study of 17 horses showing neurological signs including ataxia and other CNS disorders, although nine of the 17 were found to be naturally infected with EHV-4, no direct relationship between the infection and neurological disease was confirmed (Borchers, in press).

4.3.3 Equine Herpesviruses-2

Based on genome sequence analysis, EHV-2 has been placed in the Gammaherpesvirinae, therefore it is different from all the other herpesviruses under consideration in this chapter; but, like the above, it is also widespread among the horse population although its precise role as pathogen is still uncertain (for reviews see BROWNING and STUDDERT 1988; AGIUS and STUDDERT 1994). It has been implicated in upper respiratory tract infections, immunosuppression, anorexia and lymphadenopathy but most infections probably remain subclinical. However, it remains an important candidate as a cofactor which may up-regulate other infections, e.g. EHV-1 (PUREWAL et al. 1992; WELCH et al. 1992).

Based on this hypothesis, we conducted a study recently that showed a significantly higher prevalence of EHV-2 among horses with respiratory clinical signs, including ataxia and abortions, than in a control group of healthy animals (BORCHERS et al. 1997c). Furthermore, EHV-2 was isolated from foals with keratoconjunctivitis (THEIN and BÖHM 1976; THEIN 1978; Borchers et al., in press).

To date, very little is known about the nature of EHV-2 pathogenesis. Notably, EHV-2 was detected by a nested PCR in post-mortem specimens of trigeminal ganglion and brain. In addition, however, lymphoid tissues were also PCR-positive (RIZVI et al. 1997; BORCHERS et al. 1998b). Because of the high sensitivity of the PCR employed, the authors considered the possibility that the positive neural tissues could have been contaminated with infected blood cells. Although this cannot be ruled out absolutely, on a quantitative basis this explanation appears to be extremely unlikely.

4.4 Bovine Herpesviruses

4.4.1 Bovine Herpesvirus-1

As for EHV-1 and EHV-4 above, BHV-1 is an alphaherpesvirus and is widely distributed among cattle world-wide. The clinical manifestations are usually upper respiratory tract disease often associated with conjunctivitis. In neonates a generalised, fatal infection can occur (LUDWIG 1983; WYLER et al. 1989) and sometimes infection of pregnant cows results in abortion. A rarer, but possibly more ancient form, of disease is genitally transmitted, producing infectious pustular vulvovaginitis in females and balanoposthitis in males.

It is very well established that BHV-1 establishes classical neuronal latency analogous to HSV-1. It is of interest that this virus was the first to be shown to undergo expression of RNA during latency with production of LATs. Using experimentally infected rabbits, BHV-1 LATs derived from the IE gene homologous to HSV-1 ICP0 were detected by in situ hybridisation (ROCK et al. 1986, 1987; KUTISH et al. 1990). There is a feature of BHV-1 latency that currently appears to contrast with those infections discussed above, especially HSV-1. This is the apparent production, during latency, of a virus-induced latency-associated protein (HOSSAIN et al. 1995; SCHANG et al. 1996). The significance of this observation and role of this protein in latency, if any, remains to be elucidated.

An encephalitic form of BHV-1 infection was first described in 1962 as a sporadic disease of calves in Australia (FRENCH 1962; JOHNSTON et al. 1962). Since that time, this type of disease has been reported in several countries, although it appears to be very rare. However, because, the neurovirulent strains of BHV-1 revealed antigenic and genetic differences when compared with non-neurovirulent strains, it has since been reclassified as BHV-5 (ROIZMAN et al. 1996); indeed, the infection has been implicated in epidemics of uniformly fatal encephalitis in calves (CARILLO et al. 1983).

Notwithstanding the above CNS complications associated with BHV-5 infection, there is some evidence of CNS manifestations of the true BHV-1 producing neurological sequelae: In 1992 a bovine herpesvirus meningoencephalitis associated with infectious BHV vaccine occurred in Japan, and approximately 20 Holstein-Fresian calves showed neuroparalysis and died. The necropsy findings included a few haemorrhages and malacic lesions in the cortical to subcortical area of the cerebrum (FURUOKA et al. 1995). Milder lesions were observed in the cerebellum and brain stem. Basophilic intranuclear inclusion bodies were found in the nuclei of neurons and astrocytes. Viral antigen was detected in the cerebellum, midbrain, pons, medulla oblongata or spinal cord, whilst in the trigeminal ganglia positively stained cells were absent. Therefore, it was concluded that the infectious route to the cerebellum was through the olfactory bulb or via the meninges. The strong staining of the cell body and dendrites of the neurons indicated direct axonal transport of BHV-1 and suggested that cell to cell propagation has an important pathogenic role. The haematogenous spread of BHV-1 into the brain seemed to be relatively unlikely, since the lesions were not widely disseminated as multiple foci. Studies on experimental intranasal infections, however, suggested that the trigeminal nerve was the pathway of viral spread to the brain, based on successful virus isolations of infectious virus.

In the UK, the infection is being controlled by surveillance using serological tests and the registration of closed BHV-1-free herds. Vaccines have been developed but are of uncertain efficacy.

4.4.2 Bovine Herpesvirus-2

Bovine mamillitis, caused by BHV-2, causes ulcerative lesions on the teats and udders of infected cows (LUDWIG 1983). The virus is interesting because it is closely related to HSV-1 and it is assumed (although to our knowledge has not been formally shown to date) to establish latency in the ganglionic neurons (CASTRUCCI et al. 1982). However, experimental reactivation is difficult to demonstrate (CASTRUCCI et al. 1980, 1983) and, to date, there has been no association with encephalitis or other forms of neurological disease.

4.5 Canine Herpesvirus-1 and Feline Herpesvirus-1

Feline herpesvirus-1 causes respiratory disease in cats which can be severe, especially in neonates, and abortion is common in pregnant cats. Like the other alphaherpesvirus infections, both FHV-1 and CHV-1 establish latency in ganglia, from which it can be recovered by cocultivation (GASKELL et al. 1985) or detected by PCR. For FHV-1 LATs are encoded by the HSV-1 homologue ICP4 (OHMURA et al. 1993; KAWAGUCHI et al. 1994). The trigeminal ganglia may serve a similar function as for EHV-4 – a reservoir from where virus can be reactivated and nasally shed. During an acute infection, tonsils and nasal turbinates contain the greatest concentration of FHV-1, and by PCR different other tissues of the head were shown to be virus-positive. Both FHV-1 and CHV-1 can produce CNS disease in their respective species (especially following perinatal infection), e.g. puppies sometimes suffer a necrotising retinitis after primary infection (ALBERT et al. 1976) and the pathogenic signs of FHV-1 are dendritic ulcers on the cornea. However, there are no other known neurological syndromes associated with these infections in adults.

4.6 Marek's Disease Virus

The final herpesvirus to be considered in this chapter is Marek's disease of chickens. For many years the virus was considered to be a member of the Gammaherpesvirinae and only recently, upon genome analysis including sequencing, was it reclassified as being a member of the Alphaherpesvirinae. We include it because the virus appears to be neuropathogenic, although lymphoproliferation is the dominant feature of the pathogenesis in infected birds.

MDV infection is characterised by a mononuclear infiltrate within the peripheral nerves and other tissues or organs. MDV is one of the few tumour viruses against which a successful live vaccine has been employed using the closely related, nononcogenic herpesvirus of turkeys (HVT).

MDV-transformed cells obtained from infected birds were found to contain latent MDV DNA. The status of MDV DNA in the transformed cells has been investigated with contradictory results. However, using Southern blot hybridisation, Gardella gel electrophoresis (GARDELLA et al. 1984) and in situ hybridisation of methaphase and interphase chromosomes, DELECLUSE and HAMMERSCHMIDT (1993) unexpectedly found that MDV is integrated chromosomally in different cell lines. MDV LATs belong to a family of spliced RNAs that are antisense to the homologous ICP4 gene of HSV-1 (SUGAYA et al. 1990; CANTELLO et al. 1994).

Although integration into lymphocyte DNA has been demonstrated for the gammaherpesvirus, Epstein-Barr virus, the only indication of an integrated alphaherpesvirus came from studies on MDV. Here, in transformed T and B cell lines, besides episomes, integrated viral DNA was also detected by conventional Southern blot hybridisation, by Gardella gel electrophoresis and, additionally, by in situ hybridisation of metaphase and interphase chromosomes (DELECLUSE and HAMMERSCHMIDT 1993). All of these studies were done on model systems, indicating the difficulty of studying the physical state of latent herpesviruses under natural conditions.

Depending on the impaired nerve, asymmetric and progressive paralysis of the legs, wings or neck, can be observed in infected chickens. It is claimed that this results from the infiltration of proliferating lymphoblastic cells into peripheral nerves and, in turn, this causes ataxia and macroscopic enlargement of the peripheral nerves. So far this condition appears to be unique for this chicken virus. However, in the future, comparative studies may show analogous changes in herpesviruses of other species, although less prominent.

Table 1 is an overview of the nature of LATs for the different herpesviruses and their tissue distribution as well as the genome conformation of human and animal herpesviruses described in the text.

5 The Role of Latency in Neurological Disease Caused by Members of Virus Families Other than Herpes

While HSV and many other members of Alphaherpesvirinae are undoubtedly highly neurotropic and establish latency within neurons, their precise role in chronic disease of the CNS still remains difficult to define, with many conflicting and controversial data having been published in this area. For other virus infections the situation is extremely difficult to define. For example, it is known that many individuals contract the papovaviruses BK and JC. The occasional development of

Table 1. Latency-associated transcripts (LATs) of human and animal neurotropic herpesviruses, their tissue tropism and viral genome conformation

Virus	LAT	Localisation	Conformation
HSV-1	ICP0	TGG	Nonintegrated
HSV-2	ICP0	TGG	–
VZV	Genes 21, 29, 62, 63	TGG	–
EHV-1	ICP0	TGG	–
EHV-4	ICP0, ICP4	TGG	Nonintegrated
PRV	ICP4	TGG	Nonintegrated
FHV-1	ICP4	TGG	–
BHV-1	ICP0	TGG	–
MDV	ICP4	T cell lymphomas	Integrated

progressive multifocal leukoencephalopathy (PML), caused by JC in immunosuppressed patients, demonstrates the potential this virus has for causing neurological disease. However, there is little evidence that papovavirus latency has a role in the pathogenesis of PML or, if latency exists, that it might have a role in any other human conditions. Similar arguments could be applied to the RNA viruses, e.g. picorna viruses, including poliomyelitis, or the negative strand viruses such as influenza, which has recently been postulated to have had a role in cases of neurological disease associated with the 1981 influenza pandemic including encephalitis lethargica (J.S. Oxford, personal communication), or bornavirus, which has been linked to psychiatric disease in humans (see elsewhere in this volume). In none of these cases is there a firm foundation for experimental latency producing a model that satisfies the criteria defined in our Introduction, and in none of these cases does the virus appear to benefit from the latency as a means of perpetuating itself in the population.

Theiler's murine encephalomelitis virus is a picornavirus which can cause an acute polio-like disease or a chronic demyelinating disease, depending on the strain. The latter disease in some respects resembles multiple sclerosis and therefore has become an important model system. The acute infection is followed by a virus persistence in glia cells and macrophages which most likely induces an immune response leading to demyelination mediated by cytokines and lymphokines released by activated T cells and macrophages (RODRIGUEZ et al. 1996).

6 Conclusions

In summary, neurological disorders have been studied in naturally and experimentally infected animals leading to more insight into the mechanism of pathogenesis of neurological disorders and latency. In some cases natural infection causes severe CNS pathology (e.g. B virus in humans), while neurotropic viruses, such as PRV, are valuable tools for the study of developmental neurobiology.

Thus, direct and indirect involvement of virus infections of the nervous system depend on the virus type and genetic, humoral and immunological factors of the host. For herpesviruses, the sensory nerves and peripheral and trigeminal ganglia (and olfactory nerve and olfactory bulb in some cases) represent important routes by which the virus translocates from a mucosal site of infection. Lymphotropism and endotheliotropism may also play a role in distributing the virus and the respective cells may also serve as sites of latency.

Among the key questions that remain is the functional significance of the presence of these viruses in the nervous system, once latency has been established in neurons. Firstly, can the latent virus provoke an on-going pathological response, either directly, or by invoking host immune or other responses to its presence? Second, can the latent virus produce infection and consequent damage to the nervous system following reactivation, as appears to happen in some cases of herpes encephalitis? It is particularly intriguing to consider the possibility that occult viruses which produce no specific clinical signs to reveal their identity may yet be covert agents of neurological disease and even of psychiatric illnesses or behavioural changes. It is particularly difficult to ascribe this to viruses, including many of the herpesviruses of animals and humans that are ubiquitous among the normal population. In some cases the answers to these questions will come with the development of highly effective antiviral agents and vaccines to enable the complete eradication of the infection from the host population.

References

Agius CT, Studdert MJ (1994) Equine herpesviruses 2 and 5: comparisons with other members of the subfamily Gammaherpesvirinae. Adv Virus Res 44:357–379
Albert DM, Lahav M, Carmichael LE, Percy DH (1976) Canine herpes-induced retinal dysplasia and associated ocular anomalies. Invest Ophthalmol 15:267–278
Aleneus S, Oberg B (1978) Comparison of the therapeutic effects of five antiviral agents on cutaneous herpesvirus infections in guinea pigs. Arch Virol 58:277–288
Allen GP, Bryans JT (1986) Molecular epizootiology, pathogenesis and prophylaxis of equine herpesvirus-1 infections. Prog Vet Microbiol Immunol 2:78–144
Anderson JR, Field HJ (1983) The distribution of herpes simplex type 1 antigen in mouse central nervous system after different routes of inoculation. J Neurol Sci 60:181–195
Bartels T, Steinbach F, Hahn G, Ludwig H, Borchers K (1998) In situ study on the pathogenesis and immune reaction of equine herpesvirus type 1 (EHV-1) infections in mice. Immunology 93: 329–334
Bascunana CR, Bjornerot L, Ballagi-Pordany A, Robertsson JA, Belak S (1997) Detection of pseudorabies virus genomic sequences in apparently uninfected 'single reactor' pigs. Vet Microbiol 55:37–47
Baxi MK, Efstathiou S, Lawrence G, Whalley JM, Slater JD, Field HJ (1995) The detection of latency-associated transcripts of equine herpesvirus 1 in ganglionic neurons. J Gen Virol 76:3113–3118
Baxi MK, Borchers K, Bartels T, Schellenbach A, Field HJ (1996) Molecular studies of the acute infection, latency and reactivation of equine herpesvirus-1 (EHV-1) in the mouse model. Virus Res 40: 33–45
Becker Y (1995) HSV-1 brain infection by the olfactory nerve route and virus latency and reactivation may cause learning and behavioral deficiencies and violence in children and adults: a point of view. Virus Genes 10:217–226
Beffert U, Bertrand P, Champagne D, Gauthier S, Poirier J (1998) HSV-1 in brain and risk of Alzheimer's disease. Lancet 351:1330–1331
Bennett AM, Harrington L, Kelly DC (1992) Nucleotide sequence analysis of genes encoding glycoproteins D and J in simian herpes B. J Gen Virol 73:2963–2967
Bernstein DI, Schleupner CJ, Evans TG, Blumberg DA, Bryson Y, Grafford K, Broberg P, Martin-Munley S, Spruance SL (1997) Effect of foscarnet cream on experimental UV radiation-induced herpes labialis. Antimicrob Agents Chemother 41:1961–1964
Block TM, Hill JM (1997) The latency associated transcripts (LAT) of herpes simplex virus: still no end in sight. J Neurobiol 3:313–321
Blyth WA, Hill TJ, Field HJ, Harbour DA (1976) Reactivation of herpes simplex infection by ultraviolet light and possible involvement of prostaglandins. J Gen Virol 33:547–550
Borchers K, Ludwig H (1991) Simian agent 8 – a herpes simplex-like monkey virus. Comp Immun Microbiol Infect Dis 14:125–132
Borchers K, Weigelt W, Buhk HJ, Ludwig H, Mankertz J (1991) Conserved domains of glycoprotein B (gB) of the monkey virus, simian agent 8, identified by comparison with herpesvirus gBs. J Gen Virol 72: 2299–2304
Borchers K, Ludwig H (1997a) B Virus. In: Darai G, Handermann M, Hinz E, Sonntag H-G (eds) Lexikon der Infektionskrankheiten des Menschen. Springer, Berlin Heidelberg, pp 35–37

Borchers K, Wolfinger U, Lawrenz B, Schellenbach A, Ludwig H (1997b) Equine herpesvirus 4 DNA in trigeminal ganglia of naturally infected horses detected by direct in situ PCR. J Gen Virol 78:1109–1114

Borchers K, Wolfinger U, Goltz M, Broll H, Ludwig H (1997c) Distribution and relevance of equine herpesvirus type 2 (EHV-2) infections. Arch Virol 142:917–928

Borchers K, Wolfinger U, Schellenbach A, Lawrenz B, Glitz F, Ludwig H (1998a) Equid herpesvirus type 1 (EHV-1) and trigeminal ganglia of naturally infected horses: detection of DNA and latency associated transcripts (LATs) by RT-PCR. Equine Inf Dis VIII (in press)

Borchers K, Wolfinger U, Ludwig H, Thein P, Baxi S, Field HJ, Slater J (1998b) Virological and molecular biological investigations into equine herpesvirus type 2 (EHV-2) experimental infections. Virus Res 55:101–106

Browning GF, Studdert MJ (1988) Equine herpesvirus 2 (equine cytomegalovirus). Vet Bull 58:775–790

Cantello JL, Anderson AS, Morgan RW (1994) Identification of latency-associated transcripts that map antisense to the ICP4 homolog gene of Marek's disease virus. J Virol 68:6280–6290

Cantin EM, Hinton DR, Chen J, Openshaw H (1995) Gamma interferon expression during acute and latent nervous system infection by herpes simplex virus type 1. J Virol 69:4898–4905

Carillo BJ, Ambrogi A, Schudel AA, Vazquez M, Dahme E, Pospischil A (1983) Meningoencephalitis caused by IBR virus in calves in Argentina. Zentralbl Veterinärmed Reihe B 30:327–332

Caserta MT, Hall CB, Schnabel K, McIntyre K, Long C, Costanzo M, Dewhurst S, Insel R, Epstein LG (1994) Neuroinvasion and persistence of human herpesvirus 6 in children. J Inf Dis 170:1586–1589

Castrucci G, Frigeri F, Chilli V, Tesei B, Arush AM, Pedini B, Ranucci S, Rampichini L (1980) Attempts to reactivate bovid herpesvirus-2 in experimentally infected calves. Am J Vet Res 41:1890–1893

Castrucci G, Ferrari M, Frigeri F, Ranucci S, Chilli V, Tesei B, Rampichini L (1982) Reactivation in calves of bovid herpesvirus 2 latent infection. Arch Virol 72:75–81

Castrucci G, Chilli V, Frigeri F, Ferrari M, Ranucci S, Rampichini L (1983) Reactivation of bovid herpesvirus 1 and 2 and parainfluenza-3 virus in calves latently infected. Comp Immunol Microbiol Inf Dis 6:193–199

Charlton RM, Mitchell D, Girard A, Corner AH (1976) Meningoencephalitis in horses associated with equine herpesvirus 1 infection. Vet Path 13:59–68

Chesters PM, Allsop R, Purewal A, Edington N (1997) Detection of latency-associated transcripts of equid herpesvirus 1 in equine leucocytes but not in trigemnial ganglia. J Virol 71:3437–3443

Cheung AK (1989) Detection of pseudorabies virus transcripts in latently infected swine. J Virol 63:2908–2913

Cheung AK (1991) Cloning of the latency gene and the early protein 0 gene of pseudorabies virus. J Virol 65:5260–5271

Chors RJ, Srock K, Barbour MB, Owens G, Mahalingam R, Devlin ME, Wellish M, Gilden DH (1994) Varicella-zoster virus (VZV) transcription during latency in human ganglia: construction of a cDNA library from latently infected human trigeminal ganglia and detection of a VZV transcript. J Virol 68:7900–7908

Chors JR, Barbour M, Gilden DH (1996) Varicella-zoster virus (VZV) transcription during latency in human ganglia: detection of transcripts mapping to genes 21, 29, 62, and 63 in a cDNA library enriched for VZV RNA. J Virol 70:2789–2796

Cinque P, Marenzi R, Ceresa D (1997) Cytomegalovirus infections of the nervous system. Intervirology 40:85–97

Crabb BS, Studdert MJ (1995) Equine herpesvirus 4 (equine rhinopneumonitis virus) and 1 (equine abortion virus). Adv Virus Res 45:153–189

Craig CP, Nahmias AJ (1973) Different patterns of neurologic involvement with herpes simplex virus types 1 and 2: isolation of herpes simplex virus type 2 from the buffy coat of two adults with meningitis. J Infect Dis 127:365–372

Croen KD, Ostrove JM, Dragovic LJ, Straus SE (1988) Patterns of gene expression and sites of latency in human nerve ganglia are different for varicella-zoster and herpes simplex viruses. Proc Natl Acad Sci USA 85:9773–9777

Delecluse H-J, Hammerschmidt W (1993) Status of Marek's Disease Virus in established lymphoma cell lines: herpesvirus integration is common. J Virol 67:82–92

Dennet C, Cleator GM, Klapper PE (1997) HSV-1 and HSV-2 in herpes simplex encephalitis: a study of sixty-four cases in United Kingdom. J Med Virol 53:1–3

Deshmane SL, Fraser NW (1989) During latency, herpes simplex virus type 1 DNA is associated with nucleosomes in a chromatin structure. J Virol 63:2179–2190

Dinn JJ (1980) Transolfactory spread of virus in herpes simplex encephalitis. Br Med J 281:1392

Eberle R, Zhang M, Black DH (1993) Gene mapping and sequence analysis of the unique region of the simian herpesvirus SA8 genome. Arch Virol 130:391–411

Echevarria JM, Casas I, Martinez-Martin P (1997) Infections of the nervous system caused by varicella-zoster virus: a review. Intervirology 40:72–84

Ecob-Prince MS, Preston CM, Rixon FJ, Hassan K, Kennedy PG (1993) Neurons containing latency-associated transcripts are numerous and widespread in dorsal root ganglia following footpad inoculation of mice with herpes simplex virus type 1 mutant in 1814. J Gen Virol 74:985–994

Ecob-Prince MS, Hassan K, Denheen MT, Preston CM (1995) Expression of beta-galactosidase in neurons of dorsal root ganglia which are latently infected with herpes simplex virus type 1. J Gen Virol 76:1527–1532

Edington N, Bridges CG, Huckle A (1985) Experimental reactivation of equid herpesvirus 1 (EHV-1) following the administration of corticosteroids. Equine Vet J 17:369–372

Edington N, Brigdes CG, Patel JR (1986) Endothelial cell infection and thrombosis in paralysis caused by equid herpesvirus-1: equine stroke. Arch Virol 90:111–124

Efstathiou S, Minson AC, Field HJ, Anderson JR, Wildy P (1986) Detection of herpes simplex virus-specific DNA sequences in latently infected mice and in humans. J Virol 57:446–455

Efstathiou S, Kemp S, Darby G, Minson AC (1989) The role of herpes simplex virus type 1 thymidine kinase in pathogenesis. J Gen Virol 70:869–879

Esiri M (1982) Herpes simplex encephalitis – an immunohistochemical study of the distribution of viral antigen within the brain. J Neurol Sci 54:209–266

Field HJ, Anderson JR, Efstathiou S (1984) A quantitative study of the effects of several nucleoside analogues on established herpes encephalitis in mice. J Gen Virol 65:707–719

Field HJ, Brown GA (1989) Animal models for antiviral chemotherapy. Antiviral Res 12:165–180

Field HJ, Thackray AM (1997) Can herpes simplex virus latency be prevented using conventional nucleoside analogue chemotherapy? Antiviral Chem Chemotherapy 8:59–66

Fraser NW, Block TM, Spivack JG (1992) The latency-associated transcripts of herpes simplex virus: RNA in search of function. Virology 191:1–8

French EL (1962) A specific virus encephalitis in calves: isolation and characterization of the causal agent. Aust Vet J 38:216–221

Furuoka H, Izumida N, Horiuchi M, Osame S, Matsui T (1995) Bovine herpesvirus meningoencephalalitis association with infectious bovine rhinotracheitis (IBR) vaccine. Acta Neuropathol 90:656–671

Garcia-Blanco MA, Cullen BR (1991) Molecular basis of latency in pathogenic human viruses. Science 254:815–820

Gardella T, Medveczky P, Sairenji T, Mulder C (1984) Detection of circular and linear herpesvirus DNA molecules in mammalian cells by gel electrophoresis. J Virol 50:248–254

Gaskell RM, Dennis PE, Goddard LE, Cocker FM, Wills JM (1985) Isolation of felid herpesvirus 1 from the trigeminal ganglia of latently infected cats. J Gen Virol 66:391–394

Gilden DH, Vafai A, Shtram Y, Becker Y, Devlin M, Wellish M (1983) Varicella-zoster virus DNA in human sensory ganglia. Nature 306:478–480

Gilden DH, Dueland AN, Devlin ME, Mahalingam R, Cohrs R (1992) Varicella-zoster virus reactivation without rash. J Infect Dis 166:30–34

Goodpasture EW, Teague O (1923) Transmission of the virus of herpes fibrilis along nerves in experimentally infected rabbits. J Med Res 44:139–184

Gosztonyi G, Falke D, Ludwig H (1992) Axonal and transsynatic (transneuronal) spread of Herpesvirus simiae (B virus) in experimentally infected mice. Histol Histopath 7:63–74

Halford WP, Gebhardt BM, Carr DJJ (1996) Persistent cytokine expression in trigeminal ganglion latently infected with herpes simplex virus type 1. J Immunol 157:3542–3549

Halford WP, Gebhardt BM, Carr DJ (1997) Acyclovir blocks cytokine gene expression in trigeminal ganglia latently infected with herpes simplex virus type 1. Virology 238:53–63

Hayashi K, Niwayama S (1993) Effects of gangliosides on the growth of herpes simplex virus type 1-infected cells derived from neurons and on viral replication. Intervirology 36:134–143

Hill TJ, Blyth WA, Harbour DA (1978) Trauma to the skin causes recurrence of herpes simplex in the mouse. J Gen Virol 39:21–28

Hilliard JK, Black D, Eberle R (1989) Simian alphaherpesviruses and their relationship to the human herpes simplex viruses. Arch Virol 109:83–102

Ho DY (1992) Herpes simplex virus latency: molecular aspects. In: Melnick JL (ed) Prog Med Virol Karger, Basel, 39:76–115

Hossain A, Schang LM, Jones C (1995) Identification of gene products encoded by the latency-related gene of bovine herpesvirus 1. J Virol 69: 5345–5352

Itzhaki RF, Lin WR (1998) Herpes simplex virus type I in brain and the type 4 allele of the apolipoprotein E gene are a combined risk factor for Alzheimer's disease. Biochem Soc Trans 26:273–277

Itzhaki RF, Lin WR, Wilcock GK, Faragher B (1998) HSV-1 and risk of Alzheimer's disease [letter] Lancet 352: 238

Jackson TA, Osburn BI, Cordy DR (1977) Equine herpesvirus-1 infection of horses: studies in the experimentally induced neurologic disease. Am J Vet Res 38:709–719

Johnson RT (1964) The pathogenesis of herpes simplex encephalitis. 1. Virus pathways to the nervous system of suckling mice demonstrated by fluorescent antibody staining. J Exp Med 119:343–356

Johnston LAY, Simmons GC, McGavin MD (1962) A viral meningoencephalitis in calves. Aus Vet J 38:207–215

Kawaguchi Y, Maeda K, Miyazawa T, Ono M, Kai C, Mikami T (1994) Nucleotide sequence and characterization of the feline herpesvirus type 1 immediate early gene. Virology 204:430–435

Keeble SA (1960) B virus infection in monkeys. Ann NY Acad Sci 85:960–969

Kennedy PGE, Grinfeld E, Gow JW (1998) Latent varicella-zoster virus is located predominantly in neurons in human trigeminal-ganglia. 95:4658–4662

Knotts FB, Cook ML, Stephens JG (1973) Latent herpes simplex virus in the central nervous system of rabbits and mice. J Exp Med 138:740–744

Kohn CW, Fenner WR (1987) Equine herpes myeloencephalopathy. Equine Practice 3:405–419

Kondo K, Nagafuji H, Hata A, Tomomori C, Yamanishi K (1993) Association of human herpesvirus 6 infection of the central nervous system with recurrence of febrile convulsions. J Infect Dis 167:1197–1200

Kosz-Vnenchak M, Jcobson J, Coen DM, Knipe DM (1993) Evidence for a novel regulatory pathway for herpes simplex virus gene expression in trigeminal ganglion neurons. J Virol 67:5383–5393

Kramer MF, Coen DM (1995) Quantification of transcripts from the ICP4 and thymidine kinase genes in mouse ganglia latently infected with herpes simplex virus. J Virol 69:1389–1399

Kristensson K, Svennerholm B, Persson L, Vahlne A, Lycke E (1979) Latent herpes simplex virus trigeminal ganglionic infection in mice and demyelination in the central nervous system. J Neurol Sci 43:253–264

Kritas SK, Pensaert MB, Mettenleiter TC (1994) Role of envelope glycoproteins gI, gp63 and gIII in the invasion and spread of Aujesjky's disease virus in the olfactory nervous pathway of the pig. J Gen Virol 75:2319–2327

Kritas SK, Nauwynck HJ, Pensaert MB (1995) Dissemination of wild-type and gC-, gE- and gI-deleted mutants of Aujeszky's disease virus in the maxillary nerve and trigeminal ganglion of pigs after intranasal inoculation. J Gen Virol 76:2063–2066

Kutish G, Mainprize T, Rock D (1990) Characterization of the latency-related transcriptionally active region of the bovine herpesvirus 1 genome. J Virol 64: 5730–5737

Kwon BS, Gangarosa LP, Burch KD, deBack J, Hill JM (1981) Induction of ocular herpes simplex virus shedding by iontophoresis of epinephrine into rabbit cornea. Invest Ophthalmol Vis Sci 21: 442–449

Lachmann RH, Efstathiou S (1997) Utilization of the herpes simplex virus type 1 latency-associated regulatory region to drive stable reporter gene expression in the nervous system. J Virol 71:3197–207

Leib DA, Coen DM, Bogerd CL, Hicks KA, Yager DR, Knipe DM, Tyler KL, Schaffer PA (1989a) Immediate-early regulatory gene mutants define different stages in the establishment and reactivation of herpes simplex virus latency. J Virol 63:755–768

Leib DA, Bogard CL, Kosz-Vnenchak M, Hicks KA, Coen DM, Knipe DM, Schaffer PA (1989b) A deletion mutant of the latency-associated transcript of herpes simplex virus type 1 reactivates from the latent state with reduced frequency. J Virol 63:2893–2900

Lekstrom-Himes JA, Pesnicak L, Straus SE (1998) The quantity of latent viral DNA correlates with the relative rates at which herpes simplex virus types 1 and 2 cause recurrent genital herpes outbreaks. J Virol 72:2760–2764

Li D-S, Pastorek J, Zelnik V, Smith GD, Ross LJN (1994) Identification of novel transcripts complementary to the Marek's disease virus homologue of the ICP4 gene of herpes simplex virus. J Gen Virol 75:1713–1722

Liedtke W, Trübner K, Schwechheimer K (1995) On the role of human herpesvirus 6 in viral latency in nervous tissue and in cerebral lymphoma. J Neurological Science 134:184–188

Lin WR, Graham J, MacGowan SM, Wilcock GK, Itzhaki RF (1998) Alzheimer's disease, herpes virus in brain, apolipoprotein E4 and herpes labialis. Alzheimer Reports 1:1–6

Little PB, Thorsen J (1976) Disseminated necrotizing myeloencephalitis: a herpes-associated neurological disease of horses. Vet Pathol 13:161–167

Lokensgard JR, Thawley DG, Molitor TW (1990) Pseudorabies virus latency: restricted transcription. Arch Virol 110:129–136

Loomis MR, O'Neill T, Bush M, Montali RJ (1981) Fatal herpesvirus infection in patas monkeys and a black and white colobus monkey. J Am Vet Med Assoc 179:1236–1239

Ludwig H, Pauli G, Gelderblom H, Darai G, Koch HG, Flügel RM, Norrild B, Daniel MD (1983) B virus (Herpesvirus simiae). In: Roizman B (ed) The herpesviruses, vol 2. Plenum, New York, pp 385–428

Ludwig H (1983) Herpesviruses of bovidae. In: Roizman B (ed) The herpesviruses, vol 2. Plenum, New York, pp 135–214

Ludwig H, Thein P, Chowdhury SI (1988) Herpesvirus infections of equine animals. In: Darai G (ed) Virus diseases in laboratory and captive animals. Martinus Nijhoff, Boston, pp 283–297

Luppi M, Barozzi P, Maiorana A, Marasca R, Torelli G (1994) Human herpesvirus 6 infection in normal human brain tissue. J Infect Dis 169:943–944

Lycke J, Svennerholm B, Hjelmquist E, Frisen L, Badr G, Andersson M, Vahlne A, Andersen O (1996) Acyclovir treatment of relapsing-remitting multiple sclerosis. A randomized, placebo-controlled, double-blind study. J Neurol 243:214–224

Malherbe H, Harvin R (1958) Neurotropic virus in African monkeys. Lancet II:530

Mahalingam R, Wellish M, Chors R, Debrus S, Piette J, Rentier B, Gilden D (1996) Expression of protein encoded by varicella-zoster virus open reading frame 63 in latently infected human ganglionic neurons. Proc Natl Acad Sci USA 93:2122–2124

McCracken RM, McFerran JB, Dow C (1973) The neuronal spread of pseudorabies virus in calves. J Gen Virol 20:17–28

McLennan JL, Darby G (1980) Herpes simplex virus latency: the cellular location of virus in dorsal root ganglia and the fate of the infected cell following virus activation. J Gen Virol 51: 233–243

Mehta A, Maggioncalda J, Bagasra O, Thikkavarapu S, Saikumari P, Valyi-Nagy T, Fraser NW, Block TM (1995) In situ PCR and RNA hybridization detection of herpes simplex virus sequences in trigeminal ganglia of latently infected mice. Virology 206:633–640

Meier JL, Straus SE (1992) Comparative biology of latent varicella-zoster virus and herpes simplex virus infections. J Infect Dis 166:13–23

Mellerick DM, Fraser NW (1987) Physical state of the latent herpes simplex virus genome in a mouse model system: evidence suggesting an episomal state. Virol 158:267–275

Meyers NL, Nash AA (1994) The neurobiology of alphaherpesvirus infections. Acta Veterinaria Hungarica 42:263–275

Mumford JA (1994) Equid herpesvirus 1 and 4 infections. In: Coetzov JAW, Thomson GR, Tustin RC (eds) Infections diseases of livestock. 2nd edn, Oxford University Press, Oxford, pp 911–925

Nesburn AB, Willey DE, Trousdale MD (1983) Effect of intensive acyclovir therapy during artificial reactivation of latent herpes virus. Proc Soc Exp Biol Med 172:316–323

Nicholls SM, Blyth WA (1989) Quantification of herpes simplex virus infection in cervical ganglia of mice. J Gen Virol 70:1779–1788

Ohlinger VF, Heck R, Behrens P, Rziha H-J, Weiland F, Wittmann G (1987) Die Infektion von Zellen des hämatopoetischen Systems durch das Aujeszkyvirus – Neue Aspekte für Immunbiologie und Praxis? Tierärztliche Umschau 3:210–219

Ohmura Y, Ono E, Matsuura T, Kida H, Shimizu Y (1993) Detection of feline herpesvirus 1 transcripts in trigeminal ganglia of latently infected cats. Arch Virol 129:341–347

Okuda Y, Ishida K, Hashimoto A, Yamaguchi T, Fukushi H, Hirai K, Carmichael LE (1993) Virus reactivation in bitches with a medical history of herpesvirus infection. Am J Vet Res 54:551–554

O'Leary JJ, Chetty R, Graham AK, McGee JO'D (1996) In situ PCR: pathologist's dream or nightmare? J Pathology 178:11–20

Patel JR, Edington N, Mumford JA (1982) Variation in cellular tropism between isolates of equine herpesvirus-1 in foals. Arch Virol 74:41–51

Palmer AE (1987) B virus, herpesvirus simiae: historical perspective. J Med Primatol 16:99–130

Preston CM, Russell J, Harris RA, Jamieson DR (1994) Herpes simplex virus latency in tissue culture cells. Gene Ther 1: S49–50

Purewal AS, Smallwood AV, Kaushal A, Adegboye D, Edington N (1992) Identification and control of the cis-acting elements of the immediate early gene of equid herpesvirus type 1. J Gen Virol 73:513–519

Quinn JP, McGregor RA, Fiskerstrand CE, Davey C, Allan J, Dalziel RG (1998) Identification of a novel multifunctional structural domain in the herpes simplex virus type 1 genome: implications for virus latency. J Gen Virol 79:2529–2532

Rimon R, Halonen P (1969) Herpes simplex virus infections and depressive illness. Dis Nerv System 30:338–340
Rizvi SM, Slater JD, Wolfinger U, Borchers K, Slade AJ (1997) Detection and distribution of equine herpesvirus 2 DNA in the central and peripheral nervous systems of ponies. J Gen Virol 78:1115–1118
Rock DL, Fraser NW (1983) Detection of HSV-1 genome in central nervous system of latently infected mice. Nature 302:523–525
Rock DL, Fraser NW (1985) Latent herpes simplex virus type 1 DNA contains two copies of the virion DNA joint region. J Virol 55:849–852
Rock DL, Hagemoser WA, Osorio FA, Reeds DE (1986) Detection of bovine herpesvirus type 1 RNA in trigeminal ganglia of latently infected rabbits by in situ hybridization. J Gen Virol 67:2515–2520
Rock DL, Nesburn AB, Ghiasi H, Ong J, Lewis TL, Lokensgard JR, Wechsler SL (1987) Detection of latency-related viral RNAs in trigeminal ganglia of rabbits latently infected with herpes simplex virus type 1. J Virol 61:3820–3826
Rock DL (1990) Latent infection with alpha herpesviruses. Equine Inf Disease VI:175–180
Rodriguez M, Pavelko KD, Njenga MK, Logan WC, Wettstein PJ (1996) The balance between persistent virus infection and immune cells determines demyelination. J Immunol 157:5699–5709
Roizman B (1996) Herpesviridae. In: Fields BN, Knipe DM, Howley PM, Chanock RM, Melnick JL, Monath TP, Roizman B, Straus SE (eds) Virology. Lippincott-Raven, Philadelphia, pp 2221–2230
Rollinson E (1998) Therapeutic vaccines: a novel approach to the management of genital herpes. International Antiviral News 6:125–127
Rziha HJ, Mettenleitner TC, Ohlinger V, Wittmann G (1986) Herpesvirus (pseudorabies virus) latency in swine: occurrence and physical state of viral DNA in neuronal tissues. Virology 155:600–613
Sawtell NM, Thompson RL (1992) Rapid in vivo reactivation of herpes simplex virus in latently infected murine ganglionic neurons after transient hyperthermia. J Virol 66:2150–21546
Sawtell NM (1998) The probability of in vivo reactivation of herpes simplex virus type 1 increases with the number of latently infected neurons in the ganglia. J Virol 72:6888–6892
Sawtell NM, Poon D, Tansky CS, Thompson RL (1998) The latent herpes simplex virus type 1 genome copy number in individual neurons is virus strain specific and correlates with reactivation. J Virol 72:5343–5350
Schang LM, Hossain A, Jones C (1996) The latency-related gene of bovine herpesvirus 1 encodes a product which inhibits cell cycle progression. J Virol 70:3807–3814
Scheck AC, Wigdahl B, Rapp F (1989) Transcriptional activity of the herpes simplex virus genome during establishment, maintenance, and reactivation of in vitro virus latency. Intervirology 30:121–136
Schirm J, Mulkens PS (1997) Bell's palsy and herpes simplex. APMIS 105:815–823
Scriba M (1975) Herpes simplex infection of guinea pigs: an animal model for studying latent and recurrent herpes simplex infection. Infect Immun 12:162–165
Shearer ML, Finch SM (1964) Periodic organic psychosis associated with recurrent herpes simplex. N Eng J Med 271:494–557
Shimeld C, Whiteland JL, Nicholls SM, Easty DL, Hill TJ (1996) Immune cell infiltration in corneas of mice with recurrent herpes simplex virus disease. 77:977–985
Sinclair J, Sissons P (1997) Latent and persistent infections of monocytes and macrophages. Intervirology 39:293–301
Sinzger C, Jahn G (1997) Human cytomegalovirus cell tropism and pathogenesis. Intervirology 39:302–319
Slater JD, Gibson JS, Barnett KC, Field HJ (1992) Choriorethinopathy associated with neuropathology following infection with equine herpesvirus-1. Vet Rec 131:237–239
Slater J, Thackray A, Borchers K, Field HJ (1994) The trigeminal ganglion is a location for equine herpesvirus 1 (EHV-) latency and reactivation in the horse. J Gen Virol 75:2007–2016
Slomka MJ, Brown DW, Clewely JP, Bennett AM, Harrington L, Kelly DC (1993) Polymerase chain reaction for detection of herpesvirus simiae (B virus) in clinical specimens. Arch Virol 131:89–99
Smith J (1997) Strategies against herpes simplex virus 1. The role of vaccines. International Antiviral News 5:223–225
Spruance SL (1993) Prophylactic chemotherapy with acyclovir for recurrent herpes simpelx labialis. J Med Virol 1:27–32
Stanberry LR (1991) Evaluation of herpes simplex virus vaccines in animals: the guinea pig vaginal model. Rev Infect Dis 13:920–923
Stevens JG, Cook MI (1971) Latent herpes simplex virus in spinal ganglia of mice. Science 173:843–845
Stroop WG, Schaefer DC (1986) Production of encephalitis restricted to the temporal lobe by experimental reactivation of herpes simplex virus. J Infect Dis 153:721–731

Sugaya K, Bradley G, Nonoyama M, Tanaka A (1990) Latent transcripts of Marek's disease virus are clustered in the short and long repeat regions. J Virol 64:5773–5782

Tenser RB (1998) Trigeminal neuralgia: mechanisms of treatment. Neurology 51:17–19

Thackray AM, Field HJ (1998) Famciclovir and valaciclovir differ in the prevention of herpes simplex virus latency in mice: a quantitative study. Antimicrob Ag Chemother 42:1555–1562

Thein P, Böhm D (1976) Äthiologie und Klinik einer virusbedingten Keratokonjunktivitis beim Fohlen. Zentralbl Veterinarmed [B] 23:507–519

Thein P (1978) The association of EHV-2 infection with keratoconjunctivitis in horses and research on the occurrence of equine coital exanthema (EHV-3) in Germany. In: Bryans JT, Gerber HE (eds) Equine infectious diseases IV. Vet Publ, Princeton, pp 33–43

Ueda Y, Togaya J, Shiroki K (1968) Immunological relationship between herpes simplex virus and B virus. Arch Ges Virusforsch 24:231–244

Vizoso AD (1975) Latency of herpes simiae (B virus) in rabbits. Br J Exp Pathol 56:489–494

Wagner EK, Bloom DC (1997) Experimental investigation of herpes simplex virus latency. Clin Microbiol Rev 10:419–443

Welch HM, Bridges CG, Lyon AM, Griffiths L, Edington N (1992) Latent equid herpesviruses 1 and 4: detection and distinction using the polymerase chain reaction and co-cultivation from lymphoid tissues. J Gen Virol 73:261–268

Whitewell KE, Blunden AS (1992) Pathological findings in horses dying during an outbreak of the paralytic form of equid herpesvirus type 1 (EHV-1) infection. Equine Vet J 24:13–19

Whitley RJ, Soong SJ, Hirsch MS (1981) Herpes simplex encephalitis. Vidarabin therapy and diagnostic problems. New Eng J Med 304:313–318

Whitley RJ (1986) Therapeutic advances for severe and life-threatening herpes simplex virus infections. In: Lopez C, Roizman B (eds) Human herpesvirus infections. Raven, New York, pp 153–164

Wigdahl B, Scheck AC, Ziegler RJ, De Clercq E, Rapp F (1984a) Analysis of the herpes simplex virus genome during in vitro latency in human diploid fibroblasts and rat sensory neurons. J Virol 49:205–213

Wigdahl B, Smith CA, Traglia HM, Rapp F (1984b) Herpes simplex virus latency in isolated human neurons. Proc Natl Acad Sci USA 81:6217–6221

Wilcox CL, Johnson EM (1987) Nerve growth factor deprivation results in the reactivation of latent herpes simplex virus in vitro. J Virol 61:2311–2315

Wildy P, Field HJ, Nash AA (1982) Classical herpes latency revisited. In: Mahy BWJ, Minson AC, DaMay GD (eds) Virus persistence. Symposium 33. Soc Gen Microbiol, Cambridge University Press, pp 133–168

Wyler R, Engels M, Schwyzer M (1989) Infectious bovine rhinotracheitis/vulvovaginitis (BHV-1) In: Wittmann G (ed) Herpesvirus diseases of cattle, horses and pigs. Developments in veterinary virology. Kluwer, Dordrecht, pp 1–72

Yamanishi K, Okun T, Shiraki K, Takahashi M, Kondo T, Asano Y, Kurata T (1988) Identification of human herpesvirus-6 as a causal agent for exanthem subitum. Lancet 1:1065–1067

Programmed Cell Death in Virus Infections of the Nervous System

J.K. Fazakerley and T.E. Allsopp

1	Introduction	95
1.1	Apoptosis Is the Discrete Elimination of Cells	95
1.2	Viruses and Advantageous Cell Suicide	96
1.3	Advantageous Cell Suicide and the CNS	97
2	Apoptosis Is an Active, Gene-Dependent Program of Cell Death	98
2.1	Cell Suicide	98
2.2	Death Receptors	99
2.3	Death Regulators	99
2.4	Caspases	102
2.5	Viral Regulators of Apoptosis	102
3	Apoptosis of CNS Cells	103
3.1	Neural Cell Death in Development	103
3.2	Neuronal Cell Apoptosis and Disease	104
4	Apoptosis in Virus Infections of Neural Cells	106
4.1	Alphaviruses and CNS Cell Death	106
4.2	Rabies	109
4.3	Human Immunodeficiency Virus	109
4.4	Herpesviruses	110
4.5	Other Viruses Inducing Neural Cell Apoptosis In Vitro	110
4.6	Other Viruses Inducing Neural Cell Apoptosis In Vivo	111
5	Summary and Hypothesis	111
5.1	Multiple Factors Regulating Competence to Die	112
5.2	A Modified Altruistic Hypothesis	112
5.3	Consequence of Death Failure Following Infection	113
	References	114

1 Introduction

1.1 Apoptosis Is the Discrete Elimination of Cells

Apoptosis is the term originally coined by KERR, WYLLIE and CURRIE (1972) to describe the 'silent' disposal of cells in tissues. It is a description of the morphology adopted by cells dying in normal physiological contexts, a process evolved for the

---

Laboratory for Clinical and Molecular Virology and Fujisawa Institute of Neuroscience, University of Edinburgh, Edinburgh, UK
e-mail: John.Fazakerley@ed.ac.uk

removal of superfluous, unwanted and damaged cells during embryogenesis but also in mature tissues. Apoptosis occurs in nearly all cell types at one stage or another during the embryogenesis of multicellular organisms. The processes of involution and regression require apoptosis and classical examples are the death of interdigital mesenchymal cells during vertebrate limb development and the formation of kidney glomeruli. Apoptosis is essential in mature organ systems, for example during spermatogenesis, in the regression of the corpus luteum post-ovulation, in the turnover of specialised secretory epithelial cells lining the small intestine and in the resolution of inflammatory responses.

When cells undergo apoptosis they appear to detach themselves from surrounding cells, condense and shrink their cytoplasm and nucleus. Chromatin also condenses and marginalises to the periphery of the nuclear membrane. Shrinkage continues and terminates in a breakdown of the cell into small fragments. DNA is cleaved into regular and precise fragments during the late stages of this process. The final demise of the cell is its clearance by phagocytosis. Apoptosis is clearly distinct from necrosis, in which the damaged or diseased cell lyses, releasing its contents into the surrounding milieu and triggering an inflammatory response. Necrotic cells do not display the characteristic 'pyknotic' features of apoptotic cells and they generally occur en masse whereas, at least in developmental and physiological processes, apoptosis is typically observed in single cells. Furthermore, the DNA in necrotic cells is digested at stages after the release of lysosomal nucleases which cleave at irregular sites distinct from those cleaved during apoptosis.

Apoptosis can be triggered by a large number of physiological or pathological stimuli in a cell type- and stimulus-dependent manner. These include the presence or absence of specific hormones, e.g. glucocorticoids, growth or trophic factors, e.g. transforming growth factor (TGF)-β, nerve growth factor (NGF). Such factors can trigger cell death in certain cells but only at defined stages of tissue morphogenesis and this is dependent on the cell expressing appropriate receptors. Exposure to oxidants, chemotherapeutic agents, UV-/γ-irradiation, bacterial and viral infection and cytotoxic T cells (THOMPSON 1995) are just a small number of the pathological stimuli that have been shown to trigger apoptosis of albeit healthy cells.

1.2 Viruses and Advantageous Cell Suicide

According to the classical concepts of virology and many of today's textbooks, viruses kill the cells they infect and this cell death or cytopathic effect is the pathological process underlying many symptoms of disease. An example often given is poliovirus infection and the subsequent death of anterior horn cells in the spinal cord, the loss of which contributes to the characteristic paralysis of poliomyelitis. It is now clear however that following virus infection many cell types die by apoptosis. How did this process evolve? Have selective pressures on viruses led to the evolution of viral mechanisms to tap into and activate the cellular suicide pathways and, if so, what is the advantage to the virus and what are the selective pressures that have driven this? One possibility is that cell death allows virion release. This might be particularly important for non-enveloped viruses but would seem unnecessary for many enveloped viruses. An alternative is that it is not the viruses which have evolved to trigger the suicide response but the cells which have evolved mechanisms to commit suicide upon infection. Cells have evolved pathways to detect viral infections and may have linked these into cell death pathways. But why should a virally infected cell want to kill itself? In a multicellular organism, rapid suicide upon infection can be seen as an altruistic response. If it occurs before the virus has completed replication and assembly it might be a highly effective strategy for limiting spread of the infection. If the cell dies before infectious virions are produced, infection of other cells is effectively prevented. In this way suicide of virally infected cells can be considered as a first line of defence against infection.

In more advanced vertebrates, the fate of many virus-infected cells is nevertheless death, which is mediated by cytotoxic T cells; however this must await the priming and expansion of specific immune responses, which can take a few days. During this time the infection may have produced tens of thousands of new virions and resulted in the infection of many more cells. Suicide of the original infected cells could obviate the requirement for this later and more extensive fratricide by cytotoxic T cells. That suicide upon infection is a highly effective anti-viral mechanism evolved by the cell is supported by the presence of anti-apoptotic genes which have now been found in many, and may even be present in most, large viruses. These genes presumably function to keep the infected cell alive long enough for these large viruses, which have correspondingly long replication times, to produce infectious progeny. As mentioned previously an alternative strategy would be to complete replication and assembly of stable infectious virions very rapidly before cell death; such viruses would not have a requirement for anti-apoptotic genes and cell death may benefit or even be necessary for the virus by facilitating its release from the cell. It is likely that many of the small RNA viruses adopt this approach. An added complication and one that may be an important determinant of productive viral replication is the rate of individual cell suicide, which may differ between cell types and between differentiation and activation states.

1.3 Advantageous Cell Suicide and the CNS

The 'altruistic or advantageous cell suicide' concept would only apply to non-vital cells and cell types that are able to remain mitotic. The cells of the mature central nervous system (CNS), particularly post-mitotic differentiated neurons are likely to be different from those in other tissues. Suicide of vital irreplaceable neurons, within say the respiratory centre, would have very different consequences than suicide of skin cells. Thus we might on theoretical grounds alone anticipate that CNS neurons would have evolved not to trigger a suicide response upon infection. This would be similar to the lack of MHC I expression on functional CNS neurons which prevents surveillance of virus infection and death induction by cytotoxic T cells. Presumably, selective pressures have dictated that a live functional neuron, even a virus-infected one, is, in the balance of all considerations, better than a dead one. A few dead neurons in a vital centre may be sufficient to compromise the survival of the whole organism. Taking this argument a step further, what then are the consequences of virus infection in a mature neuron which can neither commit suicide nor be destroyed by cytotoxic T cell surveillance? One possible outcome is viral persistence, neuronal dysfunction and disease. The inability of neurons to die following many virus infections may explain the large number of viruses capable of persisting in these cells. On the other hand, the immature neurons of the developing nervous system are highly susceptible to apoptosis. What then is the consequence for virus infection of cells of the immature nervous system, suicide? Before we progress further along these roads we will first discuss the pathways of apoptosis and known points at which viruses interact with this.

2 Apoptosis Is an Active, Gene-Dependent Program of Cell Death

2.1 Cell Suicide

Apoptosis is a description of physiological or programmed cell death and can be viewed as death by suicide (or fratricide in the case of cytotoxic T cell-instigated apoptosis of target cells). Once a responsive cell receives the appropriate signal then apoptosis involves an internal program of gene function. Apoptosis regulation may require the initial expression or repression of certain genes in order to complete the cell death program. For example, fibroblasts can be sensitised to undergo apoptosis in response to tumour necrosis factor (TNF)-α if they are also treated with the protein synthesis inhibitor cycloheximide. Immature but differentiated neurons can be inhibited from undergoing apoptosis in response to trophic factor deprivation if they are exposed to protein synthesis inhibitors. Other cells appear ready and primed to undergo apoptosis, for example, peripheral T cells expressing CD95 rapidly undergo apoptosis when exposed to CD95L; blocking protein synthesis has little affect. Apoptosis is also an energy-dependent program – ATP is required for the function of key enzymes during apoptosis. This energy requirement is another distinction to necrosis.

A working scheme conveniently divides apoptosis into four distinct phases (EARNSHAW 1995; KROEMER et al. 1997). *Induction* involves the signal reception and the convergence of cell type- and/or stimulus-specific biochemical pathways and their coupling inside cells into a series of common *effector* steps. These effector points might be represented at the level of the mitochondria or endoplasmic reticulum, where protein interactions can regulate or determine the condemnation of cells to apoptosis. Once committed, cells enter an *execution* phase, mediated by enzymes whose irreversible action indicates there is no retrieval point. Unlike the periods of induction and effection, during which cells may take hours or even days to become committed, the execution phase is rapid (30–60min). Apoptotic cells in situ in the stages of effection or execution are also readily recognised and engulfed by neighbouring cells or macrophages in the *disposal* phase. Many gene products have been indirectly or directly associated with the regulation of apoptosis, and there are several detailed reviews on the pathways involved (CRYNS and YUAN 1998; RAFF 1998; ASHKENAZI and DIXIT 1998; GREEN and REED 1998; THORNBERRY and LAZEBNIK 1998; EVAN and LITTLEWOOD 1998; ADAMS and CORY 1998). At the time of writing, this is a fast moving field in which a huge amount of new information is constantly emerging. Figure 1 summarises our understanding of some of the key events to date.

2.2 Death Receptors

These receptors engage the apoptosis pathway through a region of intracellular homology designated the 'death domain'. Principle members of this burgeoning family are the tumour necrosis factor receptor-1 (TNFR-1) and CD95 (Fas or APO-1). The recent identification of new family members suggests that they may act to signal death in a highly cell-type-specific manner (CHINNAIYAN et al. 1996). The expression of 'decoy' receptors that lack the intracellular signalling domain (MACFARLANE et al. 1997) adds a degree of complexity to this system. Activation of the death signal by TNFR-1 and CD95 following receptor ligation involves the recruitment of death domain-containing adapter molecules that bind to the death domain of the receptor to form a protein-signalling complex. Both of these activated receptor types recruit FADD (Fas-associated death-domain-containing molecule; also called MORT-1), while the activated TNFR-1 in addition can bind TRADD (TNFR death domain protein), TRAF-2 (TNFR-associated factor-2), RIP (receptor-interacting protein) and RAIDD (RIP-associated Ich-1/ced-3 homologous protein with a death domain). Negative regulation of the ability of death receptors to initiate cell death when unligated has been shown recently for the naturally occurring protein SODD (silencer of death domains) (JIANG et al. 1999). Whereas the death receptors can signal for cell death, this is not the only possible outcome of their activation. There are pathways, which can also lead to survival and activation. For example, ligation of TNFR1 by TNF-α can induce cell death but it can also signal survival and production of interleukin (IL)-1β. The outcome for the cell depends upon the balance of factors such as the adapter molecules and molecules which bind to them, in this case TRADD, TRAF1,2, cIAPs (cellular inhibitors of apoptosis), RIP, RAIDD and CARDIAK (CARD-containing IL-1β converting enzyme (ICE)-associated kinase). The balance of these factors in a cell is one point at which the cell's propensity to commit suicide is regulated.

2.3 Death Regulators

The Bcl-2 family consists of proteins with either an apoptosis-suppressing (Bcl-2, Bcl-xL, Bcl-w, Mcl-1, A1), or an apoptosis-promoting function (Bax, Bak, Bok,

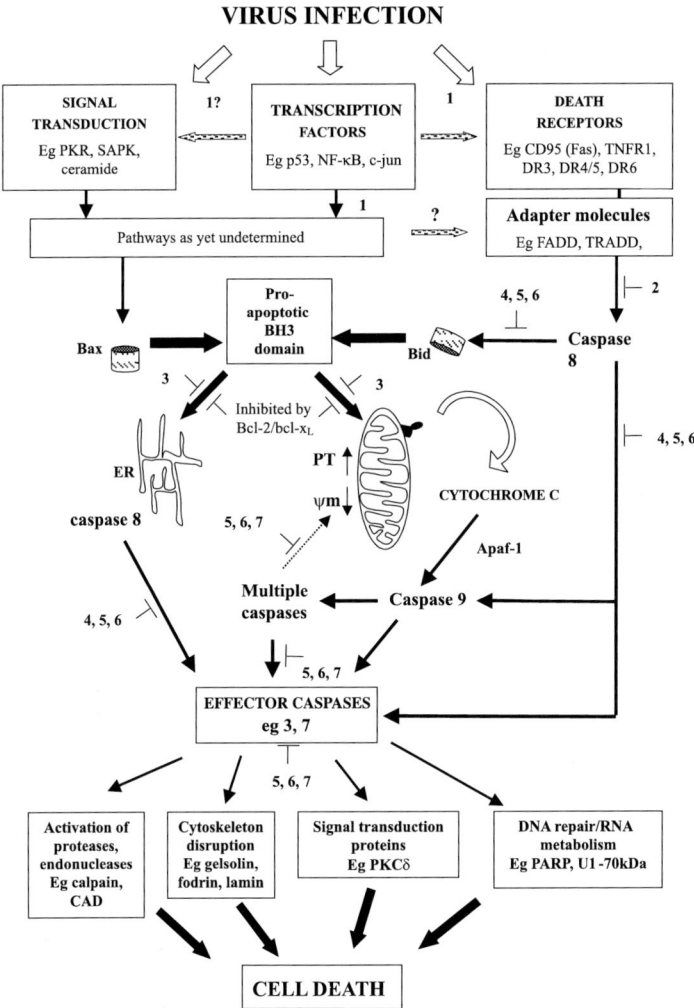

Bad, Bik/Nbk, Bim, Hrk, Bid) (KROEMER 1997). Protein domain and interaction studies have indicated that the death-suppressing members often contain a BH4 domain (Bcl-2 homology domain), although possession of a BH4 domain does not always indicate functional death suppression (INOHARA et al. 1998). In the case of Bcl-2/Bcl-xL the BH4 domain is thought to regulate the interaction with Raf-1 (kinase), calcineurin (phosphatase) and Bag-1. In addition biochemical studies in cell lines have shown that Bcl-2/Bcl-xL can bind Apaf-1 (apoptosis accelerating factor-1) thereby suppressing its function in activating apoptotic proteases. Structural and artificial phospholipid bilayer studies have previously suggested that Bcl-2, Bcl-xL and probably Bax can form membrane channels for ionic conductances. Evidence has favoured a role for anti-apoptotic proteins in preventing the release of mitochondrial cytochrome C and other apoptogenic factors (i.e. apop-

Fig. 1. Summary of known and putative intrinsic activation pathways of apoptosis and their regulation in neural cells following virus infection. A combination of in vitro and in vivo genetic evidence suggests that the core components of apoptosis regulation in death-susceptible neurons are identical to those identified in many other cell types (MICHAELIDIS et al. 1996; MOTOYAMA et al. 1995; KNUDSON et al. 1995; KUIDA et al. 1996, 1998; HAKEM et al. 1998). Central to the regulation are Bcl-2- and caspase-family members which function at intracellular membrane sites (*ER*, endoplasmic reticulum, mitochondria). *Initiation*: Cellular studies have indicated that the pro-apoptotic proteins possessing Bcl-2 homology domain (BH)-3, including Bax, Bak and Bid (ADAMS and CORY 1998), translocate upon stimulation to caspase activation sites. These proteins initiate mitochondrial changes, caspase activation and possibly modulate other organelle changes. The BH-3-domain proteins can be antagonised by Bcl-2/Bcl-xL. As a result of neural cell infection, BH-3-domain protein translocation may be indirectly initiated either: (1) subsequent to a kinase cascade requiring stress-activated protein kinases (SAPK) or protein kinase R (PKR). PKR is activated during viral replication by the formation of double-stranded RNA (WILLIAMS 1995). (2) As a result of increased gene expression by transcription factors such as p53 and c-Jun whose functional activation has been shown to correlate with apoptosis in stressed neurons; (3) Following 'death receptor' (TNFR1, DR3/4/5/6) activation of apoptosis initiator caspases. It is possible that infected neurons show enhanced sensitivity to 'death ligands' as a result of increased receptor expression. Following infection of neighbouring cells, and not indicated in the figure, inflammatory cytokines will be produced by the action of pro-inflammatory caspases (caspases 1, 4, 5). Some of these such as IL-1β, IL-18, and TNF-α may be death ligands and provoke bystander apoptosis following infection. Sensitivity to TNF-α in cell lines can be increased in a p53-dependent manner (KLEFSTROM et al. 1997). *Regulation*: Several possibilities for how Bcl-2/Bcl-xL might function include: sequestration of Apaf-1 (apoptosis accelerating factor), direct formation of pores/channels in membranes, prevention of the release of apoptogenic mitochondrial factors such as cytochrome C, caspases, AIF (apoptosis-inducing factor) or interaction with the mitochondrial voltage-dependant anion channel (VDAC) or the permeability transition pore complex. The mechanism of cytochrome C release in mammalian cells is still obscure. The figure is not intended to indicate the precise function of Bcl-2/Bcl-xL proteins nor distinguish between a direct function for Bid in causing apoptogen release or a Bid/Bax interaction in order for this to be achieved. Caspase 8-mediated bid cleavage has been shown in certain non-neuronal cells to amplify the CD95 (FasL) death trigger. *Activation*: Caspase 9 activates multiple caspases including effector caspases and these can exacerbate the increase in mitochondrial permeability transition (*PT*) and rundown of transmembrane potential (Ψ_m). Translocation of BH-3-domain proteins to the ER may facilitate the activation of caspase 8 which is associated with an ER-dependent protein complex in certain cells (NG et al. 1998). Downstream of 'death receptors' and activation of caspase 8 is the direct activation of effector caspases. The collective action of effector caspases results in either the functional inhibition or activation of proteins required to carry out apoptosis (*CAD*, caspase-activated DNase; *PKC*, protein kinase C; *PARP*, poly(ADP-ribose) polymerase). *Inhibition*: Viruses inhibit apoptosis by preventing: p53 transcriptional activity (*1* e.g. adenovirus E1B-55kD); 'death receptor' adapter molecule activation of initiator caspases (*2* e.g. γ-herpesvirus FLIPs); the pro-apoptotic function of Bax (*3* e.g. Epstein-Barr virus BHRF1); inhibition of pro-inflammatory caspases (*4* e.g. poxvirus cytokine response modulator A (crmA) protein); and general caspase activity (*5* e.g. baculovirus p35 protein, *7* viral inhibitor of apoptosis proteins – IAPs). Caspases at the apex and effector stages of apoptosis in non-neural cells can also be experimentally inhibited during infection by peptide inhibitors, for example zVADfmk (*6*)

tosis-inducing factor (AIF), proteases). In yeast Bax and Bak have individually been shown to interact with the voltage-dependent anion channel (VDAC) in mitochondria (SHIMIZU et al. 1999). This interaction increases the permeability of the channel for cytochrome C. Bcl-xL appears to antagonise this interaction. Whether this is also the case in mammalian cells remains unclear.

A common element of all the death-promoting members is the possession of a BH3 domain that may serve as an intracellular apoptotic ligand (Fig. 1). BH3-containing family members may thus work either by acting as ligands to modify the normal function of mitochondrial proteins (e.g. VDAC or the adenine nucleotide transporter, ANT), or as dimerising ligands that antagonise the physiological function of Bcl-2/Bcl-xL. Of interest is that Bcl-2/Bcl-xL also interacts with other membranous organelles (endoplasmic reticulum, nucleus) arguing that other modulators/targets of these proteins may exist in non-mitochondrial compartments.

2.4 Caspases

Caspases (cysteine-containing, aspartate proteases) reside as inactive zymogens in cells, possess an essential active-site cysteine residue and are themselves activated by proteolysis during apoptosis at aspartate-containing sites. Crystallographic studies on caspases 1 and 3 indicate that the active enzyme consists of a heterotetramer formed by two large (containing the active site) and two small subunits. Caspases play a prominent role in apoptosis as either the apical initiators of a caspase-mediated proteolytic cascade (caspases 8, 9, 10) or the terminal effectors of such a cascade (caspases 3, 6, 7) (THORNBERRY and LAZEBNIK 1998). Caspases 8 and 10 are recruited to, and activated at, the signalling complex of death receptors, although other sites of activation may exist (NG and SHORE 1998; Fig. 1). Caspase 9 is activated by the binding of Apaf-1 in combination with cytochrome C (released from mitochondria) and ATP (Fig. 1). The effector caspases degrade specific protein substrates of the cytoskeleton, RNA/DNA metabolism and signal transduction pathways. All the caspases possess an NH_2-terminal prodomain that varies in size and sequence depending on caspase identity. The large prodomain of the caspases 2, 8 and 9 in particular, has been shown to be important for determining the localisation and subsequent activation of the enzymes (LI et al. 1997a; BUTT et al. 1998; MUZIO et al. 1998).

2.5 Viral Regulators of Apoptosis

Emerging studies have indicated that viruses may use one or more of many general strategies to inhibit the cell suicide response to infection. There is insufficient space to review these here (for reviews see TEODORO and BRANTON 1997; O'BRIEN 1998; HARDWICK 1998). Such strategies are indicated in Fig. 1 and include the inhibition of transcription factors, inhibition of caspase activation by 'death receptor' ligation, suppression of death induction by pro-apoptotic BH-3 domain-containing proteins and direct inhibition of active caspases. The normal, pro-apoptotic function of the tumour suppressor and transcription factor protein p53 is targeted by adenovirus (E1B-55K), human papillomavirus (E6), SV40 (T antigen), hepatitis B virus (pX) and Epstein-Barr virus (EBV) (BZLF1). Some gammaherpesviruses encode an inhibitor, v-FLIP (FLICE-like inhibitory proteins) that prevents the recruitment and activation of caspase 8 by the FADD adapter molecule. Several viruses have been shown to encode Bcl-2 homologues. These are present in adenovirus (E1B 19K), African swine fever virus (5-HL) and the gammaherpesviruses, EBV (BHRF1), human herpesvirus-8 (KSbcl-2), herpesvirus saimiri (ORF16) and the murine gammaherpesvirus-68 (ORF M11).

Viruses can also directly inhibit activated caspases. Adenovirus (E3 14.7K protein) is able to inhibit caspase 2 and 8 activation (SHISLER and GOODING 1998). Baculovirus encodes a caspase inhibitor (p35 protein) that is very effective against apoptosis initiator and effector caspases (caspases 2, 3, 6, 7, 8, 9, 10). In contrast, cowpox virus (Crm A), and vaccinia virus (SPI-2) encode inhibitors that are more selective for the caspases involved in producing inflammatory cytokines (caspases 1, 4, 5, 11). Baculovirus also encodes a protein, IAP (inhibitor of apoptosis), which is a member of a large, evolutionary conserved protein family. Some of these, cIAP-2 and X-linked IAP (XIAP), have been shown to be direct inhibitors of effector caspases (ROY et al. 1997; DEVERAUX et al. 1997).

Some viruses have evolved to suppress apoptosis at more than one stage. For example, the gammaherpesviruses which, in addition to the suppression of p53 activity (BZLF1), inhibition of caspase 8 (vFLIP) and the presence of a Bcl-2 homologue (BHRF1), also promote survival via the LMP1 protein and the B-cell A20 protein. It is likely that these reflect strategies to inhibit the death mechanism that is regulated differently in diverse cell types, at distinct stages of the virus life cycle and at different stages of cellular differentiation.

3 Apoptosis of CNS Cells

3.1 Neural Cell Death in Development

Upon development of the vertebrate nervous system into discrete cell populations, as many as 50% of the neurons generated may die as they differentiate to form functional connections. Prior to this stage of target-dependent apoptosis, from the early stages of nervous system morphogenesis, naturally occurring cell death is also likely to regulate the size of precursor pools of neurons and glia (reviewed by BUREK and OPPENHEIM 1998). It is generally assumed that an extracellular influence that affects the magnitude of this naturally occurring cell death is the limited availability of soluble trophic factors. In addition it is clear that the intrinsic state of activity of central neurons in particular may affect viability in that deafferentation or blockade of afferent activity leads to increased apoptosis (KELLEY et al. 1997). Depolarisation may increase survival by making neurons more responsive to soluble factors or by enhancing the synthesis of endogenous autocrine-acting factors. Recently, evidence has accumulated for a role of soluble pro-apoptotic factors in regulating cell numbers during development. NGF and bone morphogenetic protein 4 (BMP-4) are two such factors. NGF produced by microglia in the chick retina positively regulates apoptosis of the low-affinity NGF-receptor-expressing (p75) neurons (FRADE and BARDE 1998), while BMP-4 induces apoptosis of rhombencephalic neural crest cells (GRAHAM et al. 1994).

A large number of studies have indicated a role for Bcl-2 family proteins and caspases in the regulation of naturally occurring apoptosis during neuronal development (reviewed in PETTMAN and HENDERSON 1998). The major apoptotic pathways that probably operate in immature neurons downstream of transduced death signals or in the absence of survival signals are summarised in Fig. 1. Genetic studies have indicated that Bcl-xL is essential for normal CNS development whereas Bcl-2 regulates the post-natal survival of certain peripheral and central neurons (MICHAELIDIS et al. 1996; MOTOYAMA et al. 1995). Animals homozygous null ($^{0/0}$) for the pro-apoptotic protein Bax display increased numbers of both peripheral and central neurons but otherwise undergo normal development (WHITE et al. 1998). Variability in the regulation of viability of diverse neuronal populations is indicated by the fact that bcl-xL$^{0/0}$ and $bax^{0/0}$ double-knockout animals show selective neuronal protection against the loss generated in bcl-xL$^{0/0}$ animals. In the rodent, bcl-2 is widely expressed during CNS development but is only present at low levels in the adult brain (MERRY et al. 1994). Virtually all neonatal neurons are bcl-x positive in the neonatal rat but expression is progressively restricted during post-natal development. In the adult expression is only observed in a few areas, including a few hypothalamic nuclei, the olfactory bulb, the hippocampus and some nuclei in the brain stem and the cerebellum (ALONSO et al. 1997). In the adult squirrel monkey brain, bcl-2 is expressed in microglia and subsets of neurons in the amygdala, hippocampus, hypothalamus, limbic cortices and striatum (BERNIER and PARENT 1998).

In the developing nervous system, a most striking neuronal phenotype is observed following deletion of the caspase 3 or caspase 9 gene. These each lead to apoptosis failure and result in widespread perturbations of cortical and forebrain development (KUIDA et al. 1996; 1998; HAKEM et al. 1998). The related brain defects observed in both of these genotypes are mimicked in the phenotype of the apoptosis-associated factor-1 ($apaf$-1)$^{0/0}$ animals (CECCONI et al. 1998; YOSHIDA et al. 1998). Apaf-1 is a factor originally identified from in vitro cell studies which, in the presence of cytochrome C and ATP, activates a caspase 9- and caspase 3-dependant pathway (LI et al. 1997b). Taken together these studies indicate that an Apaf-1 and caspase 9-mediated cascade are required for caspase 3 activation during cell death in the developing nervous system. However, these studies have not eliminated a role for a caspase 9-induced activation of effector caspases other than caspase 3 as a fraction of caspase 9 $^{0/0}$ animals remain viable post-natally.

3.2 Neuronal Cell Apoptosis and Disease

Apoptosis has prominent significance in CNS disease. It leads to cell loss in acute injury (MACMANUS and LINNIK 1997). Indicators of apoptosis including caspase 3 activation have reliably been shown in animal models of stroke and the infarct volume can be reduced by experimental suppression of apoptosis (MARTINOU et al. 1994; LODDICK et al. 1996). Cell death in ischaemia and acute head injury is thought to result from the accumulation of high local concentrations of excitatory amino acids, for example glutamate. In glutamate receptor-expressing neurons this leads to calcium influx and cell death (predominantly by necrosis). The release of low amounts of glutamate from dying cells, or an inhibition of uptake by glial cells, may sensitise neurons to apoptosis. Increased expression of death receptors and their ligands, free radical-induced damage of mitochondria or deprivation from trophic support may each lead to propagation of the lesion by apoptosis.

Apoptosis in the nervous system can be triggered experimentally by a wide variety of insults including axotomy, toxins, free radicals and irradiation. In certain experimental systems the triggering of apoptosis has been shown to be dependent on the transcription factor p53. Cerebellar granule neurons from $p53^{0/0}$, but not wild-type mice are resistant to apoptosis induced by γ-rays (WOOD and YOULE 1995) but not methyl azo-oxymethanol. Many other types of $p53^{0/0}$ neurons in the CNS (hippocampal, cortical, amygdala) are protected from death induced by kainic acid or glutamate exposure (MORRISON et al. 1996; XIANG et al. 1996). The mechanism of action of p53 in these pathological circumstances may involve the increased expression of ligands and receptors of the 'death receptor' system and pro-apoptotic BH-3-domain containing Bcl-2 family members; p53 can act as a transcription factor for the expression of these components.

Apoptosis is also likely to be important in chronic neurodegenerative disorders such as amyotrophic lateral sclerosis (ALS, motor neuron disease), Parkinson's, Huntington's and Alzheimer's diseases and the transmissible spongiform encephalopathies. The majority of clinical cases of ALS, Parkinson's and Alzheimer's are sporadic, with unidentified, crucial causes. A fraction of the cases are due to the inheritance of familial forms. For example, the majority of the familial cases of Alzheimer's disease represent mutations in the genes for presenilins 1 and 2 which have been implicated in the regulation of apoptosis (WOLOZIN et al. 1996; KIM et al. 1997; but see BURSZTAJIN et al. 1998). Damage caused by free radicals, perturbations in cellular redox state and mitochondrial dysfunction have been proposed to contribute to pathology in both Alzheimer's and Parkinson's disease (KELLER and MATTSON 1998). It is conceivable that some, if not many of those, components identified in regulating neural cell viability during development (Fig. 1) may also be involved in the death of neurons in the mature nervous system, but a precise role for apoptotic regulators and effectors in determining cell death in chronic neurodegeneration is so far not available. Recently however various approaches have indicated that the Alzheimer's proteins, amyloid precursor protein and presenilins, can be caspase substrates (GERVAIS et al. 1999; WALTER et al. 1999). Inhibiting caspase 1 slows disease progression in a mouse model of Huntington's disease (ONA et al. 1999) and caspase 8 is required for cell death induced by toxic polyglutamine repeats (SANCHEZ et al. 1999). Abnormal expression of p53 and CD95 (Fas) have been detected in Alzheimer's disease (DE LA MONTE et al. 1997). A number of studies based on TUNEL staining have reported that neuronal loss in both natural and experimental spongiform encephalopathies is apoptotic (KRETZCHMAR et al. 1997; WILLIAMS et al. 1997; see also the chapter by Glese and Kretzschmar).

Whereas in most disease situations involving apoptotic death of neural cells it is likely that the trigger for this is derived from within the nervous system, the possibility that it is initiated by other systemic events, involving for example the neuroendocrine system, should not be overlooked. For example, adrenalectomy in rats results in a highly specific apoptotic loss of hippocampal dentate gyrus granule cells (SLOVITER et al. 1993).

4 Apoptosis in Virus Infections of Neural Cells

4.1 Alphaviruses and CNS Cell Death

A number of virus infections of neural cells in culture and in vivo have been examined but perhaps the best studied are the alphaviruses Semliki Forest virus (SFV) and Sindbis virus (SV) (for reviews on these viruses see GRIFFIN and HARDWICK 1997; GRIFFIN 1998; ATKINS et al. 1999). Infection of primary cultures of mouse embryonic sensory neurons with the A7(74) or L10 strain of SFV results in death by apoptosis (ALLSOPP et al. 1998). The morphological changes of surface pitting and neurite fragmentation are identical to the changes observed during apoptotic death following withdrawal of the survival factor NGF and both processes are inhibited by the caspase peptide inhibitors zVADfmk and DEVD-aldehyde. Interestingly, cell death in these primary neuronal cultures varies according to the age of the animal from which the cultures are derived. Following infection with the A7(74) strain of SFV, embryonic neuronal cultures die more rapidly than primary neuronal cultures from neonatal mice, and neurons from adult mice do not die following infection (T. Allsopp and J.K. Fazakerley, unpublished observations). An age-related change in the susceptibility of sensory neurons is also observed with SV. Infection of rat embryonic dorsal root ganglia explants with the AR339 strain of SV results in death within 1–4 days whereas explants allowed to mature in culture for 2 weeks before infection survive for more than 14 days after infection (LEVINE et al. 1993). In contrast, infection with the L10 strain of SFV results in death of primary sensory neuronal cultures by apoptosis irrespective of the age of the animal from which the cultures were derived (T. Allsopp and J.K. Fazakerley, unpublished observations). In another series of studies (GLASGOW et al. 1997), primary neuronal and glial cell cultures prepared from neonatal rat brain were infected with the A7 or SFV4 strains of SFV. The cells were infected after 7 or 12 days in culture and death was observed over the following 4 days. Cerebellar granule cell cultures survived the infection but cultured glial cells died by apoptosis. The difference between these two studies (ALLSOPP et al. 1998; GLASGOW et al. 1997) may reflect the difference between the neuronal cells used, sensory (DRG) neurons versus cerebellar granule cells, the culture conditions, the time in culture prior to infection or the strains of virus.

Primary neural cell cultures can be difficult to obtain and maintain and a number of studies have been done with continuous neural cell lines. SV infection of rat PC12 pheochromocytoma cells results in death by apoptosis as characterised by chromatin changes and poly (ADP-ribose) polymerase cleavages (UBOL et al. 1996). In these cells nuclear changes characteristic of apoptosis are cell-cycle-dependent (G_0/G_1 but not S phase cells) and death does not involve or require an increase in intracellular calcium. Death of these cells can be delayed by expression of dominant inhibitory *ras* (JOE et al. 1996). However, it should be noted that PC12 cells are transformed cell lines and may not therefore be a good model of post-mitotic neurons in vivo.

There are also many studies on non-neural, vertebrate continuous cell lines which have examined events in alphavirus-induced cell death (LEVINE et al 1993, 1996, 1998; UBOL et al. 1994; SCALLAN et al. 1997; NAVA et al. 1998; GRANDGIRARD et al. 1998; LIANG et al. 1998). All studies indicate that alphavirus infection results in apoptotic cell death with activation of caspases along the classical pathways set out in Fig. 1. Peptide inhibitors of caspases or over-expression of *bcl*-2, *bcl*-xL, IAP or beclin can block or delay this death, at least with some strains of virus (e.g. AR339 strain of SV) if not with others (e.g. neuro-adapted SV). Interestingly, there is also evidence that infection can result in the proteolytic cleavage of *bcl*-2. The alphaviruses are generally transmitted by mosquitoes and SV infection of non-vertebrate *Aedes albopictus* cells also results in apoptosis (KARPF and BROWN 1998).

The mechanism(s) that trigger apoptosis in neural or non-neural cells infected with alphaviruses remain unclear. Indeed, they remain unclear for all RNA virus infections of all cell types. SFV- and SV-based vectors have been developed as systems for gene expression. These vectors, which often contain only the non-structural genes of the virus, mediate apoptosis upon infection or transfection into a number of cell lines. These observations suggest that the structural viral genes are not required to trigger apoptosis (GLASGOW et al. 1998). In contrast, in one study expression of the transmembrane domains of the SV envelope glycoproteins were found to be sufficient to induce cell death (JOE et al. 1998). Bcl-2-mediated protection of adenocarcinoma cells (AT3) from SV apoptosis maps to amino acid 55 of the E2 envelope glycoprotein (UBOL et al. 1994). As with probably all productive virus infections, infection of the cell and virus replication perturb or activate a number of cellular processes several of which have the potential to act as triggers of apoptosis. These include activation of the interferon (IFN) system, activation of gene expression via NF-κB and activation of the cell stress response (e.g. p38 mitogen-activated protein kinase, c-Jun NH_2-terminal kinases and heat shock proteins). These have all been reported to occur in alphavirus infections (LIN et al. 1998; NAKATSUE et al. 1998).

In the animal, CNS infection of neonatal or suckling mice with all strains of SFV, SV or other alphaviruses is invariably fatal. Following intranasal infection of neonatal mice the A7(74) strain of SFV tracks along neuronal connections into and within the olfactory bulb and then out into secondary and tertiary olfactory nuclei (OLIVER and FAZAKERLEY 1998). The olfactory pathways are well-defined and thus the course and sequence of the areas infected are known. Double immunostaining of brains sampled at various time points after infection for virus and apoptosis (TUNEL) demonstrates that in the olfactory bulb a wave of infection is followed by a wave of apoptosis. Interestingly, although many of the apoptotic cells are virally infected, there are also a significant number that do not appear to be so. It is likely that some cells die following loss of contacts with other cells or perhaps as a result of high levels of cytopathic cytokines. Infection with SFV A7(74) of suckling transgenic mice which over-express *bcl-2* in the CNS results in reduced spread of the virus in the CNS and a smaller proportion of apoptotic cells (J.K. Fazakerley et al., unpublished data). These mice also show reduced developmental cell death and have larger than normal brains (DUBOIS-DAUPHIN et al. 1994).

In a series of studies a number of potentially anti-apoptotic genes have been genetically engineered into the TE strain of SV (double subgenomic SV vector) so that the protein products of these genes are expressed in SV-infected cells. Using this elegant approach *bcl-2*, *crm*A and *beclin* have been shown to prevent SV-induced neuronal cell death in vivo and to protect suckling mice from lethal encephalitis (LEVINE et al. 1996; NAVA et al. 1998; LIANG et al. 1998). As with the NSE-bcl-2 transgenic mice, expression of these genes in infected cells results in a reduction in both CNS virus titres and the number of foci of infection, again demonstrating that these anti-apoptotic genes not only affect cell death but also virus replication and spread.

Whereas all strains of alphaviruses are virulent in neonatal and suckling mice, strains differ in virulence in adult mice. For example, the L10 strain of SFV is virulent whereas the A7(74) strain is avirulent, similarly the TE strain of SV is virulent and the 633 strain avirulent. Infection of suckling mice with the 633 strain results in widespread infection and apoptosis of infected cells in the brain and spinal cord, whereas infection of 2-week-old animals results in reduced infection and minimal apoptosis (LEWIS et al. 1996). In contrast, TE infection (virulent for adults) results in widespread apoptosis in both suckling and adult mouse CNS. These two viruses differ by only a single amino acid in the E2 envelope glycoprotein. How this difference mediates the difference in apoptosis is not clear. Similarly, the Trinidad Donkey strain of Venezuelan equine encephalitis virus is virulent in both suckling and adult mice and infection of adult mice results in widespread neuronal apoptosis (JACKSON and ROSSITER 1997a). In contrast, infection of the adult rat brain with the A7 and SFV4 strains of SFV results in death of infected neuronal cells in the forebrain and olfactory system by necrosis (SAMMIN et al. 1999). However, infection of cells of the rostral migratory stream resulted in death by apoptosis. Apoptotic death in these cells of the adult rat brain is particularly interesting since these cells are immature and retain the ability to differentiate into mature neurons.

Despite, all the studies demonstrating that expression of anti-apoptotic genes in infected CNS cells can protect suckling mice from nominally virulent alphavirus encephalitides, it is not clear how this is mediated. The effect may result from the prolonged survival of infected cells expressing these anti-apoptotic genes or the reduced viral load and infected cell number which results from the effects of these genes on viral replication. Furthermore, it remains to be determined whether the age-related virulence of alphaviruses reflects maturational changes in CNS cell susceptibility to apoptosis or perhaps other age-related events such as a severe stress response or cytokine production (TRGOVICICH et al. 1997; LIANG et al. 1999).

Of perhaps greater direct relevance to human health is the related Rubella virus. This virus of the *Rubivirus* genus is, like the alpha-viruses, a member of the

Togaviridae. Rubella virus has been shown to induce apoptosis in cultured cell lines and, as with the alphaviruses, in some cell lines this can be prevented by over expression of *bcl*-2 (PUGACHEV and FREY 1998; DUNCAN et al. 1999). It is highly likely that many of the congenital abnormalities produced by this virus result from apoptosis of infected cells in the developing embryo. Developmental changes in the susceptibility of different cell populations results in a variety of malformations that depend upon the time of gestation at which infection occurs.

4.2 Rabies

Intracerebral inoculation of adult mice with the CVS strain of rabies virus results in apoptosis of infected neurons within 4 days. This is particularly marked in the pyramidal neurons of the hippocampus and neurons scattered in the cerebral cortex, but other areas including the basal ganglia, thalamus and brain stem are also involved (JACKSON and ROSSITER 1997b). Cells can be double-labelled for viral antigens and apoptosis (TUNEL) and increased immunostaining for the Bax protein is present in the areas of infection. Interestingly, although many cerebellar Purkinje cells are infected these do not show morphological changes. Immunosuppression studies suggest that the apoptosis observed in rabies-infected adult mouse brains is independent of cellular immune responses (THEERASURAKARN and UBOL 1998). A similar but more rapid and more extensive apoptosis is observed in suckling mice (JACKSON and PARK 1998; THEERASURAKARN and UBOL 1998). Again, viral protein and TUNEL-positive neurons are readily observed. As in adult mice, Purkinje cells are infected but are not apoptotic. A notable difference to adult mice is infection and apoptosis of hippocampal dentate gyrus cells. Neuronal apoptosis was detectable within 25h, before inflammatory responses or upregulation of IL-1β, TNF-α or IFN-γ transcription. Apoptotic, TUNEL-positive, neurons in the hippocampus and brain stem have been described in a human brain from a patient with rabies and AIDS (ADLE-BIASSETTE et al. 1996). Rabies virus has also been shown to induce apoptosis in cultured neuroblastoma cells (THEERASURAKARN and UBOL 1998).

4.3 Human Immunodeficiency Virus

The neuropathology of HIV encephalitis includes brain atrophy, reactive gliosis, demyelination, microglial nodules, multinucleate giant cells and neuronal loss. The neuronal loss is from discrete brain areas including the neocortex and basal ganglia. Using TUNEL a number of studies have observed apoptotic neurons in these areas (PETITO and ROBERTS 1995; ADLE-BIASSETTE et al. 1995; GELBARD et al. 1995; SHI et al. 1996) and gel electrophoresis shows DNA fragmentation. Apoptotic astrocytes and, rarely, apoptotic multinucleate giant cells are also observed. A comparative study of HIV-1 positive pre-AIDS patients, patients with AIDS but no neurological illness and patients with HIV encephalitis observed apoptotic cells in some patients in each group but the frequency of positive cases increased through this series (AN et al. 1996). In a study of pediatric patients with HIV-1 encephalitis and progressive encephalopathy, a spatial association was observed between HIV-infected inflammatory cells and apoptotic neurons (GELBARD et al. 1995). However, in most studies, the degree of neuronal loss appears to be more extensive than the infection and there is no spatial correlation between the two. Given this difference in location and given that HIV infects predominantly microglial cells and rarely neurons, neuronal apoptosis is generally considered to result from some indirect mechanism. Soluble viral products such as the surface envelope glycoprotein gp120 or the viral Tat protein or soluble cellular products such as cytokines, particularly TNF-α or nitric oxide (NO), have been postulated to mediate this death. In primary human fetal brain cell cultures, HIV, gp120 and Tat are each able to induce apoptotic death of neurons, and TNF-α can potentiate the effect of Tat (LANNUZEL et al. 1997; NEW et al. 1997; SHI et al. 1996, 1998). There is evidence that gp120 induces apoptosis following its binding to cell surface chemokine receptors (HESSELGESSER et al. 1998). Simian immunodeficiency virus infection of macaque brains also results in apoptosis of neuronal, endothelial and glial cells (ADAMSON et al. 1996).

4.4 Herpesviruses

The outcome of infection of cultured cells with herpes simplex virus (HSV) varies; although frequently destructive this death is often necrotic not apoptotic. The virus carries at least three genes ($\gamma(1)34.5$, Us3 and ICP27) which when deleted result in an increase in apoptosis upon infection, at least in some cell types (CHOU and ROIZMAN 1992; LEOPARDI et al. 1997; AUBERT and BLAHO 1999). For example, the $\gamma(1)34.5$ gene suppresses apoptosis in human neuroblastoma cells but not Vero or HEp-2 cells. In vivo, experimental studies have demonstrated neuronal apoptosis following HSV infection. Following intravitreal infection in the mouse, HSV (strain F) spreads to the CNS resulting in neuronal apoptosis. Apoptosis is more extensive in IFN-$\gamma^{0/0}$ than control mice (GEIGER et al. 1997). Skin injection and dissemination of HSV-2 (186) into mouse dorsal root ganglia results in infection and apoptosis of cells in both ganglia and spinal cord (OZAKI et al. 1997).

Like rubella virus, human cytomegalovirus (CMV) can produce neurological birth defects. Murine CMV infection of the developing mouse brain results in apoptosis. As in HIV, this is likely to result from an indirect mechanism since in this case the apoptotic cells appear not to be virally infected (KOSUGI et al. 1998).

4.5 Other Viruses Inducing Neural Cell Apoptosis In Vitro

Japanese encephalitis (JE) is a major cause of human encephalitis and is most devastating in children. Infection of N18 murine neuroblastoma cells and human NT2 neural progenitor cells with JE virus (JEV) results in death by apoptosis (LIAO et al. 1998). Intriguingly, in contrast to its effect on SV, expression of *bcl*-2 in N18 cells fails to block JEV-induced apoptosis. Infection of neuronal cells by the related dengue virus also results in apoptosis (DESPRES et al. 1996; MARIANNEAU et al. 1998). Infection of N18 neuroblastoma cells with La Crosse virus results in apoptosis which can be inhibited by *bcl*-2 (PEKOSZ et al. 1996). The BeAn strain of Theiler's virus has been shown to infect cells of the monocyte/macrophage lineage, including CNS microglial cells. Cultured macrophages can be infected and susceptibility to apoptosis depends upon their differentiation state (JELACHICH et al. 1999).

4.6 Other Viruses Inducing Neural Cell Apoptosis In Vivo

By double-labelling techniques, a study of human brains from three cases of subacute sclerosing panencephalitis demonstrated apoptosis of neurons, oligodendrocytes, microglia and lymphocytes in all of three brains studied but not in controls (McQUAID et al. 1997). Following intracerebral injection into neonatal mouse brain, La Crosse virus results in infection and apoptotic death of CNS cells (PEKOSZ et al. 1996) as does the T3D (Dearing) strain of reovirus type 3 (OBERHAUS et al. 1997) and human isolates of dengue virus (DESPRES et al. 1998). Infection of the mouse CNS with vesicular stomatitis virus results in apoptosis which is particularly striking in the olfactory bulb (BI et al. 1995). Theiler's virus infection of mice causes apoptosis of both neurons and oligodendrocytes (TSUNODA et al. 1997). The virulent GDVII strain results in a more widespread infection and greater apoptosis than does the avirulent DA strain.

5 Summary and Hypothesis

Apoptosis is a normal physiological process in the developing nervous system and serves to sculpt neuronal circuits. In the developing nervous system, at least in the rodent, infection with many neurotropic viruses is highly virulent and results in widespread apoptosis. This has now been observed for SFV, SV, rabies virus, La Crosse virus, reovirus and dengue virus and it is likely to occur in many other neurotropic virus infections. In the developing nervous system many neurons are susceptible to apoptosis and many that do not form correct connections are eliminated by this process. It is likely that immature developing glial cells are also highly susceptible to apoptosis and that numbers of these cells are also regulated by this process. As argued earlier, apoptosis of a replaceable cell upon infection may be an altruistic response designed to limit virus spread. Death of infected immature cells in the developing nervous system can be seen in this context. However, not all neurons in the developing mouse brain are equally susceptible; although many rabies-infected neurons die by apoptosis in the suckling mouse brain, Purkinje cells become infected but do not die (JACKSON and PARK 1998), and neuronal susceptibility to infection and apoptosis following SFV infection changes in different neuronal populations at different times between birth and 2 weeks of age (OLIVER and FAZAKERLEY 1998; J.K. Fazakerley et al., unpublished data). It is also worth noting that not all neurotropic viruses are virulent in the developing rodent nervous system. Infection of neonatal mice with lymphocytic choriomeningitis virus (LCMV) results in a life-long persistent infection. It remains unclear why this simple RNA virus does not induce apoptosis in these immature neurons.

5.1 Multiple Factors Regulating Competence to Die

With maturation most neurons may increase their resistance to apoptosis, though it is likely that this varies substantially between neuronal populations. Similar changes may also occur in glial cells. Susceptibility to apoptosis depends on the distinct cell type expression pattern of regulatory proteins and their relative expression levels. These include death receptors (e.g. CD95, TNFR) and endogenous inhibitors of their activation, signal transduction components (e.g. FADD, TRADD, TRAF, RIP), apoptotic regulators (e.g. Bax, Bak, Bad, Bid, Bcl-2, Bcl-xL, Bcl-w) and effector molecules such as the caspases and their inhibitors. For most of these, developmental changes in neural cell expression or differential expression between neural cell populations remains unknown. Where it has been looked at, there are major changes in expression of *bcl*-2 and *bcl*-x with development and there are clear differences in expression of these anti-apoptotic regulators between different neuronal populations in the adult. However, the age-related developmental changes observed are from high to low expression and are the reverse of those that might be expected given the relative resistance of adult neurons to apoptosis. This resistance is likely therefore to be mediated by up- or down-regulation of other apoptotic gene(s) or by variation in one or more of the other components of the death pathway. One candidate would be beclin as this anti-apoptotic regulator is expressed in the mature nervous system (LIANG et al. 1998). An alternative would be that mature neurons have to acquire a 'competence to die' before they become susceptible to typical death-inducing stimuli; a threshold that needs to be exceeded to trigger death. Experimental evidence indicates that uncharacterised factors are required in order for NGF-deprived, mature sympathetic neurons to gain death competence in response to intracellular cytochrome C injection (DESHMUKH and JOHNSON Jr 1998).

5.2 A Modified Altruistic Hypothesis

In contrast to the situation in the developing nervous system, there are fewer studies of viral induction of apoptosis in the mature nervous system and the situation seems to vary between viruses and strains. Two studies on the SFV4 strain of SFV in the mature rodent brain report that any neuronal cell death with this virulent strain is by necrosis not apoptosis (GLASGOW et al. 1997; SAMMIN et al. 1999), and following infection with the 633 strain of SV, few apoptotic cells are observed in the brains of mice 2 weeks or more of age (LEWIS et al. 1996). However, the situation cannot be generalised since in the adult rodent brain infected neurons of the rostral migratory stream (SAMMIN et al. 1999) and the olfactory bulb (J.K. Fazakerley, unpublished results) die by apoptosis. Interestingly, in the otherwise mature adult brain these are both renewable immature neuronal populations.

The situation in the olfactory bulb is particularly interesting. Neurons in the olfactory bulb are formed late in development and continue to turn over in the adult. The reason for this is not clear – it may be linked to a requirement for ongoing neuronal plasticity. However, an additional possibility is that, as the neurons of the olfactory nerve provide direct access to the CNS from the environment, olfactory bulb cells might remain highly susceptible to apoptosis to prevent spread of viruses into the CNS along this route. If correct this would be a clear example of the protective effect of an altruistic suicide response to infection in CNS cells.

In contrast to the situation described above, many neurotropic viruses do trigger neuronal apoptosis in the adult CNS; this is the case for the TE strain of SV, the Trinidad Donkey strain of Venezuelan equine encephalitis virus, the F strain of herpes simplex virus and vesicular stomatitis virus (LEWIS et al. 1996; JACKSON and ROSSITER 1997a; GEIGER et al. 1997; BI et al. 1995). Neuronal cell death in these and other infections could be an active viral function that overcomes the expression of endogenous cellular factors which function to maintain cell viability and designed to facilitate viral spread. Alternatively, this cell death could result from some aspect of these infections which exceeds a threshold required to trigger cell death.

On the basis of all the data to date, we can modify our previously stated altruistic cell suicide hypothesis to predict that cell suicide is triggered in immature cells by many, if not all, viruses but is only triggered in valuable, highly differentiated mature cells by specific conditions present only during infection with some viruses. According to this modified hypothesis, immature replaceable cells would rapidly undergo apoptosis so as to limit spread of the infection, whereas mature irreplaceable cells would only die under tightly regulated circumstances that were dominant over protective mechanisms.

5.3 Consequence of Death Failure Following Infection

In the absence of immune responses, the avirulent alphaviruses can persist in the mature rodent CNS, apparently without causing damage and essentially for life (LEVINE et al. 1991; AMOR et al. 1996). In this situation mature CNS cells fail to initiate a suicide response and in the absence of antibodies virus persists. The situation is essentially similar to persistent LCMV infection and is in marked contrast to the widespread destruction produced by these same alphaviruses in the developing nervous system. Many other viruses including herpes simplex, varicella zoster, HIV, HTLV-I, measles, mumps, rubella, Borna virus and picornaviruses have been reported to persist in the mature rodent or human CNS. Some of these are DNA viruses which can initiate latency but many are RNA viruses which, with the exception of retroviruses, cannot go latent and must therefore persist in some other form. It seems likely that the relative resistance of irreplaceable post-mitotic neurons to apoptosis upon infection is one factor which allows some RNA viruses to persist. According to our modified altruistic cell suicide hypothesis, the virus infects these neurons but the stringent conditions for suicide are not met, the cells survive and in the absence of an effective immune response the virus persists. The requirements for apoptosis could continue to be absent in a persistently infected cell or with time a threshold could eventually be reached triggering the suicide response. This could lead to a slow but progressive loss of neurons which may only be observed by careful quantitation over time. Measles virus persistence in neurons and oligodendrocytes in subacute sclerosing panencephalitis may be an example of RNA virus persistence in the relatively apoptosis-resistant cells of the human CNS. Just as mature irreplaceable neurons do not readily commit suicide upon infection, lack of MHC class I expression ensures that they do not participate in surveillance and destruction by cytotoxic T cell responses. Antibody responses, however, are highly effective in reducing virus infections of neurons (LEVINE et al. 1991; AMOR et al. 1996).

The understanding that non-latent viruses may be able to persist for long periods in mature CNS cells which do not readily undergo apoptosis may be important in understanding neurological diseases suggested to have a viral etiology. These include post-polio syndrome, motor neuron disease, multiple sclerosis, Alzheimer's disease and Parkinsonian conditions. Whether this non-latent viral persistence in the mature CNS might lead to impaired neuronal function and possibly neuropsychiatric disease is an intriguing question.

References

Adams JM, Cory S (1998) The Bcl-2 protein family: arbiters of cell survival. Science 281:1322–1326
Adamson DC, Dawson TM, Zink MC, Clements JE, Dawson VL (1996) Neurovirulent simian immunodeficiency virus infection induces neuronal, endothelial, and glial apoptosis. Molec Med 2:417–428
AdleBiassette H, Levy Y, Colombel M, Poron F, Natchev S, Keohane C, Gray F (1995) Neuronal apoptosis in HIV-infection in adults. Neuropath Appl Neurobiol 21:218–227
AdleBiassette H, Bourhy H, Gisselbrecht M, Chretien F, Wingertsmann L, Baudrimont M, Rotivel Y, Godeau B, Gray F (1996) Rabies encephalitis in a patient with AIDS: a clinicopathological study. Acta Neuropath 92:415–420
Allsopp TE, Scallan MF, Williams A, Fazakerley JK (1998) Virus infection induces neuronal apoptosis: A comparison with trophic factor withdrawal. Cell Death Differentiation 5:50–59
Alonso G, Guillemain I, Dumoulin A, Privat A, Patey G (1997) Immunolocalization of Bcl-x(L/S) in the central nervous system of neonatal and adult rats. Cell Tiss Res 288:59–68
Amor S, Scallan MF, Morris MM, Dyson H, Fazakerley JK (1996) Role of immune responses in protection and pathogenesis during Semliki Forest virus encephalitis. J Gen Virol 77:281–291
An SF, Giometto B, Scaravilli T, Tavolato B, Gray F, Scaravilli FJ (1996) Programmed cell death in brains of HIV-1-positive AIDS and pre-AIDS patients. Acta Neuropathol 91:169–173

Ashkenazi A, Dixit VM (1998) Death receptors: signaling and modulation. Science 281:1305–1308
Atkins GJ, Sheahan BJ, Liljestrom P (1999) The molecular pathogenesis of Semliki Forest virus: a model made useful. J Gen Virol 80:2287–2297
Aubert M, Blaho JA (1999) The herpes simplex virus type 1 regulatory protein ICP27 is required for the prevention of apoptosis in infected human cells. J Virol 73:2803–2813
Bernier PJ, Parent A (1998) The anti-apoptosis bcl-2 proto-oncogene is preferentially expressed in limbic structures of the primate brain. Neurosci 82:635–640
Bi ZB, Quandt P, Komatsu T, Barna M, Reiss CS (1995) IL-12 promotes enhanced recovery from vesicular stomatitis-virus infection of the central-nervous-system. J Immunol 155:5684–5689
Butt AJ, Harvey NL, Parasivam G, Kumar S (1998) Dimerisation and autoprocessing of the nedd2 (caspase 2) precursor requires both the prodomain and the carboxy-terminal regions. J Biol Chem 273:6763–6768
Burek MJ, Oppenheim RW (1998) Cellular interactions that regulate programmed cell death in the developing vertebrate nervous system. In: Koliatsos V, Ratan R (eds) Cell death and diseases of the nervous system. Human Press, Totowa
Bursztajn S, DeSouza R, McPhie DL, Berman SA, Shioi J, Robakis NK, Neve RL (1998) Overexpression in neurons of human presenilin-1 or a presenilin-1 familial Alzheimer disease mutant does not enhance apoptosis. J Neurosci 18:9790–9799
Cecconi F, Alvarez-Bolado G, Meyer BI, Roth KA, Gruss P (1998) Apaf1 (CED-4 homologue) regulates programmed cell death in mammalian development. Cell 94:727–737
Chinnaiyan AM, O'Rourke K, Yu G-L, Lyons RH, Garg M, Duan DR, Xing L, Gentz R, Ni J Dixit VM (1996) Signal transduction by DR3, a death domain-containing receptor related to TNFR-1 and CD95. Science 274:990–992
Chou J, Roizman B (1992) The gamma-134.5 Gene of herpes-simplex virus-1 precludes neuroblastoma-cells from triggering total shutoff of protein-synthesis characteristic of programmed cell-death in neuronal cells. Proc Natl Acad Sci USA 89:3266–3270
Cryns V, Yuan J (1998) Proteases to die for. Genes Dev 12:1551–1570
De la Monte SM, Sohn YK, Wards JR (1997) Correlates of p53 and Fas(CD95)-mediated apoptosis in Alzheimer's disease. J Neurol Sci 152:73–83
Deshmukh M, Johnson Jr EM (1998) Evidence of a novel event during neuronal death: development of competence-to-die in response to cytoplasmic cytochrome C. Neuron 21:695–705
Despres P, Frenkiel MP, Ceccaldi PE, DosSantos CD, Deubel V (1998) Apoptosis in the mouse central nervous system in response to infection with mouse-neurovirulent dengue viruses. J Virol 72:823–829
Deveraux QL, Takahashi R, Salvesen GS, Reed JC (1997) X-linked IAP is a direct inhibitor of cell-death proteases. Nature 388:300–304
Dubois-Dauphin M, Frankowski H, Tsujimoto Y, Huarte J, Martinou J-C (1994) Neonatal motoneurons overexpressing the bcl-2 protooncogene in transgenic mice are protected from axotomy-induced cell death. Proc Natl Acad Sci USA 91:3309–3313
Duncan R, Muller J, Lee N, Esmaili A, Nakhasi HL (1999) Rubella virus-induced apoptosis varies among cell lines and is modulated by Bcl-X-L and caspase inhibitors. Virology 255:117–128
Earnshaw WC (1995) Apoptosis: lessons from in vitro systems. Trends Cell Biol 5:217–220
Evan G, Littlewood T (1998) A matter of life and cell death. Science 281:1317–1322
Frade JM, Barde Y-A (1998) Microglia derived nerve growth factor causes cell death in the developing retina. Neuron 20:35–41
Geiger KD, Nash TC, Sawyer S, Krahl T, Patstone G, Reed JC, Krajewski S, Dalton D, Buchmeier MJ, Sarvetnick N (1997) Interferon-gamma protects against herpes simplex virus type 1-mediated neuronal death. Virology 238:189–197
Gelbard HA, James HJ, Sharer LR, Perry SW, Saito Y, Kazee AM, Blumberg BM, Epstein LG (1995) Apoptotic neurons in brains from pediatric-patients with HIV-1 encephalitis and progressive encephalopathy. Neuropath Appl Neurobiol 21:208–217
Gervais FG, Xu D, Robertson GS, Vaillancourt JP, Zhu Y, Huang J, LeBlanc A, Smith D, Rigby M, Shearman MS, Clarke EE, Zheng H, Van der Ploeg LHT, Ruffolo SC, Thornberry NA, Xanthoudakis S, Zamboni RJ, Roy S, Nicholson DW (1999) Involvement of caspases in proteolytic cleavage of Alzheimer's amyloid-ss precursor protein and amyloidogenic Ass peptide formation. Cell 97:395–406
Glasgow GM, Mcgee MM, Sheahan BJ, Atkins GJ (1997) Death mechanisms in cultured cells infected by Semliki Forest virus. J Gen Virol 78:1559–1563
Glasgow GM, McGee MM, Tarbatt CJ, Mooney DA, Sheahan BJ, Atkins GJ (1998) The Semliki Forest virus vector induces p^{53}-independent apoptosis. J Gen Virol 79:2405–2410

Graham A, Francis-West P, Brickell P, Lumsden A (1994) The signalling molecule BMP-4 mediates apoptosis in the rhombencephalic neural crest. Nature 372:684–686

Grandgirard D, Studer E, Monney L, Belser T, Fellay I, Borner C, Michel MR (1998) Alphaviruses induce apoptosis in Bcl-2-overexpressing cells: evidence for a caspase-mediated, proteolytic inactivation of Bcl-2. Embo J 17:1268–1278

Green DR, Reed JC (1998) Mitochondria and apoptosis. Science 281:1309–1312

Griffin DE, Hardwick JM (1997) Regulators of apoptosis on the road to persistent alphavirus infection. Ann Rev Microbiol 51:565–592

Griffin DE (1998) A review of alphavirus replication in neurons. Neurosci Biobehav Rev 22:721–723

Hakem R, Hakem A, Duncan GS, Henderson JT, Woo M, Soengas MS, Elia A, de la Pompa JL, Kagi D, Khoo W, Potter J, Yoshida R, Kaufman SA, Lowe SW, Penninger JM, Mak TW (1998) Differential requirement for caspase 9 in apoptotic pathways in vivo. Cell 94:339–352

Hardwick JM (1998) Viral interference with apoptosis. Semins Cell Devel Biol 9:339–349

Hesselgesser J, Taub D, Baskar P, Greenberg M, Hoxie J, Kolson DL, Horuk R (1998) Neuronal apoptosis induced by HIV-1 gp120 and the chemokine SDF-1 alpha is mediated by the chemokine receptor CXCR4. Curr Biol 8:595–598

Inohara N, Ekhterae D, Garcia I, Carrio R, Merino J, Merry A, Chen S, Nunez G (1998) Mtd, a novel bcl-2 family member activates apoptosis in the absence of heterodimerisation with bcl-2 and bcl-xL. J Biol Chem 273:8705–8710

Jackson AC, Rossiter JP (1997a) Apoptotic cell death is an important cause of neuronal injury in experimental Venezuelan equine encephalitis virus infection of mice. Acta Neuropath 93:349–353

Jackson AC, Rossiter JP (1997b) Apoptosis plays an important role in experimental rabies virus infection. J Virol 71:5603–5607

Jackson AC, Park H (1998) Apoptotic cell death in experimental rabies in suckling mice. Acta Neuropath 95:159–164

Jelachich ML, Bramlage C, Lipton HL (1999) Differentiation of M1 myeloid precursor cells into macrophages results in binding and infection by Theiler's murine encephalomyelitis virus and apoptosis. J Virol 73:3227–3235

Jiang Y, Woronicz JD, Liu W, Goeddel DV (1999) Prevention of constitutive TNF receptor-1 signaling by silencer of death domains. Science 283:543–546

Joe AK, Ferrari G, Jiang HH, Liang XH, Levine B (1996) Dominant inhibitory ras delays Sindbis virus-induced apoptosis in neuronal cells. J Virol 70:7744–7751

Joe AK, Foo HH, Kleeman L, Levine B (1998) The transmembrane domains of Sindbis virus envelope glycoproteins induce cell death. J Virol 72:3935–3943

Karpf AR, Brown DT (1998) Comparison of Sindbis virus-induced pathology in mosquito and vertebrate cell cultures. Virology 240:193–201

Keller JN, Mattson MP (1998) Roles of lipid peroxidation in modulation of cellular signalling pathways, cell dysfunction and death in the nervous system. Rev Neurosci 9:105–116

Kelley MS, Lurie DI, Rubel EW (1997) Rapid regulation of cytoskeletal proteins and their mRNAs following afferent deprivation in the avian cochlear nucleus. J Comp Neurol 389:469–483

Kerr JFR, Wyllie AH, Currie AR (1972) Apoptosis: A basic biological phenomenon with wide-ranging implications for tissue kinetics. Br J Cancer 26:239–257

Kim T-W, Pettingell WH, Jung Y-K, Kovacs DM, Tanzi RE (1997) Alternative cleavage of Alzheimer-associated presenilins during apoptosis by a caspase –3 family protease. Science 277:373–376

Klefstrom J, Arighi E, Littlewood T, Jaattela M, Saksela E, Evan GI, Alitalo K (1997) Induction of TNF-sensitive cellular phenotype by c-Myc involves p53 and impaired NF-kappaB activation. Embo J 16:7382–7392

Knudson CM, Tung KS, Tourtellotte WG, Brown GA, Korsmeyer SA (1995) Bax-deficient mice with lymphoid hyperplasia and male germ cell death. Science 270:96–99

Kosugi I, Shinmura Y, Li RY, AibaMasago S, Baba S, Miura K, Tsutsui Y (1998) Murine cytomegalovirus induces apoptosis in non-infected cells of the developing mouse brain and blocks apoptosis in primary neuronal culture. Acta Neuropath 96:239–247

Kretzschmar HA, Giese A, Brown DR, Herms J, Keller B, Schmidt B, Groschup M (1997) Cell death in prion disease. J Neural Transm Suppl 191–210

Kroemer G, Zamzami N, Susin SA (1997) Mitochondrial control of apoptosis. Immunol Today 18: 44–51

Kroemer G (1997) The proto-oncogene bcl-2 and its role in regulating apoptosis. Nat Med 3:614–620

Kuida K, Zheng TS, Na S, Kuan C-YI, Yang D, Karasuyama H, Rakic P, Flavell RA (1996) Decreased apoptosis in the brain and premature lethality in CPP32-deficient mice. Nature 384:368–372

Kuida K, Haydar TF, Kuan CY, Gu Y, Taya C, Karasuyama H, Su MSS, Rakic P, Flavell RA (1998) Reduced apoptosis and cytochrome C-mediated caspase activation in mice lacking caspase 9. Cell 94:325–337

Lannuzel A, Barnier JV, Hery C, VanTan H, Guibert B, Gray F, Vincent JD, Tardieu M (1997) Human immunodeficiency virus type 1 and its coat protein gp120 induce apoptosis and activate JNK and ERK mitogen-activated protein kinases in human neurons. Ann Neurol 42:847–856

Leopardi R, VanSant C, Roizman B (1997) The herpes simplex virus 1 protein kinase U(s)3 is required for protection from apoptosis induced by the virus. Proc Natl Acad Sci USA 94:7891–7896

Levine B, Hardwick JM, Trapp BD, Crawford TO, Bollinger RC, Griffin DE (1991) Antibody-mediated clearance of alphavirus infection from neurons. Science 254:856–860

Levine B, Huang Q, Isaacs JT, Reed JC, Griffin DE, Hardwick JM (1993) Conversion of lytic to persistent alphavirus infection by the bcl-2 cellular oncogene. Nature 361:739–742

Levine B, Goldman JE, Jiang G, Griffin D, Hardwick JM (1996) bcl-2 protects mice against fatal alphavirus encephalitis. Proc Natl Acad Sci USA 93:4810–4815

Levine B, Rosen A, Veliuona MA, Clem RJ, Hardwick JM (1998) Sindbis virus induces apoptosis through a caspase-dependent crmA-sensitive pathway. J Virol 72:452–459

Lewis J, Wesselingh SL, Griffin DE, Hardwick JM (1996) Alphavirus induced apoptosis in mouse brains correlates with neurovirulence. J Virol 70:1828–1835

Li H, Bergeron L, Cryns V, Pasternack MS, Zhu H, Shi L, Greenberg A, Yuan J (1997) Activation of caspase 2 in apoptosis. J Biol Chem 272:21010–21017

Li P, Nijhawan D, Budihardjo I, Srinivasula SM, Ahmad M, Alnemri ES, Wang X (1997) Cytochrome C and dATP-dependent formation of apaf-1/caspase 9 complex initiates an apoptotic protease cascade. Cell 91:479–489

Liang XH, Kleeman LK, Jiang HH, Gordon G, Goldman JE, Berry G, Herman B, Levine B (1998) Protection against fatal Sindbis virus encephalitis by Beclin, a novel Bcl-2-interacting protein. J Virol 72:8586–8596

Liang XH, Goldman JE, Jiang HH, Levine B (1999) Resistance of interleukin-1 beta-deficient mice to fatal Sindbis virus encephalitis. J Virol 73:2563–2567

Liao CL, Lin YL, Shen SC, Shen JY, Su HL, Huang YL, Ma SH, Sun YC, Chen KP, Chen LK (1998) Antiapoptotic but not antiviral function of human bcl-2 assists establishment of Japanese encephalitis virus persistence in cultured cells. J Virol 72:9844–9854

Lin KI, DiDonato JA, Hoffmann A, Hardwick JM, Ratan RR (1998) Suppression of steady-state, but not stimulus-induced NF-kappa B activity inhibits alphavirus-induced apoptosis. J Cell Biol 141:1479–1487

Loddick SA, MacKenzie A, Rothwell NJ (1996) An ICE inhibitor, z-VAD-DCB attenuates ischaemic brain damage in the rat. Neuroreport 7:1465–1468

MacFarlane M, Ahmad M, Srinivasula SM, Fernandes-Alnemri T, Cohen GM, Alnemri ES (1997) Identification and molecular cloning of two novel receptors for the cytotoxic ligand TRAIL. J Biol Chem 272:25417–25420

MacManus JP, Linnik MD (1997) Gene expression induced by cerebral ischemia: an apoptotic perspective. J Cereb Blood Flow Metab 17:815–832

Marianneau P, Flamand M, Courageot MP, Deubel V, Despres P (1998) Apoptotic cell death triggered by dengue virus infection: implications in viral pathogenesis. Ann Biol Clin (Paris) 56:395–405

Martinou J-C, Dubois-Dauphin M, Staple JK, Rodriguez I, Frankowski H, Missotten M, Albertini P, Talabot D, Catsicas S, Pietra C, Huarte J (1994) Overexpression on Bcl-2 in transgenic mice protects neurons from naturally occurring cell death and experimental ischemia. Neuron 13:1017–1030

McQuaid S, McMahon J, Herron B, Cosby SL (1997) Apoptosis in measles virus-infected human central nervous system tissues. Neuropath Appl Neurobiol 23:218–224

Merry DE, Veis DJ, Hickey WF, Korsmeyer SJ (1994) Bcl-2 protein expression is widespread in the developing nervous system and maintained in the adult PNS. Development 120:301–311

Michaelidis TM, Sendtner M, Cooper JD, Airaksinen MS, Holtmann B, Meyer M, Thoenen H (1996) Inactivation of bcl-2 results in progressive degeneration of motoneurons, sympathetic and sensory neurons during early postnatal development. Neuron 17:75–89

Morrison RS, Wenzel HJ, Kinoshita Y, Robbins CA, Donehower LA, Schwartzkroin PA (1996) Loss of the p53 tumour suppressor gene protects neurons from kainate-induced cell death. J Neurosci 16:1337–1345

Motoyama N, Wang F, Roth KA, Sawa H, Nakayama K, Negishi I, Senju S, Zhang Q, Fuji S, Loh DY (1995) Massive cell death of immature hematopoietic cells and neurons in bcl-x-deficient mice. Science 267:1506–1510

Muzio M, Stockwell BR, Stennicke HR, Salvesen GS, Dixit VM (1998) An induced proximity model for caspase 8 activation. J Biol Chem 273:2926–2930

Nakatsue T, Katoh I, Nakamura S, Takahashi Y, Ikawa Y, Yoshinaka Y (1998) Acute infection of Sindbis virus induces phosphorylation and intracellular translocation of small heat shock protein HSP27 and activation of p38 MAP kinase signaling pathway. Biochem Biophys Res Comm 253:59–64

Nava VE, Rosen A, Veliuona MA, Clem RJ, Levine B, Hardwick JM (1998) Sindbis virus induces apoptosis through a caspase-dependent, CrmA-sensitive pathway. J Virol 72:452–459

New DR, Ma MH, Epstein LG, Nath A, Gelbard HA (1997) Human immunodeficiency virus type 1 Tat protein induces death by apoptosis in primary human neuron cultures. J Neurovirol 3:168–173

Ng FWH, Shore GC (1998) Bcl-xL cooperatively associates with the bap31 complex in the endoplasmic reticulum dependent on procaspase 8 and ced-4 adaptor. J Biol Chem 273:3140–3143

O'Brien V (1998) Viruses and apoptosis. J Gen Virol 79:1833–1845

Oberhaus SM, Smith RL, Clayton GH, Dermody TS, Tyler KL (1997) Retrovirus infection and tissue injury in the mouse central nervous system are associated with apoptosis. J Virol 71:2100–2106

Oliver KR, Fazakerley JK (1998) Transneuronal spread of Semliki Forest virus in the developing mouse olfactory system is determined by neuronal maturity. Neurosci 82:867–877

Ona VO, Mingwei L, Vonsattel JPG, Andrews LJ, Khan SQ, Chung WM, Frey AS, Menon AS, Li X-J, Stieg PE, Yuan J, Penney JB, Young AB, Cha J-H J, Friedlander RM (1999) Inhibition of caspase 1 slows disease progression in a mouse model of Huntington's disease. Nature 399:263–267

Ozaki N, Sugiura Y, Yamamoto M, Yokoya S, Wanaka A, Nishiyama Y (1997) Apoptosis induced in the spinal cord and dorsal root ganglion by infection of Herpes simplex virus type 2 in the mouse. Neurosci Letts 228:99–102

Pekosz A, Phillips J, Pleasure D, Merry D, Gonzalez-Scarano FJ (1996) Induction of apoptosis by La Crosse virus infection and role of neuronal differentiation and human bcl-2 expression in its prevention. J Virol 70:5329–5335

Petito CK, Roberts B (1995) Evidence of apoptotic cell-death in HIV encephalitis. Am J Path 146: 1121–1130

Pettman B, Henderson CE (1998) Neuronal cell death. Neuron 20:633–647

Pugachev KV, Frey TK (1998) Rubella virus induces apoptosis in culture cells. Virology 250:359–370

Raff M (1998) Cell suicide for beginners. Nature 396:119–122

Roy N, Deveraux QL, Takahashi R, Salvesen GS, Reed JC (1997) The c-IAP-1 and c-IAP-2 proteins are direct inhibitors of specific caspases. The EMBO J 16:6914–6925

Sammin DJ, Butler D, Atkins GJ, Sheahan BJ (1999) Cell death mechanisms in the olfactory bulb of rats infected intranasally with Semliki Forest virus. Neuropath Appl Neurobiol (in press)

Sanchez I, Xu C-J, Juo P, Kakizaka A, Blenis J, Yuan J (1999) Caspase-8 is required for cell death induced by expanded polyglutamine repeats. Neuron 22:623–633

Scallan MF, Allsopp TE, Fazakerley JK (1997) bcl-2 acts early to restrict Semliki Forest virus replication and delays virus-induced programmed cell death. J Virol 71:1583–1590

Shi B, DeGirolami U, He JL, Wang S, Lorenzo A, Busciglio J, Gabuzda D (1996) Apoptosis induced by HIV-1 infection of the central nervous system. J Clin Invest 98:1979–1990

Shi B, Raina J, Lorenzo AE, Busciglio J, Gabuzda D (1998) Neuronal apoptosis induced by HIV-1 Tat protein and TNF-alpha: potentiation of neurotoxicity mediated by oxidative stress and implications for HIV-1 dementia. J Neurovirol 4:281–290

Shimizu S, Narita M, Tsujimoto Y (1999) Bcl-2 family proteins regulate the release of apoptogenic cytochrome C by the mitochondrial channel VDAC. Nature 399:483–486

Shisler JL, Gooding LR (1998) Adenoviral inhibitors of the apoptotic cascade. Trends Microbiol 6: 337–339

Sloviter RS, Sollas AL, Dean E, Neubort S (1993) Adrenalectomy-induced granule cell degeneration in the rat hippocampal dentate gyrus – characterization of an in vitro model of controlled neuronal death. J Comp Neurol 330:324–336

Teodoro JG, Branton PE (1997) Regulation of apoptosis by viral gene products. J Virol 71:1739–1746

Theerasurakarn S, Ubol S (1998) Apoptosis induction in brain during the fixed strain of rabies virus infection correlates with onset and severity of illness. J Neurovirol 4:407–414

Thompson CB (1995) Apoptosis in the pathogenesis and treatment of disease. Science 267:1456–1462

Thornberry NA, Lazebnik Y (1998) Caspases: enemies within. Science 281:1312–1316

Trgovcich J, Ryman K, Extrom P, Eldridge C, Aronson JF, Johnston RE (1997) Sindbis virus infection of neonatal mice results in a severe stress response. Virology 227:234–238

Tsunoda I, Kurtz CB, Fujinami RS (1997) Apoptosis in acute and chronic central nervous system disease induced by Theiler's murine encephalomyelitis virus. Virology 228:388–393

Ubol S, Tucker PC, Griffin DE, Hardwick JM (1994) Neurovirulent strains of alphavirus induce apoptosis in bcl-2-expressing cells – role of a single amino-acid change in the E2 glycoprotein. Proc Natl Acad Sci USA 91:5202–5206
Ubol S, Suk P, Budihardjo I, Desnoyers S, Montrose MH, Poirier GG, Kaufmann SH, Griffin DE (1996) Temporal changes in chromatin, intracellular calcium, and poly(ADP-ribose) polymerase during Sindbis virus-induced apoptosis of neuroblastoma cells. J Virol 70:2215–2220
Walter J, Schindzielorz A, Grunberg J, Haass C (1999) Phosphorylation of presenilin-2 regulates its cleavage by caspases and retards progression of apoptosis. Proc Natl Acad Sci USA 96:1391–1396
White FA, Keller-Peck CR, Knudson CM, Korsmeyer SJ, Snider WD (1998) Widespread elimination of naturally occurring neuronal death in *Bax*-deficient mice. J Neurosci 18:1428–39
Williams A, Lucassen PJ, Ritchie D, Bruce M (1997) Prp deposition, microglial activation and neuronal apoptosis in murine scrapie. Exp Neurol 144:433–438
Williams BRG (1995) The role of the dsRNA-activated kinase, PKR, in signal transduction. Semin Virol 6:191–202
Wolozin B, Iwasaki K, Vito P, Ganjei JK, Lacana E, Sunderland T, Zhao B, Kusiak JW, Wasco W, D'Adamio L (1996) Participation of presenilin 2 in apoptosis enhanced basal activity conferred by an Alzheimer mutation Science 274:1710–1713
Wood KA, Youle RJJ (1995) The role of free radicals and p53 in neuron apoptosis in vivo. J Neurosci 15:5851–5857
Xiang H, Hochman DW, Saya H, Fujiwara T, Schwartzkroin PA, Morrison RS (1996) Evidence for p53-mediated modulation of neuronal viability. J Neurosci 16:6753–6765
Yoshida H, Kong Y-Y, Yoshida R, Elia AJ, Hakem A, Hakem R, Penninger JM, Mak TW (1998) Apaf-1 is required for mitochondrial pathways of apoptosis and brain development. Cell 94:739–750

Interactions of Viral Proteins with Neurotransmitter Receptors May Protect or Destroy Neurons

G. Gosztonyi[1] and H. Ludwig[2]

1 Introduction... 121
2 Virus Receptors... 122
3 Receptor Affinities of Rabies Virus.......................... 122
4 Receptor Affinities of Borna Disease Virus................... 128
5 HIV-1 gp120 Protein and the NMDA Receptor................... 135
6 Conclusions... 138
References.. 139

1 Introduction

Virus infection, as a rule, is associated with the development of cytopathic changes in the host cell both in vivo and in vitro. The study of persistent virus infections, however, has led to the recognition that this cytopathic change may evolve slowly and may even be minimal. At the end of the 1970s it became evident that some viruses may alter cell function even without causing any kind of morphological damage to the host cell (OLDSTONE 1984). During this process some of the differentiated functions of the cell suffered, while the vital functions could remain intact, securing survival of the cell. This phenomenon was first noted in vitro in murine neuroblastoma cell lines exhibiting a disturbance of acetylcholine (ACh) metabolism when persistently infected with lymphocytic choriomeningitis virus (LCMV) (OLDSTONE et al. 1977). In vivo infection of mice induced disturbance in growth hormone (GH) production and glucose metabolism; however, there were no lytic changes in the GH-producing somatotrophic pituitary cells, although replication of LCMV was restricted to this cell type of the anterior pituitary (OLDSTONE et al. 1982, 1984; RODRIGUEZ et al. 1983). Another interesting observation emerged from

---

[1] Abteilung für Neuropathologie, Freie Universität Berlin, Universitätsklinikum Benjamin Franklin, Hindenburgdamm 30, 12200 Berlin, Germany
e-mail: gegos@zedat.fu-berlin.de
[2] Institut für Virologie, Freie Universität Berlin, Königin-Luise-Str. 49, 14195 Berlin, Germany

experimental infection of mice with canine distemper virus, which induced severe obesity and other hormonal abnormalities. Cell destruction, however, could not be observed (LYONS et al. 1982).

Such disturbances arise from an interference of viral nucleic acids and proteins with host cell functions. If there is an interference with the vital functions of the cell, the result is in most cases a lytic infection. Conversely, interference with a differentiated cell function results in specific functional disturbances without cell lysis. Differentiated cell functions are often individual features of definite organs, tissues or cell groups. Therefore, not infrequently, a parallelism exists between tissue tropism of a virus and the disturbance of such functions.

2 Virus Receptors

Differentiated cell functions are often regulated by signal molecules that attach to surface receptors, which therefore have a crucial role in the biology of the cell. At the same time, some of these receptors may also serve as virus receptors, enabling attachment and entry of the virus into the cell. Viral structures that secure this attachment, the so-called viral attachment proteins, occur also in a soluble form and, having a high affinity to specific cell surface receptors, can bind to them and may prevent the attachment of their natural ligands.

Accordingly, cell surface receptors that are specific for a definite tissue or cell type determine the selective affinity of a virus to these tissues or cells. As to neural tissue, those are the neurotransmitter receptors that are unique for neurons and indispensable for their function. Therefore, it is conceivable that viruses that under natural conditions replicate exclusively or almost exclusively in neural tissue may have affinities to receptors that occur solely in this tissue. This consideration prompted us to postulate that neurotransmitter receptors are the substrates of viral neurotropism (GOSZTONYI and LUDWIG 1984b). Indeed, among the neurotropic viruses, an affinity to neurotransmitter receptors has been proven for those viruses exhibiting the highest degree of neurotropism. These viruses are Borna disease virus (BDV), with an affinity to glutamate receptors (GOSZTONYI and LUDWIG 1995), and rabies virus (RV), with an affinity to ACh receptors (LENTZ et al. 1982). Therefore, as to interactions between viruses and neurotransmitter receptors, these two agents will preferentially be considered in this chapter.

3 Receptor Affinities of Rabies Virus

Enveloped viruses attach to cell surface receptors by their envelope glycoproteins. Thus, it has been shown that RV attaches to ACh receptors on myotubes and peripheral nerve fibers and this attachment can be greatly reduced by α-bungarotoxin and d-tubocurarin, ligands of the ACh receptor (LENTZ et al. 1982). This affinity to ACh receptors, however, is not a decisive factor in the spread of rabies virus within the central nervous system (CNS), since there its distribution is not restricted to cholinergic systems (REAGAN and WUNNER 1985; KUCERA et al. 1985; LAFAY et al. 1991; GOSZTONYI et al. 1993; CHARLTON 1994). This fact, however, does not jeopardize the ACh receptor theory, since some viruses can bind to more than one receptor (HAYWOOD 1994). The spread of RV within the CNS is determined by other virus/receptor interactions than in the periphery. In these interactions within the CNS, the viral glycoproteins do not necessarily play the decisive role. It has been established that in the front-line of RV infection the trans-synaptic spread of infection takes place in the form of viral nucleocapsids and full viruses will be formed only 24–48h after the initial arrival of infection (GOSZTONYI 1978a; GOSZTONYI et al. 1993). Accordingly, epitopes on the viral ribonucleoprotein (RNP) must have a decisive role in the virus/receptor interactions that decide attachment and entry of infectious viral RNPs during their spread within the CNS. As to the nature of this receptor, the distribution pattern of RV antigens within the CNS may give important clues. There are certain peculiarities in this distribution pattern in the brains of rats and mice infected with the CVS strain of RV. The distribution pattern of RV antigens is rather diffuse, with only a few, but characteristic exceptions. In the hippocampal formation, the pyramidal cells and some interneurons are heavily laden with RV antigens, but the granule cells of the dentate gyrus are almost completely spared (JACKSON and REIMER 1989, FU et al. 1993). A further peculiarity is that two structures built up of adrenergic neurons, the locus coeruleus in the brain stem (LAFAY et al. 1991; ASTIC et al. 1993) and the superior cervical ganglion (TSIANG et al. 1983; LAFAY et al. 1991), are not infected by RV. Following inoculation of RV into the anterior chamber of the eye the virus did not spread along adrenergic, only along cholinergic and trigeminal sensory fibers (KUCERA et al. 1985).

There are controversies as to the *cellular localization* of RV in the nervous system. The majority of observers regard RV as an exclusively neuronal pathogen (see CHARLTON 1994). However, authors of some electron microscopic studies reported occasional infection of astrocytes (MATSUMOTO 1963; IWASAKI et al. 1973; IWASAKI and CLARK 1975; BAER et al. 1980), and immunohistochemical studies described occurrence of RV antigens in glial cells (SCHNEIDER 1975; TIRAWATNPONG et al. 1989), astrocytes and oligodendrocytes (FEIDEN et al. 1985, 1988).

Comparison of this peculiar distribution pattern with that of known neurotransmitter receptors might shed light on the receptor affinities of RV within the CNS. The majority of neurotransmitter receptors may be sorted out as RV receptor candidates, since they have a rather limited differential distribution. The very widespread, almost ubiquitous, regional distribution of RV parallels most of all with the distribution pattern of glutamate receptors. The latter form a family, encompassing at least 21 members, classified in four groups: N-methyl-D-aspartate (NMDA), α-amino-3-hydroxy-5-methylisoxazole propionic acid (AMPA), kainate and metabotropic (NAKANISHI 1992; NICHOLLS 1994; WHETSELL 1996). These receptor subunits have again a differential distribution in various brain areas, the comparison of which with RV distribution may bring us closer to the determination of the RV receptor within the CNS.

The idea that a glutamate receptor might be a RV receptor arose first in a tissue culture study of TSIANG et al. (1991). In rat primary cortical neurons a non-competitive NMDA antagonist, dizocilpine (MK-801), inhibited RV replication. The results remained, however, controversial, since MK-801 was effective only in the millimolar, and not in the micromolar, range. Therefore, the authors concluded that MK-801, although exerting a selective inhibition on RV replication, does not operate through a high-affinity binding site mechanism. Probably due to these controversies the NMDA hypothesis has been abandoned.

The hippocampal formation has a unique architecture, in so far as various neuronal systems terminate in it in a very regular, laminar pattern and in the termination fields the various neurotransmitter receptors also have a laminar distribution. This formation, therefore, is most suitable for the study of receptor affinities of neurotropic viruses. As to RV infection, it was hoped that the sparing of the dentate gyrus (Fig. 1) might bring us closer to determining the receptor affinities of this agent. When analyzing the peculiar protection of the dentate gyrus from RV infection, the possibility emerges first that in dentate gyrus neurons the surface specialization is missing, which may serve as RV receptor within the CNS. Therefore, we analyzed the distribution of various neurotransmitter receptors and

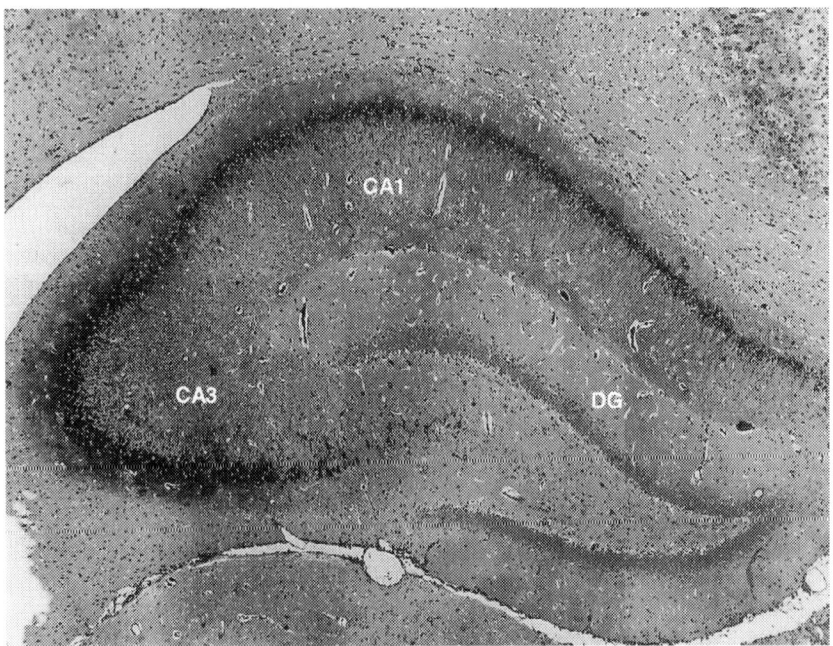

Fig. 1. Hippocampus of a rat infected with the CVS strain of rabies virus (RV). Immunohistochemical demonstration of the distribution of RV N-protein. The granule neurons of the dentate gyrus are spared from infection. *CA1, CA3*, subfields of the cornu ammonis (hippocampus proper); *DG*, dentate gyrus

associated proteins in the CA1, CA3 subfields of the hippocampus proper and in the dentate gyrus on the basis of the data in the literature. This analysis was enhanced by the significant increase of our knowledge on the distribution of various glutamate receptors in the brain in the last 15 years. In particular, the distribution of the NMDA receptors has been elucidated in several meticulous studies (MONAGHAN and COTMAN 1985; GREENAMYRE et al. 1985; MORIYOSHI et al. 1991; MONYER et al. 1992; KUTSUWADA et al. 1992; MONYER et al. 1994; MORI and MISHINA 1995). Through the survey of these studies it has become clear that the four NMDAR2 subtypes (NR2A, NR2B, NR2C and NR2D) do not come into question as RV receptor candidates, since they all have a too-restricted topographic distribution. Only NR1, with a most widespread distribution, is in line with a similarly diffuse distribution of RV in the brain.

However, the distribution pattern of the NMDA receptors does not give an explanation for the sparing of the dentate gyrus. All NMDA receptors that are present in the hippocampal formation are equally expressed in the CA1 and CA3 subfields, just as in the dentate gyrus. Furthermore, no other known neurotransmitter receptor has been found that was present in the CA subfields and was missing in the dentate gyrus. Only one single protein was found with a markedly differential distribution within the CA subfields and the dentate gyrus, α-actinin-2, a putative NMDA-receptor-anchoring protein. This protein was present only in the dentate gyrus and in the narrow CA2 subfield (WYSZINSKI et al. 1998). Thus, the presence of this protein might be related to the resistance of the dentate gyrus to RV infection. However, the distribution of this protein in other brain areas contradicts this assumption, since α-actinin-2 is present also in the striatum and the substantia nigra, brain areas that harbor abundantly RV antigens.

In order to understand the sparing of the dentate gyrus from RV infection, the arrival and subsequent distribution of RV into the hippocampal formation has been studied following nasal inoculation of rats with the CVS strain (Gosztonyi, Dietzschold and Koprowski, unpublished observations). As the earliest sign of infection, scattered hippocampal pyramidal cells expressed RV antigen in both the CA1 and CA3 subfields. Simultaneously, a few interneurons in both CA subfields and in the hilus of the dentate gyrus also became immunoreactive. RV spreads in the early phase of infection almost exclusively by retrograde axonal transport. There are only two exceptions: its spread along the olfactory nerve following nasal inoculation and its spread from the spinal and trigeminal sensory ganglion cells along their central processes in an anterograde fashion after peripheral inoculation. It is, therefore, conceivable that the pyramidal cells and interneurons became infected by the retrograde axonal spread of RV. The source of infection is the lateral septum pellucidum, to which a few, randomly distributed, hippocampofugal CA1 and CA3 pyramidal cells project (LERANTH and FROTSCHER 1989). The scattered, RV-immunoreactive interneurons in all subfields of the hippocampus are identical with GABAergic interneurons that project to the medial septum pellucidum (TÓTH et al. 1993; FREUND and BUZSÁKI 1996). This pattern of primary infection of the hippocampal formation documents clearly that this formation is infected by the RV only by a retrograde axonal spread of the agent. The rich afferent connections of the hippocampal formation cannot be exploited for this spread, since RV cannot use the anterograde axonal transport for its conveyance. That is also the reason why the dentate gyrus is not infected by the massive afferent input, the perforant path. This system is the hippocampopetal projection of neurons of the entorhinal cortex, whose axons terminate on the dendrites of granule cells in the molecular layer of the dentate gyrus.

If RV uses retrograde axonal transport, it could spread from the CA3 pyramidal cells to the dentate gyrus along the mossy fiber system. The latter arises from the axons of dentate granule cells, moves to the CA3 subfield and establishes multiple synaptic contacts with the proximal dendritic segments of CA3 pyramidal cells in the stratum lucidum. These rich contacts could give ample opportunities for RV to be taken up and carried retrogradely to the cell bodies of dentate granule cells, but this uptake does not take place. The dentate granule cell/CA3 pyramidal cell connection is a glutamatergic one, it has, however, a peculiarity in its receptor architecture. This synaptic subfield (stratum lucidum of the CA3 area), in contrast to all other subfields in the hippocampal formation, does not contain NMDA receptors, but has a high density of kainate and AMPA binding sites (MONAGHAN and COTMAN 1985; GREENAMYRE et al. 1985). Consequently, RV does not bind to mossy fiber terminals in the stratum lucidum, because it has no affinity towards kainate and AMPA receptors, but to NMDA receptors that, in turn, are not expressed on the mossy fiber terminals.

There are three important conclusions from these observations. First, dentate gyrus will not be infected via the hippocampopetal perforant path, because RV does not spread in the anterograde direction. Second, RV will not be taken up by the mossy fiber terminals and carried retrogradely to the dentate granule neuronal perikarya, because the latter terminals do not express NMDA, only kainate and AMPA receptors. Third, due to the latter feature and due to the fact that NR1 has a diffuse, and the four NR2 receptors a rather differential, spatial distribution, it is evident that *NR1 is the most probable RV receptor candidate* within the CNS. Summing up, protection of the neurons of the dentate gyrus against RV has two roots: first, the inability of RV to use the anterograde axonal transport in this neuronal system, second, the lack of the putative RV receptor, NR1, on the terminals of the mossy fiber system.

The exemption of the dentate gyrus from RV infection is valid only for adult animals. In newborn mice infected with RV the infection extends also to this structure, so that dentate granule cells are infected and destroyed (GOSZTONYI 1978b; JACKSON and PARK 1998). These observations indicate that in the immature nervous system non-specific uptake mechanisms may also be at work.

The observation that in the hippocampal subfields RV antigens appear simultaneously in scattered pyramidal cells and scattered interneurons indicates that RV spreads retrogradely from the septum pellucidum both along the axons of the glutamatergic pyramidal cells and along the axons of the GABAergic interneurons. A review of the various types of hippocampal interneurons reveals that there is a single group of interneurons with far-reaching projection down to the septum pellucidum and this group is GABAergic (FREUND and BUZSÁKI 1996).

With great certainty, the RV-expressing interneurons correspond to this group of GABAergic interneurons. Consequently, RV has a threefold receptor affinity: first, to the ACh receptor, predominantly in the periphery, second, to NMDAR1 (NR1) receptors in widespread areas of the CNS, and third, to GABA receptors in inhibitory neuronal systems. Studying the spread of the CVS strain of RV and its avirulent mutant AvO1, LAFAY et al. (1991) also came to the conclusion that the CVS strain could bind to several different receptors, but they did not suggest what kinds of receptors these could be. HANHAM et al. (1993) developed a monoclonal anti-ACh receptor (anti-idiotypic) antibody derived by immunization with a monoclonal antibody specific for RV glycoprotein. This antibody was bound to widespread areas in the normal mouse brain that are well-known targets of RV infection. However, not all areas and cell groups were labeled, which harbor RV antigens in the course of experimental infections in the mouse brain. Therefore, these authors concluded that the nicotinic ACh receptor serves as the RV receptor also in the CNS; however, there may be multiple receptors for RV in this organ.

The possibility that among these multiple receptors those for GABA may also represent RV receptors (see above) seems to be questioned by the RV antigen distribution in the olfactory bulb. Characteristically, in the olfactory bulb of rats infected nasally with the CVS strain of RV, the granule neurons that are GABAergic are spared and only the mitral and tufted neurons are infected (Gosztonyi, Dietzschold and Koprowski, unpublished observations). The granule cells of the olfactory bulb in mice behave in a similar way (ASTIC et al. 1993). In contrast, the Purkinje cells of the cerebellar cortex that are also GABAergic (see SCHULMAN 1983) harbor RV antigens. This controversy can perhaps be resolved by the unique structure of the granule cells of the olfactory bulb. These neurons do not have an axon, only two polar dendrites, of which the peripheral one forms reciprocal dendro-dendritic synapses with the mitral and tufted neurons (HALÁSZ and SHEPHERD 1983), which are otherwise permissive for RV infection. It may be that this unusual synaptic type is impenetrable for the spread of RV, which passes preferentially axo-somatic and axo-dendritic synapses in a retrograde way.

In these considerations of the receptor affinities of RV, the questionable infection of glial cells has not been included. As described above, virus particles and RV antigens have been found occasionally in astrocytes, oligodendrocytes and macrophages/microglial cells. In the evaluation of these results, however, caution must be recommended. Hypertrophic astrocytes bind antibodies not infrequently in a non-specific way. Macrophages may phagocytose RV particles that retain their antigenicity for some time. If glial cells harbor RV antigens, it is rather an exception than the rule. High local concentration of RV may lead to uptake in cells in a way not mediated by specific surface receptors. When comparing the wide distribution of another neurotropic agent, BDV, in almost all CNS cell types, RV still remains rather rigorously restricted to neurons. Studies of TUFFEREAU et al. (1998) also indicate that RV binds to neuronal cells. They expressed the glycoprotein of the CVS strain of RV on the surface of *Spodoptera frugiperda* cells. These cells bound specifically to neuroblastoma cells in tissue culture, but not to glioma cell lines.

Study of the distribution of RV RNA in various cell types in the CNS might answer this controversial question.

To the question of a possible replication of RV in glial cells it is very important to decide whether or not glial cells express NMDA receptors. The views concerning this question are at present not unequivocal. Some authors described the presence of NMDA receptors in astrocytes and oligodendrocytes (CONTI et al. 1996; WANG et al. 1996), others established regional variations in expression and dependence on the functional state of the glial cells (GOTTLIEB and MATUTE 1997). Taken together, NMDA receptors are, under certain conditions, expressed in glial cells; this expression, however, seems to be less constant than the expression of the glutamate receptors of the non-NMDA type, the various AMPA and kainate subunits. Further studies are required to elucidate the expression of various NMDA receptor subtypes in various glial cells, and to determine whether a differential distribution of NR1 and NR2 receptors in these cell types can be documented. Resolution of the latter question would be of crucial importance for the glial expression of RV, since a restricted or missing expression of NR1 on glial cells could offer an explanation for the rigorous neuronotropism of RV.

4 Receptor Affinities of Borna Disease Virus

Rabies virus and BDV have many biological and molecular biological features in common and are widely distributed in nature (LUDWIG and BODE 2000). They both belong to the order mononegavirales (SCHNEEMANN et al. 1995; DE LA TORRE 1994). While RV spreads in the form of ribonucleoproteins (RNPs) only in the first half of its replication cycle and in the second half already full virus particles are formed, BDV seems to be present in the CNS in the form of RNPs during the complete replication cycle (GOSZTONYI et al. 1993). The most important argument to support this view is the fact that BDV particles could never be visualized by electron microscopy in the CNS, and in vitro an accumulation of RNPs with infectivity has been shown in the nuclei of infected cells (CUBITT and DE LA TORRE 1994). However, viral proteins, filling more or less evenly the infected cells, could readily be displayed by light and electron microscopic immunocytochemistry (GOSZTONYI and LUDWIG 1995). The ability of BDV to spread intra-axonally (KREY et al. 1979; GOSZTONYI and LUDWIG 1984a; CARBONE et al. 1987) throughout the entire nervous system, most probably by RNPs (GOSZTONYI et al. 1993), points to a perfect adaptation of this agent to neural tissue (LUDWIG and BODE 2000). The distribution patterns and receptor affinities of BDV can best be studied by the demonstration of viral antigens in various structural elements of the nervous system (GOSZTONYI and LUDWIG 1995).

BDV proteins exhibit a most peculiar distribution pattern in a definite phase of infection in the hippocampal formation (GOSZTONYI and LUDWIG 1984b; LUDWIG et al. 1988; MORALES et al. 1988; GOSZTONYI and LUDWIG 1995). This pattern consists of a stratified or laminated distribution of BDV antigens specifically in the hippocampus: BDV proteins occupy the stratum oriens and stratum radiatum in the CA1 region and are absent from the stratum pyramidale and stratum lacunosum-moleculare (Fig. 2). In order to understand the significance of this unique pattern, it has to be correlated with that of termination of various afferent neuronal systems and of the neurotransmitters they are using. The strata oriens and radiatum of the CA1 region represent the termination site of the recurrent collaterals (Schaffer collaterals) of the homolateral CA3 pyramidal neurons and of the commissural projections of the contralateral CA3 pyramidal neurons. These axons all establish synaptic contacts with the basal and apical dendrites of CA1 pyramidal neurons. These synapses use glutamate and aspartate as transmitters, while systems terminating in the other two strata are non-glutamatergic. This fact has allowed us to conclude that BDV has a specific affinity to glutamatergic neurons (GOSZTONYI and LUDWIG 1995). These structural correlations in mind the spread of BDV RNPs in the hippocampus can be reconstructed as follows: BDV moves along the axons predominantly using the anterograde transport mechanism and arrives, for the most part, along the perforant path from the entorhinal cortex to the dentate gyrus neurons. The granule cells of the dentate gyrus forward the infection also anterogradely along the mossy fiber system to the CA3 neurons. The latter, along

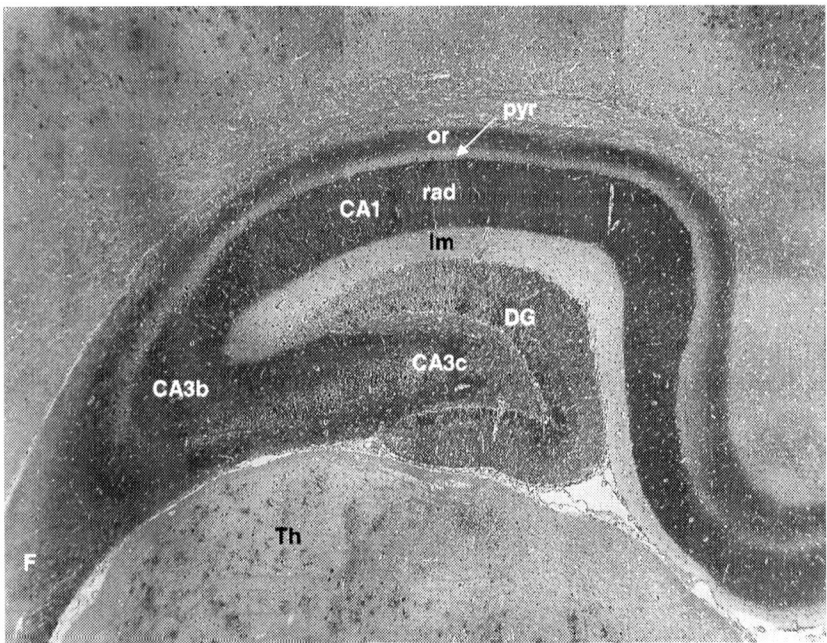

Fig. 2. Hippocampus of a rat persistently infected with Borna disease virus (BDV). Immunohistochemical demonstration of BDV antigen showing a laminar distribution in the CA3 and CA1 subfields. *or*, stratum oriens; *pyr*, stratum pyramidale; *rad*, stratum radiatum; *lm*, stratum lacunosum-moleculare; *CA1, CA3b, CA3c*, subfields of the Cornu Ammonis (hippocampus proper); *DG*, dentate gyrus; *F*, fimbria hippocampi; *Th*, thalamus

their homolateral and contralateral connections, transport viral RNPs to their presynaptic terminals in the strata oriens and radiatum of the CA1 hippocampal subfield. This region, when examined with light- and electron microscopic immunocytochemistry, shows axons, presynaptic axonal segments and synaptic boutons filled with BDV antigens, while the CA1 pyramidal neurons, on which these boutons terminate, are free (with a few exceptions) from these antigens (Fig. 3). This means that CA3 pyramidal neurons take up and replicate BDV, but when the infectious RNPs are carried to the synaptic contacts on the CA1 neurons, they will not be transferred into the latter – they are stagnating in the terminal segments of the CA3 axons. What is the basic difference between CA3 and CA1 neurons,which renders the first permissive for and the second protected against BDV infection? Both neuronal types are glutamatergic, and, in spite of that, they behave towards BDV infection in an opposite way. The basic difference between these two types of hippocampal pyramidal neurons is that CA3 neurons express the kainate 1 (KA-1) receptor, and CA1 neurons do not (WERNER et al. 1991; WISDEN and SEEBURG 1993). Accordingly, CA1 neurons are protected against BDV infection, because they do not possess the receptor that BDV RNPs most probably

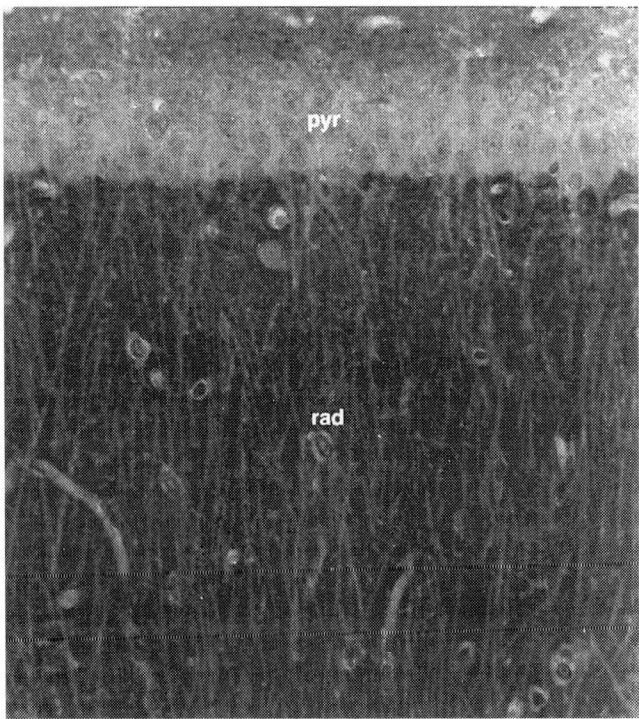

Fig. 3. Stratum pyramidale (*pyr*) and stratum radiatum (*rad*) in the CA1 subfield of the hippocampus. Immunohistochemical demonstration of Borna disease virus (BDV) antigen. Note absence of positivity in the pyramidal layer. In the stratum radiatum (*rad*) BDV antigen is found in synaptic boutons and terminal/preterminal axonal segments, but not in the dendrites of the CA1 pyramidal cells

use for their entry into these neurons. Thus, we postulate that *the KA-1 receptor represents the BDV receptor within the CNS*. The question arises whether the distribution pattern of BDV antigens is in accordance with that of KA-1 receptors also in other CNS regions? There is a high degree of conformity between the two distribution patterns, and there are significant differences to the distribution pattern of another non-NMDA glutamate receptor, KA-2. In particular, the accordance of KA-1 expression and BDV-positivity in oligodendrocytes and astrocytes has to be emphasized. This explains, why BDV, as a highly neurotropic virus, is not only neuronotropic, but also gliotropic.

These specific correlations allow the conclusion that cells without the KA-1 receptor are protected from BDV infection. No such infection can be demonstrated, e.g. in the GABAergic inhibitory neurons of the hippocampus. These internuncial neurons have dense terminations on the cell bodies of CA1 pyramidal neurons. There are no BDV-positive structures in the pyramidal layer of the CA1 region (Fig. 3). On the other hand, the KA-1 receptor renders neurons vulnerable to BDV. Beyond that, binding of viral proteins to their receptor makes that specific receptor inaccessible for its natural ligand, glutamate. In the CNS, glutamate receptors may be homomeric, i.e. all constituents of the receptor complex may be identical, or they may be heteromeric, so that the individual constituents represent different receptor proteins of a receptor family. Heteromeric receptor complexes occur more frequently (WISDEN and SEEBURG 1993). If, e.g. BDV RNPs bind to a KA-1 receptor which is a component of a receptor complex, then the viral macromolecule may prevent also the binding of glutamate to the other, non-KA-1 constituents of the receptor complex. We studied experimentally the binding of an antibody to GluR1, a high-affinity AMPA receptor in the hippocampus of normal and BDV-infected rats. Binding to GluR1 was markedly reduced in the CA3 region, which is heavily loaded with BDV proteins, as compared with the normal control hippocampus (Fig. 4A–C). Comparable results were obtained with the anti-GluR2–3 antibody directed against two other high-affinity AMPA-receptors. This phenomenon can be explained by binding of the BDV protein to its KA-1 receptor, thus preventing the binding of the anti-GluR1 and anti-GluR2–3 antibodies to their antigens during the immunohistochemical reaction, due to the close proximity of the receptor protein subunits. Thus, the consequences of binding of BDV proteins to the KA-1 receptors are more widespread than could be expected from the absolute number of these receptors. The widespread distribution of KA-1 receptors in the CNS correlates with the diffuse distribution pattern of BDV antigens. Within this pattern, however, the heavy concentration of viral antigens in the dentate gyrus and the CA3 hippocampal region has to be emphasized. These regions are also very rich in KA-1 receptors.

In the course of BDV infection almost all neurons of the dentate gyrus and the CA3 region express viral antigen. Therefore, the blockage of a great deal of neuronal glutamate receptors must have severe consequences. Processing of input impulses arriving through the perforant path and dentate gyrus suffers severe reduction and delay. The balance between glutamatergic and other neurotransmitter systems is upset in the hippocampal formation. This imbalance is reflected in

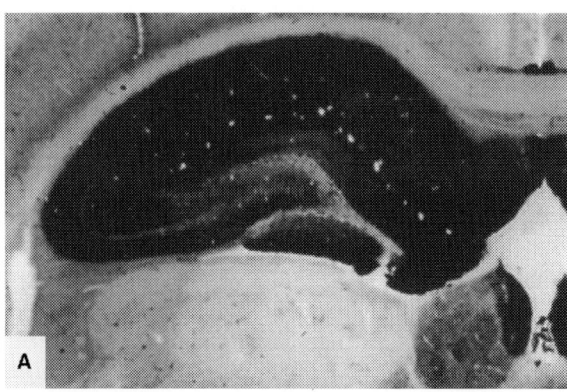
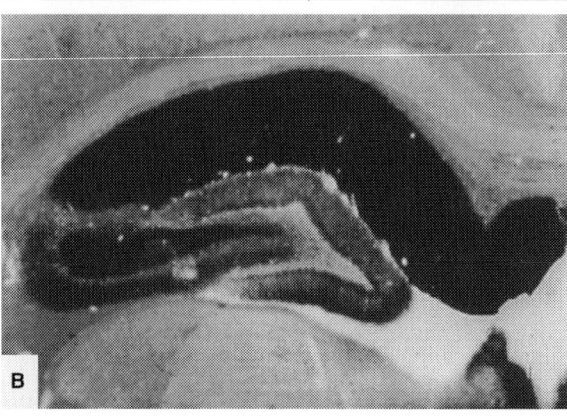
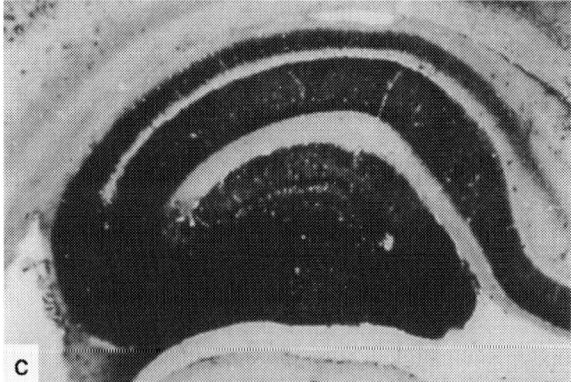

Fig. 4A–C. Immunohistochemical demonstration of the distribution of Borna disease virus (BDV) antigen, and GluR1 receptor protein in the hippocampal formation in BDV-infected and uninfected rats. **A** Uninfected rat, distribution of GluR1 protein. **B** BDV-infected rat, distribution of GluR1 protein. **C** BDV-infected rat, distribution of BDV antigens

alterations in the bioelectrical activity of the hippocampus already during early phases of infection (GIEREND 1982). It is known that the hippocampal formation has an important role in learning and memory processing. This explains why persistently BDV-infected rats show severe learning deficiencies, based on heavy accumulation of BDV antigens which interfered with hippocampal functions (DITTRICH et al. 1989). Changes in social behavior in persistently infected tree shrews which were free of clinical symptoms point to the same etiology (SPRANKEL et al. 1978).

While the blocking of glutamate receptors by BDV proteins has a deleterious effect on the function of a great number of neurons, it may have a protective effect as well. It is well-known that an increased activity of glutamatergic systems results in severe disturbances of the CNS. Excessive stimulation of glutamate gated ion channels leads to a massive influx of Na^+, Cl^- and Ca^{2+} into the cell, to decreased glutamate uptake and to extracellular glutamate accumulation, which ultimately result in damage and degeneration of the neuron. This effect of the excitatory amino acids has been called excitotoxicity (OLNEY et al. 1971) and appears in acute neurological conditions, such as ischemia, stroke, hypoglycemia, craniocerebral trauma and seizures; but excitotoxicity may also have a role in the pathogenesis of slowly evolving, degenerative CNS processes, such as Huntington's disease (HD), amyotrophic lateral sclerosis (ALS) and Parkinson's disease (PD) (COYLE and PUTTFARCKEN 1993; WHETSELL 1996). Excitotoxicity is closely related to another pathological condition, oxidative or metabolic stress. We understand under oxidative stress the cytotoxic consequences of free oxygen radicals (superoxide anion, hydroxy radical and hydrogen superoxide) that are generated as byproducts of normal and aberrant metabolic processes which utilize molecular oxygen (COYLE and PUTTFARCKEN 1993). Excitotoxicity may elicit oxidative stress, but the two processes may operate not only in a sequential, but also in an interacting manner (COYLE and PUTTFARCKEN 1993; WHETSELL 1996). In the CNS, both processes finally end in neuronal degeneration, but this may be counteracted with glutamate antagonists, in the case of excitotoxicity, and with antioxidants, in the case of oxidative stress. One way by which these processes can bring about neuronal destruction is the induction of programmed cell death (apoptosis) (FERRARI et al. 1995; see chapter by Fazakerley and Allsopp, this volume).

There are certain parallelisms between viral neurotropism, excitotoxicity and oxidative stress: all of them have a selective character. In acute ischemia, e.g., a selective decay of the CA1 pyramidal cells of the hippocampus due to the involvement of NMDA receptors is conspicuous (AUER et al. 1989; KNUCKEY et al. 1995). While many studies have been done on the role of excitotoxicity and oxidative stress in acute neurological conditions (stroke, hypoglycemia, seizures) and in chronic progressive, systemic neuronal degenerations (HD, ALS, PD), the participation of excitotoxicity and oxidative stress in neuronal damage in acute and chronic-persistent viral CNS infections has hardly been considered.

It is, however, conceivable that virus-induced metabolic changes might lead to excessive discharge of excitatory amino acids and/or increased production of oxygen radicals.

As to Borna disease, due to binding of BDV proteins to glutamate receptors the neurons are rather protected against excitotoxic mechanisms. This might be one of the reasons why cytopathic changes of neurons are absent or appear only rather late in the course of persistent infection. Besides that, BDV is known to cause no cytopathic effect in vitro. Nevertheless, since BDV proteins bind preferentially to one of the AMPA receptors, KA-1 (see above), it cannot be excluded that excitotoxic mechanisms might assert themselves by activating the NMDA receptors.

In persistent infection of rats two exceptions from this general rule, that neurons are spared from cytopathic changes, are observed: the subacute/chronic degeneration of the dentate gyrus and of the retina (LUDWIG et al. 1988, GOSZTONYI and LUDWIG 1995). Five to six weeks following inoculation of BDV, the neurons (granule cells) of the dentate gyrus degenerate and disappear, while the CA1 pyramidal neurons of the hippocampus remain completely, and the CA3 neurons in the overwhelming majority of the cases are intact (LUDWIG et al. 1988; CARBONE et al. 1991; GOSZTONYI and LUDWIG 1995). Interestingly, the septal part of the dentate gyrus is more severely involved than its temporal part. This most selective destruction of the dentate gyrus seems to be a very peculiar and specific feature during BDV infection. In a study on the adaptation of a BDV isolated from the CNS of cats suffering from a non-suppurative encephalomyelitis ("staggering disease") to newborn rats, a selective degeneration of the dentate gyrus was found in the first passage of the agent. This selective degeneration was completely identical with that found in rats persistently infected with BDV. The only difference was that in the first passage of the cat agent the immunohistochemical reactions with monoclonal anti-BDV antibodies were negative. In the second rat passage, however, viral antigens could readily be demonstrated (LUNDGREN et al. 1995). These observations indicate that some component of BDV has a most genuine affinity to a certain structural element of dentate granule neurons and this association interferes with the survival of this neuronal type. Although the nature of this association is unknown, prompted by the high selectivity of the neuronal degeneration we suspected that these neurons were targeted by an excitotoxic mechanism and the latter might have induced an oxidative stress with the production of free oxygen radicals. Starting from this working hypothesis we administered an antioxidant, *N*-acetylcysteine (NAC), through the drinking water to rats persistently infected with BDV. At the histological evaluation the degeneration of dentate granule neurons was significantly less expressed in the NAC-treated group than in the non-treated, but infected control group (GOSZTONYI et al. 2001).

The question arises why do dentate granule neurons degenerate in the course of persistent infection with BDV, while other neurons seem to be protected by the blockage of their glutamate receptors by BDV proteins? In glutamatergic synapses there is a mechanism by which the postsynaptic neuron is protected from excessive glutamate release. There are presynaptic autoreceptors that exert a negative feedback onto the presynaptic bouton. The receptor on these terminals, probably belonging to the metabotropic, mGluR subgroup, act through a protein kinase C (PKC)-mediated mechanism, inhibiting Ca^{2+} and activating K^+ channels (NICHOLLS 1994). It is highly hypothetical to involve this mechanism in BDV-induced cell death; nevertheless, it may be that a BDV protein interacts with the presynaptic autoreceptor and by blocking it the glutamate release mechanism could run out of control. These autoreceptors are diffusely distributed in glutamatergic systems; in the dentate gyrus, however, the same type of neuron occurs closely packed in high density in a laminar pattern, so that the neuronal degeneration becomes more conspicuous. In the neocortex, by contrast, randomly distributed neuronal degenerations and dropouts may escape perception. In the mossy fiber/CA3 pyramidal cell synapses these presynaptic autoreceptors have also been described (NICHOLLS 1994). And indeed, in some of the rats persistently infected with BDV a subacute/chronic, non-inflammatory degeneration of CA3 neurons could be observed (GOSZTONYI and LUDWIG 1995).

SLOVITER et al. (1989, 1993a) observed that 3–4 months after adrenalectomy, an almost complete loss of granule cells of the dentate gyrus was induced in rats. This loss was due to apoptotic cell death (SLOVITER et al. 1993a,b). This highly selective dentate gyrus degeneration could be prevented by substitution of corticosteroids (SLOVITER et al. 1989). It is, however, improbable that adrenal insufficiency could play a role in BDV-induced dentate gyrus degeneration. The latter develops in 5–6 weeks following infection, when no signs of adrenal involvement are discernible.

In the laminar structure of the retina, a chronic degeneration evolves as well in rats persistently infected with BDV (HIRANO et al. 1983; LUDWIG et al. 1988; GOSZTONYI and LUDWIG 1995). This degeneration, however, develops more slowly than the involvement of the dentate gyrus, it begins after 7–8 months and progresses into a complete disappearance of the retina. In this context, the richness of glutamatergic transmission among retinal neurons has to be emphasized.

Beside the interplay of BDV proteins with glutamate receptors, interactions with other neurotransmitter mechanisms, especially those of the dopamine system, have also been established (LIPKIN et al. 1988; SOLBRIG et al. 1995; see also chapter by Hornig et al., this volume).

As can be seen, there are multiple interactions between BDV and neurotransmitter systems. These interactions obviously lead to neurotransmitter imbalance. Behavioral changes known to occur during BDV infection of animals (SPRANKEL et al. 1978; DITTRICH et al. 1989; SOLBRIG et al. 1995), as well as mood disorders in humans associated with BDV infections (BODE 1995; BODE et al. 1996; LUDWIG and BODE 1997), could be explained on the basis of these pathogenetic mechanisms.

5 HIV-1 gp120 Protein and the NMDA Receptor

In the examples cited above the viral products acted on the same cell type, i.e. on neurons, in which the virus was replicated. In HIV infection of the brain, however, the viral products exert their deleterious effects also on cells which do not replicate the virus. Since the end of the 1980s, much attention has been devoted to the neurotoxic effect of the HIV-1 envelope protein gp120 (see also chapter by Sanders et al., this volume). There is general agreement that, during HIV infection of the brain, virus is produced in monocytoid cells (blood-derived monocytes, macrophages, resident microglia) and multinucleated giant cells, while neurons are not infected directly, or very rarely, if at all (WILEY et al. 1986; KOENIG et al. 1986; PRICE et al. 1988; BUDKA et al. 1991; SHARER 1992; GOSZTONYI et al. 1994). In spite of the lack of convincing evidence for neuronal infection a significant loss of neurons has been assessed by quantitative techniques (KETZLER et al. 1990; WEIS et al. 1993) and correlated with the presence of dementia (EVERALL et al. 1993a,b). This widespread cortical neuronal loss in the absence of direct infection by HIV has been regarded as a viral model of neurodegeneration (EVERALL et al. 1993b), the mechanism of which has to be elucidated.

In 1988 BRENNEMAN and colleagues, based on hippocampal tissue culture studies, proposed that the envelope glycoprotein gp120 shed from HIV could directly damage neurons due to an interference with endogenous neurotrophic substances. Vasoactive intestinal polypeptide (VIP) has sequence similarity with gp120, and on this basis they may compete for the same neuronal receptor. DREYER et al. (1990) established that the neurotoxic effect of gp120 is enforced by an increase in the intracellular calcium, and neurotoxicity can be abrogated by the calcium channel antagonist nimodipine. GIULIAN et al. (1990) observed that HIV-1-infected human monocytoid cells release toxic agents that destroy chick and rat neurons in culture. They suggested that these substances act by stimulating the NMDA receptor. NMDA antagonists protected against the HIV-1-induced neurotoxicity, while antagonists to non-NMDA-type glutamate receptors had no protective effect. Beside the gp120-induced activation of the NMDA-operated ion channels the activation of voltage-dependent calcium channels can also lead to a lethal influx of Ca^{2+} (LIPTON 1991). It has been documented that gp120 does not directly activate NMDA receptor-associated ion channels; its neurotoxic effect is blocked by degradation of glutamate in the neuronal cultures (LIPTON et al. 1991).

GENIS and colleagues (1992) described that cocultures of HIV-infected monocytes and astroglia released high levels of cytokines, i.e. tumor necrosis factor (TNF)-α and interleukin (IL)-1β, and arachidonate metabolites. They suggested that the neuronotoxic effect of gp120 is mediated by these substances. The studies of LIPTON (1992a,b) provided further evidence for the mechanism of HIV-induced neuronal damage. The binding of gp120 coat protein alone, in the absence of viral infection, is sufficient to trigger macrophages to release toxins that lead to NMDA receptor-mediated injury. Furthermore, Lipton's group noted that gp120 is capable of inducing the release of quinolinate from macrophages and probably also from astrocytes. Quinolinate is an endogenous glutamate agonist, which, in concert with glutamate, can elicit excessive activation of NMDA receptors. SAVIO and LEVI (1993) confirmed, by the aid of rat cerebellar granule cell cultures, the results of previous authors on the neurotoxic action of gp120, which was mediated by voltage-dependent Ca^{2+} channels and NMDA receptors and could be prevented by nifedipine and the NMDA channel antagonists, D-2-amino-5-phosphonovalerate (APV) and dizocilpine (MK-801). They suggested that gp120 might sensitize the neurons to the toxic effect of glutamate. DAWSON et al. (1993), also in tissue culture studies, revealed that nitric oxide (NO) and superoxide anions contribute to gp120 neurotoxicity. Peroxynitrite, a neurotoxic reaction product of NO and superoxide anion, mediates oxidative damage in the brains of AIDS patients with dementia (BOVEN et al. 1999). Beside NMDA antagonists a neuroprotective agent, riluzole, which inhibits the release of L-glutamate and L-aspartate, also proved to prevent gp120-induced neurodegeneration in cortical cultures (SINDOU et al. 1994).

Beside numerous tissue culture studies, some authors examined the in vivo neurotoxicity of gp120 HIV coat protein. GLOWA et al. (1992) administered gp120 into the cerebral ventricles of adult rats and observed learning impairment. HILL et al. (1993) injected neonatal rats with gp120 and observed retardation in developmental milestones associated with complex motor behaviors. BARKS et al. (1997) injected gp120 intrahippocampally into perinatal rats and observed hippocampal atrophy and loss of [^{3}H]glutamate receptor binding. Furthermore, dystrophic changes appeared in cortical pyramidal neurons. Interesting data resulted from studies on transgenic mice. In a study of TOGGAS et al. (1994) gp120 HIV-1 coat protein was expressed in astrocytes of transgenic mice. At neuropathological examination of the brains, extensive vacuolization of dendrites, decrease in synapto-dendritic complexity, loss of large pyramidal neurons and widespread reactive astrocytosis were found. These changes resembled those found in the brains of HIV-1 infected humans (SCHWENK et al. 1987; ARTIGAS et al. 1989; KETZLER et al. 1990; WILEY et al. 1991; MASLIAH et al. 1992).

In the meantime, more light has been shed on the mechanisms of glutamate release. Under physiological conditions, the overwhelming majority of the glutamate is in the intracellular compartment: its intracellular concentration is approximately 10mmol/l, and the extracellular approximately 0.6µmol/l. Excitatory amino acids are cleared from the extracellular space into neurons and astrocytes by a high-affinity uptake system (glutamate transporters). Excessive release into, or delayed removal from, the extracellular space of glutamate can lead to excitotoxic neuronal damage (LIPTON and ROSENBERG 1994). There are multiple ways by which gp120 might induce extracellular glutamate accumulation. The latter, followed by the activation of NMDA receptors and voltage-dependent Ca^{2+} channels leading to influx of Ca^{2+}, can be regarded as the *final common pathway* resulting in neuronal injury. This pathway operates not only in gp120-induced brain damage, but also in hypoxia, hypoglycemia, trauma and in several chronic neurodegenerative disorders (LIPTON and ROSENBERG 1994). In the case of AIDS, there is a complex web of interactions that ultimately concentrate onto this final common pathway of glutamate release (LIPTON 1994). These interactions, taken together, are established between monocytoid cells, astrocytes and neurons. HIV-infected or gp120-stimulated macrophages, after interacting with astrocytes, release neurotoxic factors, such as platelet activating factor (PAF), arachidonic acid and its metabolites. How these neurotoxic factors activate the NMDA receptor is not quite clear, but this activation can be prevented by the NMDA antagonists, APV and MK-108, as well as by degradation of endogenous glutamate. PAF has been shown to raise intracellular neuronal Ca^{2+} and to increase glutamate release (BITO et al. 1992). Arachidonic acid inhibits high-affinity uptake of glutamate into synaptosomes and astrocytes and increases extracellular glutamate levels (VOLTERRA et al. 1992). Cytokines TNF-α and IL-1β, released from HIV-1-infected macrophages, induce astrocytic proliferation (SELMAJ et al. 1990). IFN-γ produced in AIDS brains can induce macrophages to release quinolinate, an endogenous NMDA agonist that further increases intracellular Ca^{2+} levels (HEYES et al. 1992). Neurons, following excitation elicited by neurotoxic factors, release glutamate themselves onto second-order neurons and induce further excitotoxic damage (LIPTON 1994).

While the majority of investigators is of the opinion that gp120 induces neuronal damage by an indirect mechanism, some authors believe that a direct mode of action, without the intervention of macrophages, is also possible. WU et al. (1996) examined the question by using a nearly pure population of neurons, human NT cells. There was a severe decrease in the number of viable cells after exposure to gp120. However, APV, MK-801 or nimodipine prevented this cell loss. TOGGAS et al. (1996) treated gp120 transgenic mice with the NMDA receptor antagonist memantine. This drug prevented the appearance of the severe dendritic pathology seen in non-treated transgenic animals. MEUCCI and MILLER (1996) observed that in pure neuronal cultures deprived of the glial feeder layer a modest neurotoxicity still could be observed. They concluded that, although the indirect, macrophage- and astrocyte-mediated neurotoxic mechanism prevails, a direct toxic effect of gp120 cannot be excluded.

As to the exact mechanism of excitotoxic neuronal degeneration, it is probable that both necrotic and apoptotic cell death may occur in the same cell population (BONFOCO et al. 1995; MEUCCI and MILLER 1996). Recent studies of MEUCCI et al. (1998) suggested that gp120 can induce apoptosis in cultured hippocampal neurons directly, in the complete absence of the glial feeder layer. Hippocampal neurons possess chemokine receptors, gp120 binds to these and interferes with the normal trophic effects of the chemokines, inducing thus neuronal death by apoptosis. This represents a direct mechanism of neurotoxicity, in addition to release of neurotoxins from non-neural cells. KAUL and LIPTON (1999) established that chemokine receptors occur on brain macrophages and microglia, astrocytes and neurons. This way, apoptosis may ensue by direct interaction of gp120 with neurons, indirectly by stimulation of glia to release neurotoxic factors, or by both pathways. Accordingly, beside the indirect neurotoxicity mediated by brain macrophages and NMDA receptor activation, another, but direct, mechanism has become known that results from the interaction of gp120 and chemokine receptors, leading to neuronal apoptosis.

6 Conclusions

Viruses, while attacking the nervous system, adapt to the specific milieu of this organ and exploit the peculiar neurobiological mechanisms for their entry, spread and replication. Viral proteins, by chance or through adaptation, develop multiple interactions with neural structures. Of these, the interactions with neurotransmitter receptors are probably the most important, determining the distribution patterns in the course of virus infections within the nervous system. As typical examples, the receptor affinities of two strictly neurotropic viruses have been described here: the affinity of RV to the NR1 NMDA receptor, and that of BDV to the KA-1 receptor. An interaction between a viral protein and a neurotransmitter receptor may secure penetration of the virus and render the neuron vulnerable, on the one hand, while the lack of a specific receptor on neuronal systems may protect them from the viral attack. Virus protein–neuronal interaction may also result in a more or less limited, functional damage which, however, does not basically endanger the existence of the neuron. Blocking of receptors may cause partial or complete functional deafferentation. On the other hand, a permanent binding of viral proteins to, e.g. glutamate receptors, might protect them from excitotoxic damage. An interaction between viral proteins and axonal transport systems determines the direction of virus spread. If this transport is possible for an infectious agent in one direction only, it may secure protection for neuronal systems with limited interconnections, as in the case of the dentate gyrus in RV infection. The interplay of HIV-1 coat protein with the NMDA receptor is an example that a viral protein can interact with and even destroy cell types that are not infected by the agent itself. The understanding of several of these interactions has been made possible only in the light of novel neurobiological discoveries. Consequently, a better understanding of these virus–neuron interactions depends greatly upon further progress in neurobiology.

Acknowledgements. We are indebted to Dr. Tamás F. Freund and Dr. Liv Bode for stimulating and constructive discussions. Thanks are due to colleagues, in particular to Dr. Bernhard Dietzschold, Dr. Moujahed Kao, Dr. Ralf Dürrwald, Dr. Gerald Czech and Dr. Fedik Rantam, for supply of animal brains with natural and experimental infections. We are grateful to Dr. Peter Petrusz for technical advice and for supplying us with the anti-GluR1 and anti-GluR2-3 antibodies, and to Dr. A. Wandeler for presenting an anti-rabies virus antibody. This study has been supported by the Deutsche Forschungsgemeinschaft (DFG), by grants to Georg Gosztonyi (No. Go 426/3-1) and to Hanns Ludwig (No. Lu 142/5-1, -2, -3) as well as by a grant from the European Union (No. BMH-I-CT 94-1791).

References

Artigas J, Niedobitek F, Grosse G, Heise W, Gosztonyi G (1989) Spongiform encephalopathy in AIDS dementia complex: Report of five cases. J AIDS 2:374–381
Astic L, Saucier D, Coulon P, Lafay F, Flamand A (1993) The CVS strain of rabies virus as transneuronal tracer in the olfactory system of mice. Brain Research 619:146–156
Auer RN, Jensen ML, Whishow IQ (1989) Neurobehavioral deficit due to ischemic brain damage limited to half of the CA-1 sector of the hippocampus. J Neurosci 9:1641–1647
Baer GM, Harrison AK, Bauer SP, Shaddock JH, Murphy FA (1980) A bat rabies isolate with an unusually short incubation period. Exp Mol Pathol 33:211–222
Barks JDE, Liu XH, Sun R, Silverstein FS (1997) gp120, a human immunodeficiency virus-1 coat protein, augments hippocampal injury in perinatal rats. Neurosci 76:397–409
Bito H, Nakamura M, Honda Z, Isumi T, Iwatwubo T, Seyama Z, Sgura A, Kido Y, Schimizu T (1992) Platelet-activating factor (PAF) receptor in rat brain: PAF mobilizes intracellular Ca^{2+} in hippocampal neurons. Neuron 9:285–294

Bode L (1995) Human infections with Borna disease virus and potential pathogenetic implications. In: Koprowski H, Lipkin WI (eds) Borna Disease. Curr Top Microbiol Immunol, 190:103–130

Bode L, Dürrwald R, Rantam FA, Ferszt R, Ludwig H (1996) First isolates of infectious human Borna disease virus from patients with mood disorders. Molec Psychiat 1:200–212

Bonfoco E, Kraine D, Ancarcrona M, Nicotera P, Lipton S (1995) Apoptosis and necrosis: two distinct events induced, respectively, by mild and intense insults with N-methyl-D-aspartate or nitric oxide/superoxide in cortical cell cultures. PNAS USA 92:7162–7166

Boven LA, Gomes L, Hery C, Gray F, Verhoef J, Portegies P, Tardieu M, Nottet HSLM (1999) Increased peroxynitrite activity in AIDS dementia complex: implications for the neuropathogenesis of HIV-1 infection. J Immunol 162:4319–4327

Brenneman DE, Westbrook GL, Fitzgerald SP, Ennist DL, Elkins KL, Ruff MR, Pert CB (1988) Neuronal cell killing by the envelope protein of HIV and its prevention by vasoactive intestinal polypeptide. Nature 335:639–642

Budka H, Wiley CA, Kleihues P, et al. (1991) HIV-associated disease of the nervous system: review of the nomenclature and proposal for neuropathology-based terminology. Brain Pathol 1:143–152

Carbone KM, Duchala CS, Griffin JW, Kincaid AL, Narayan O (1987) Pathogenesis of Borna disease in rats: evidence that intra-axonal spread is the major route for virus dissemination and the determinant of disease incubation. J Virol 61:3431–3440

Carbone KM, Park SW, Rubin SA, Waltrip RW, Vogelsang GB (1991) Borna disease: association with a maturation defect in the cellular immune response. J Virol 65:6154–6164

Charlton KM (1994) The pathogenesis of rabies and other lyssaviral infections: recent studies. In: Rupprecht CE, Dietzschold B, Koprowski H (eds) Lyssaviruses. Current Top Microbiol Immunol 187:95–119

Conti F, DeBiasi S, Minelli A, Melone M (1996) Expression of NR1 and NR2 A/B subunits of the NMDA receptor in cortical astrocytes. Glia 17:254–258

Coyle JT, Puttfarcken P (1993) Oxidative stress, glutamate, and neurodegenerative disorders. Science 262:689–695

Cubitt B, de la Torre JC (1994) Borna disease virus (BDV), a nonsegmented RNA virus, replicates in the nuclei of infected cells where infectious BDV ribonucleoproteins are present. J Virol 68: 1371–1381

Dawson VL, Dawson TM, Uhl GR, Snyder SH (1993) Human immunodeficiency virus type 1 coat protein neurotoxicity mediated by nitric oxide in primary cortical cultures. PNAS USA 90:3256–3259

De La Torre JC (1994) Molecular biology of Borna disease virus: prototype of a new group of animal viruses. J Virol 68:7669–7675

Dittrich W, Bode L, Ludwig H, Kâo M, Schneider K (1989) Learning deficiencies in Borna disease virus-infected but clinically healthy rats. Biol Psychiatry 26:818–828

Dreyer EB, Kaiser PK, Offermann JT, Lipton SA (1990) HIV-1 coat protein neurotoxicity prevented by calcium channel antagonists. Science 248:364–367

Everall I, Glass J, McArthur J, Spargo E, Lantos P (1993a) Neuronal loss in the superior frontal gyrus correlates with HIV associated dementia. Clin Neuropathol 12:S10

Everall I, Luthert P, Lantos P (1993b) A review of neuronal damage in human immunodeficiency virus infection: its assessment, possible mechanisms and relationship to dementia. J Neuropathol Exp Neurol 52:561–566

Ferrari G, Yun C, Yan I, Greene LA (1995) N-Acetylcysteine (D- and L-stereoisomers) prevents apoptotic death of neuronal cells J Neurosci 5:2857–2866

Freund TF, Buzsáki G (1996) Interneurons of the hippocampus. Hippocampus 6:347–470

Genis P, Jett M, Bernton EW et al. (1992) Cytokines and arachidonic metabolites produced during human immunodeficiency virus (HIV)-infected macrophage-astroglia interactions: implications for the neuropathogenesis of HIV disease. J Exp Med 176:1703–1718

Gierend M (1982) Zur Pathogenese der Bornaschen Krankheit: Untersuchungen über die zelluläre Immunantwort, die immunosuppressive Behandlung und die Elektroencephalographie (EEG). Vet med dissertation, Berlin

Glowa JR, Panlilio LV, Brenneman DE, Gozes I, Fridkin M, Hill JM (1992) Learning impairment following intracerebral administration of the HIV envelope protein gp120 or a VIP antagonist. Brain Res 570:49–53

Gosztonyi G (1978a) Axonal and transsynaptic spread of viral nucleocapsids in fixed rabies virus encephalitis. J Neuropathol Exp Neurol 37:618

Gosztonyi G (1978b) Light and electron microscopic pathology of two virus encephalitides (in Hungarian). Thesis for the Hungarian Academy of Sciences, Budapest

Gosztonyi G, Ludwig H (1984a) Borna disease of horses. An immunohistochemical and virological study of persistently infected animals. Acta Neuropathol 64:213–221

Gosztonyi G, Ludwig H (1984b) Neurotransmitter receptors and viral neurotropism. Neuropsychiatr Clin 3:107–114

Gosztonyi G, Ludwig H (1995) Borna disease – neuropathology and pathogenesis. Curr Top Microbiol Immunol 190:39–73

Gosztonyi G, Dietzschold B, Kao M, Rupprecht CE, Ludwig H, Koprowski H (1993) Rabies and Borna disease: a comparative pathogenetic study of two neurovirulent agents. Lab Invest 68:285–295

Gosztonyi G, Artigas J, Lamperth L, Webster H deF (1994) Human immunodeficiency virus (HIV) distribution in HIV encephalitis: study of 19 cases with combined use of in situ hybridization and immunocytochemistry. J Neuropath Exp Neurol 53:521–534

Gosztonyi G, Rantam F, Ludwig H (2001) N-acetylcysteine (NAC) decreases Borna disease virus-induced selective neuronal degeneration of the dentate gyrus in persistently infected rats (in preparation)

Gottlieb M, Matute C (1997) Expression of ionotropic glutamate receptor subunits in glial cells of the hippocampal CA1 area following transient forebrain ischemia. J Cerebral Blood Flow Metab 17:290–300

Greenamyre JT, Olson JMM, Penney JB Jr, Young AB (1985) Autoradiographic characterization of N-methyl-D-aspartate-, quisqualate- and kainate-sensitive glutamate binding sites. J Pharmacol Exp Therapeutics 233:254–263

Halász N, Shepherd GM (1983) Neurochemistry of the vertebrate olfactory bulb. Neuroscience 10:579–619

Haywood AM (1994) Virus receptors: binding, adhesion strengthening, and changes in viral structure. J Virol 68:1–5

Heyes MP, Saito K, Markey SP (1992) Human macrophages convert L-tryptophan into the neurotoxin quinolinic acid. Biochem J 283:633–635

Hill JM, Mervis RF, Avidor R, Moody TW, Brenneman DE (1993) HIV envelope protein-induced neuronal damage and retardation of behavioral development in rat neonates. Brain Res 603:222–233

Hirano N, Kao M, Ludwig H (1983) Persistent, tolerant or subacute infection in Borna disease virus infected rats. J Gen Virol 64:1521–1530

Iwasaki Y, Clark HF (1977) Cell to cell transmission of rabies virus in the central nervous system. II. Experimental rabies in the mouse. Lab Invest 33:391–399

Iwasaki Y, Wiktor TJ, Koprowski H (1973) Early events of rabies virus replication in tissue cultures. An electron microscopic study. Lab Invest 28:142–148

Jackson AC, Reimer DL (1989) Pathogenesis of experimental rabies in mice: an immunohistochemical study. Acta Neuropathol 78:159–168

Jackson AC, Park H (1998) Apoptotic cell death in experimental rabies in suckling mice. Acta Neuropathol 95:159–164

Kaul M, Lipton SA (1999) Chemokines and activated macrophages in HIV gp120-induced neuronal apoptosis. PNAS USA 96:8212–8216

Ketzler S, Weis S, Haug H, Budka H (1990) Loss of neurons in the frontal cortex in AIDS brains. Acta Neuropathol 80:92–94

Knuckey NW, Palm D, Primiano M, Epstein MH, Johanson CE (1995) N-acetylcysteine enhances hippocampal neuronal survival after transient forebrain ischemia in rats. Stroke 26:305–311

Koenig S, Gendelman HE, Orenstein JM, Dal Canto MC, Pezeshkpour GH, Yungbluth M, Janotta F, Aksamit A, Martin MA, Fauci AS (1986) Detection of AIDS virus in macrophages in brain tissue from AIDS patients with encephalopathy. Science 233:1089–1093

Krey HF, Ludwig H, Rott R (1979) Spread of infectious virus along the optic nerve into the retina in Borna disease virus-infected rabbits. Arch Virol 61:283–288

Kucera P, Dolivo M, Coulon P, Flamand A (1985) Pathways of the early propagation of virulent and avirulent rabies strains from the eye to the brain. J Virol 55:158–162

Kutsuwada T, Kashiwabuchi N, Mori H, Sakimura K, Kushiya E, Araki K, Meguro H, Lafay F, Coulon P, Astic L, Saucier D, Riche D, Holley A, Flamand A (1991) Spread of CVS strain of rabies virus and of the avirulent AvO1 along the olfactory pathways of the mouse after intranasal inoculation. Virology 183:320–330

Lentz TL, Burrage TG, Smith AL, Crick J, Tignor GH (1982) Is the acetylcholine receptor a rabies virus receptor? Science 215:182–184

Leranth C, Frotscher M (1989) Organization of the septal region in the rat brain: cholinergic-GABAergic interconnections and the termination of hippocampo-septal fibers. J Comp Neurol 289:304–314

Lipkin WI, Carbone KM, Wilson MC, Duchala CS, Narayan O, Oldstone BA (1988) Neurotransmitter abnormalities in Borna disease. Brain Res 475:366–370
Lipton SA (1991) Calcium channel antagonists and human immunodeficiency virus coat protein-mediated neuronal injury. Ann Neurol 30:110–114
Lipton SA, Sucher NJ, Kaiser PK, Dreyer EB (1991) Synergistic effects of HIV coat protein and NMDA receptor-mediated neurotoxicity. Neuron 7:111–118
Lipton SA (1992a) Requirement for macrophages in neuronal injury induced by HIV envelope protein gp120. Neuro Report 3:913–915
Lipton SA (1992b) Models of neuronal injury in AIDS: Another role of the NMDA receptor? Trends Neurosci 15:75–79
Lipton SA (1994) HIV-related neuronal injury. Potential therapeutic intervention with calcium channel antagonists and NMDA antagonists. Mol Neurobiol 8:181–196
Lipton SA, Rosenberg PA (1994) Excitatory amino acids as a final common pathway for neurologic disorders. New Eng J Med 330:613–622
Ludwig H, Bode L (1997) The neuropathogenesis of Borna disease virus infections. Intervirology 40: 185–197
Ludwig H, Bode L (2000) Borna disease virus: New aspects of infection, disease, diagnosis and epidemiology. Rev Sci Tech Off Int Epiz 19:259–288
Ludwig H, Bode L, Gosztonyi G (1988) Borna disease: a persistent virus infection of the central nervous system. Prog Med Virol 35:107–151
Lundgren AL, Zimmermann W, Bode L, Czech G, Gosztonyi G, Lindberg R, Ludwig H (1995) Staggering disease in cats: isolation and characterization of the feline Borna disease virus. J Gen Virol 76:2215–2222
Lyons MJ, Faust IM, Hemmes RB, Buskirk DR, Hirsch J, Zabriskie JB (1982) A virally induced obesity syndrome in mice. Science 216:82–85
Masaki H, Kumanishi T, Srakawa M, Mishina M (1992) Molecular diversity of the NMDA receptor channel. Nature 358:36–41
Masliah E, Achim CL, Ge N, DeTeresa R, Terry RD, Wiley CA (1992) Spectrum of human immunodeficiency virus-associated neocortical damage. Ann Neurol 32:321–329
Matsumoto S (1963) Electron microscope studies of rabies virus in mouse brain. J Cell Biol 19:565–591
Meucci O, Miller RJ (1996) gp120-induced neurotoxicity in hippocampal pyramidal neuron cultures: Protective action of TGF-β1. J Neurosci 16:4080–4088
Meucci O, Fatatis A, Simen AA, Bushell TJ, Gray PW, Miller RJ (1998) Chemokines regulate hippocampal neuronal signaling and gp120 neurotoxicity. PNAS USA 95:14500–14505
Monaghan DT, Cotman CW (1985) Distribution of N-methyl-D-aspartate-sensitive L-[^{3}H] glutamate-binding sites in rat brain. J Neurosci 5:2909–2919
Monyer H, Sprengel R, Schoepfer R, Herb A, Higuchi M, Lomeli H, Burnashev N, Sakmann B, Seeburg PH (1992) Heteromeric NMDA receptors: molecular and functional distinction of subtypes. Science 256:1217–1221
Monyer H, Burnashev N, Laurie DJ, Sakmann B, Seeburg PH (1994) Developmental and regional expression in the rat brain and functional properties of four NMDA receptors. Neuron 12:529–540
Morales JA, Herzog S, Kompter C, Frese K, Rott R (1988) Axonal transport of Borna disease virus along olfactory pathways in spontaneously and experimentally infected rats. Med Microbiol Immunol 177:51–68
Moriyoshi K, Masu M, Ishii T, Shigemoto R, Mizuno N, Nakanishi S (1991) Molecular cloning and characterization of the rat NMDA receptor. Nature 354:31–37
Nakanishi S (1992) Molecular diversity of glutamate receptors and implications for brain function. Science 258:597–603
Nicholls DG (1994) Proteins, transmitters and synapses. Blackwell, Oxford
Oldstone, MBA (1984) Virus can alter cell function without causing cell pathology: disordered function leads to imbalance of homeostasis and disease. In: Notkins AL, Oldstone MBA (eds) Concepts in viral pathogenesis. Springer Verlag, New York, Heidelberg, Tokyo
Oldstone MBA, Holmstoen J, Welsh RM (1977) Alteration of acetylcholine enzymes in neuroblastoma cells persistently infected with lymphocytic choriomeningitis virus. J Cell Physiol 91:459–472
Oldstone MBA, Rodriguez M, Daughaday WH, Lampert PW (1984) Viral perturbation of endocrine function: Disorder of cell function leading to disturbed homeostasis and disease. Nature 307:278–280
Oldstone MBA, Sinha YN, Blount P, Tishon A, Rodriguez M, von Wedel R, Lampert PW (1982) Virus-induced alterations in homeostasis: alterations in differentiated functions of infected cells in vivo. Science 218:1125–1127

Olney JW, Ho OL, Rhee V (1971) Cytotoxic effects of acidic and sulphur containing amino acids on the infant mouse central nervous system. Exp Brain Res 14:61–76

Price RW, Brew B, Sidtis J, Rosenblum M, Scheck AC, Cleary P (1988) The brain in AIDS: central nervous system HIV-1 infection and AIDS dementia complex. Science 239:586–592

Reagan KJ, Wunner WH (1985) Rabies virus interaction with various cell lines is independent of the acetycholine receptor. Arch Virol 84:277–282 (1985)

Rodriguez M, von Wedel RJ, Garret RS, Lampert PW, Oldstone MBA (1983) Pituitary dwarfism in mice persistently infected with lymphocytic choriomeningitis virus (LCMV). Lab Invest 49:48–53

Sabio T, Levi G (1993) Neurotoxicity of HIV coat protein gp120, NMDA receptors, and protein kinase C: a study with rat cerebellar granule cell cultures. J Neurosci Res 34:265–272

Schneemann A, Schneider PA, Lamb RA, Lipkin WI (1995) The remarkable coding strategy of Borna disease virus: a new member of the nonsegmented negative strand RNA viruses. Virology 210: 1–8

Schneider LG (1975) Spread of virus within the central nervous system. In: Baer GM (ed) The natural history of rabies, vol I. Academic Press, New York, pp 199–216

Schulman JE (1983) Chemical neuroanatomy of the cerebellar cortex. In: Emson PC (ed) Chemical neuroanatomy. Raven, New York, pp 209–228

Schwenk J, Cruz-Sanchez F, Gosztonyi G, Cervós-Navarro J (1987) Spongiform encephalopathy in a patient with acquired immune deficiency syndrome (AIDS). Acta Neuropathol 74:389–392

Selmaj KN, Farooq M, Norton T, Raine CS, Brosman CF (1990) Proliferation of astrocytes in vitro in response to cytokines. A primary role for tumor necrosis factor. J Immunol 144:129–135

Sharer LR (1992) Pathology of HIV-1 infection of the central nervous system (A review). J Neuropathol Exp Neurol 51:3–11

Sloviter RS, Valiquette G, Abrams GM, Ronk EC, Sollas AL, Paul LA, Neubort S (1989) Selective loss of hippocampal granule cells in the mature rat brain after adrenelectomy. Science 243:535–538

Sloviter RS, Sollas AL, Dean E, Neubort S (1993a) Adrenalectomy-induced granule cell degeneration in the rat hippocampal dentate gyrus: Characterization of an in vivo model of controlled neuronal death. J Comp Neurol 330:324–336

Sloviter RS, Dean E, Neubort S (1993b) Electron microscopic analysis of adrenalectomy-induced hippocampal granule cell degeneration in the rat: apoptosis in the adult central nervous system. J Comp Neurol 330:337–351

Solbrig MV, Fallon JH, Lipkin WI (1995) Behavioral disturbances and pharmacology of Borna disease. In: Koprowski H, Lipkin WI (eds) Borna Disease. Curr Top Microbiol Immunol 190:93–99

Sprankel H, Riharz K, Ludwig H, Rott R (1978) Behavior alterations in tree shrews (*Tupaia glis*, Diard 1820) induced by Borna disease virus. Med Microbiol Immunol 165:1–18

Toggas SM, Masliah E, Rockenstein EM, Rall GF, Abraham CR, Mucke L (1994) Central nervous system damage produced by expression of the HIV-1 coat protein gp120 in transgenic mice. Nature 367:188–193

Toggas SM, Masliah E, Mucke L (1996) Prevention of HIV-1 gp120-induced neuronal damage in the central nervous system of transgenic mice by the NMDA receptor antagonist memantine. Brain Res 706:303–307

Tóth K, Borhegyi Z, Freund TF (1993) Postsynaptic targets of GABAergic hippocampal neurons in the medial septum – diagonal band of Broca complex. J Neurosci 13:3712–3724

Tsiang H, Derer M, Taxi J (1983) An in vivo and in vitro study of rabies virus infection of the rat superior cervical ganglia. Arch Virol 76:231–243

Tsiang H, Ceccaldi PE, Ermine A, Lockhart B, Guillemer S (1991) Inhibition of rabies virus infection in cultured rat cortical neurons by an N-methyl-D-aspartate noncompetitive antagonist, MK-801. Antimicrobial Agents and Chemotherapy 35:572–574

Tuffereau C, Benejean J, Roque Alfonso AM, Flamand A, Fishman MC (1998) Neuronal cell surface molecules mediate specific binding to rabies virus glycoprotein expressed by a recombinant baculovirus on the surfaces of lepidopteran cells. J Virol 72:1085–1091

Volterra A, Trotti D, Cassutti P, Tromba C, Salvaggio A, Melcangi RC, Racagni G (1992) High sensitivity of glutamate uptake to extracellular free arachidonic acid levels in rat cortical synaptosomes and astrocytes. J Neurochem 59:600–606

Wang C, Pralong WF, Schulz MF, Rougon G, Aubry GM, Pagliusi S, Robert A, Kiss JZ (1996) Functional N-methyl-D-aspartate receptors in O-2 A glial precursor cells: a critical role in regulating polysialic acid-neural cell adhesion molecule expression and cell migration. J Cell Biol 135:1565–1581

Weis S, Haug H, Budka H (1993) Neuronal damage in the cerebral cortex of AIDS brains: a morphometric study. Acta Neuropathol 85:185–189

Werner P, Voigt M, Keinänen K, Voigt M, Wisden W, Seeburg PH (1991) Cloning of a putative high-affinity kainate receptor expressed predominantly in hippocampal CA3 cells. Nature 351: 742–744

Whetsell WO Jr (1996) Current concepts of excitotoxicity. J Neuropathol Exp Neurology 55:1–13

Wiley CA, Schrier RD, Nelson JA, Lampert PW, Oldstone MBA (1986) Cellular localization of human immunodeficiency virus infection within the brains of acquired immune deficiency syndrome patients. PNAS USA 83:7089–7093

Wiley CA, Masliah E, Morey M, Lemere C, DeTeresa R, Grafe M, Hansen L, Terry R (1991) Neocortical damage during HIV infection. Ann Neurol 29:651–657

Wisden W, Seeburg PH (1993) A complex mosaic of high-affinity kainate receptors in rat brain. J Neurosci 13:3582–3598

Wu P, Price P, Du B, Hatch WC, Terwilliger EF (1996) Direct cytotoxicity of HIV-1 envelope protein on human NT neurons. Neuro Report 7:1045–1049

Wyszynski M, Kharazia V, Shanghvi R, Rao A, Beggs AH, Craig AM, Weinberg R, Sheng M (1998) Differential regional expression and ultrastructural localization of α-actinin-2, a putative NMDA receptor-anchoring protein, in rat brain. J Neurosci 18:1383–1392

Mechanisms of Virus-Induced Neuronal Damage and the Clearance of Viruses from the CNS

B. Dietzschold, K. Morimoto, and D.C. Hooper

1 Introduction	145
2 CNS Response to Infection	146
3 Virus-CNS Interaction in Rabies	147
4 Virus-CNS Interaction in Borna Disease	149
5 Mechanisms of Virus Clearance from the CNS	150
6 Outlook	153
References	153

1 Introduction

The pathogenic mechanisms underlying virus-induced neurological disease are complex but fall into two general categories, neuronal damage or dysfunction, resulting: (1) from within, as a direct consequence of the virus infection, and (2) from without, due to the indirect action of resident and invading immune/inflammatory cells responding to viral antigens. Our studies have focused on two viruses that can cause acute, lethal neurological diseases representative of these classes: rabies virus (RV) and Borna disease virus (BDV). RV infection induces significant electrophysiological changes in the CNS and sleep alterations but is accompanied by only minor histopathological changes in the CNS (Gourmelon et al. 1986, 1991). In contrast, acute Borna disease is associated with extensive neuropathology including astrogliosis, perivascular cuffing, monocytic infiltration of the brain parenchyma, and massive neuronal loss (Richt et al. 1990; Gosztonyi and Ludwig 1995; Stitz et al. 1995; Morimoto et al. 1996), which are dependent on a BDV-specific immune response. The profound difference in the pathogenesis of rabies and Borna disease is evidenced by the fact that immunosuppression either has no effect or is detrimental to the outcome of rabies infection but is therapeutic

---

Center for Neurovirology, Department of Microbiology and Immunology, Thomas Jefferson University, 1020 Locust Street, Philadelphia, PA 19107-6799, USA
e-mail: bdietzschold@reddi1.uns.tju.edu

in Borna disease (STITZ et al. 1995; MORIMOTO et al. 1996). Despite the clear differences in neuropathology between rabies and Borna disease, there are immune strategies for each that can evidently clear these viruses from the CNS thereby preventing a lethal outcome to the infections (DIETZSCHOLD et al. 1992; DIETZSCHOLD 1993; RICHT 1994). Further knowledge of the different pathogenic processes underlying rabies and Borna disease is key to the development of therapeutic strategies for diverse CNS viral diseases. In this chapter we discuss the response of CNS resident cells to infection, the link between these responses and the induction of virus-specific immunity, and the mechanisms through which virus can be cleared from the CNS.

2 CNS Response to Infection

An interesting characteristic of RV and BDV infections is an almost undetectably low replication rate in the first half of the incubation period, which may reflect the relatively quiescent metabolic state of the neuron at the initiation of infection. During this time, viral RNA can only be detected using the reverse transcriptase-polymerase chain reaction (RT-PCR) (SHANKAR et al. 1991, 1992). After this eclipse period, virus RNA levels increase exponentially (FU et al. 1993), which correlates spatio-temporally with the dramatic and specific induction of immediate-early response genes (IEGs) encoding transcription factors such as c-Fos, Jun B, and Egr-1 (FU et al. 1993). The up-regulation of expression mRNAs for these transcription factors in neurons can occur in the absence of inflammation and therefore is likely a direct consequence of the virus infection (MORIMOTO et al. 1996).

One of the possible functions of transcription factors is to induce changes in neurons that render them more competent in supporting virus replication (FU et al. 1993). On the other hand, the activation of certain transcription factors in the infected CNS may be important in initiating the expression of genes that are responsible for triggering an inflammatory response. Important in this regard are the transcription factors NF-κB, AP-1, CREB, and C/EBP, which are activated in BDV infection, as demonstrated by gel-shift assays (Fig. 1), and known to be involved in the regulation of a number of anti-viral response mechanisms (JONAT et al. 1990; BAEUERLE 1991; STEIN and BALDWIN 1993). In rabies, which does not have a strong inflammatory component like Borna disease, only the activities of NF-κB and AP-1 are enhanced (Fig. 1). Because of the differences in the extent of inflammation normally seen in Borna disease in comparison with rabies, we speculate that the activation of NF-κB and AP-1 may have different purposes and outcomes in these two diseases.

While neither RV nor BDV causes lytic infection and both up-regulate IEGs, the expression of which does not depend on protein synthesis, these viruses differ fundamentally in their effects on the transcription of late host response genes. For example, BDV infection enhances the expression in neurons of enkephalin

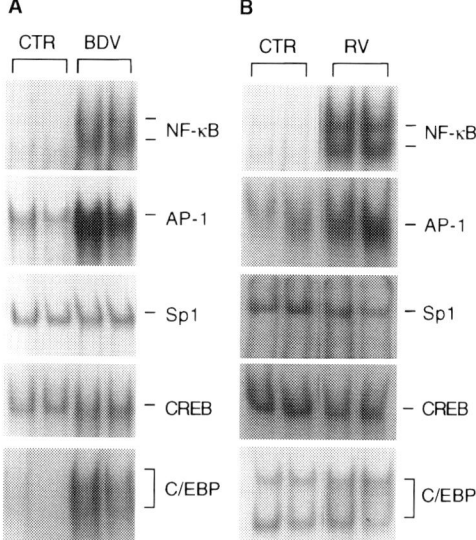

Fig. 1. Electrophoretic mobility shift assay of nuclear extract from Borna disease virus (*BDV*)-infected rat brain (**A**) and rabies virus (*RV*)-infected mouse brain (**B**). Nuclear extracts (10μg) were incubated with ^{32}P-labeled NF-κB (5'-AGTTGAGGGGACTTTCCCAGGC-3'), AP-1 (5'-CGCTTGATGACT-CAGCCGGAA-3'), Sp1 (5'-ATTCGATCGGGGCGGGGCGAGC-3'), CREB (5'-AGAGATTGCCT-GACGTCAGAGAGCTAG-3'), or C/EBP (5'-AAGTACTTTCAGTTTCATATTACTCTA-3') double-stranded consensus oligonucleotide, and the nucleotide-protein complexes were electrophorized on a 6% nondenaturing polyacrylamide gel. *CTR*, nuclear extract from normal uninfected rat or mouse brain; *BDV*, nuclear extract from BDV-infected rat brain at day 22 post-infection; *RV*, nuclear extract from rabies virus-infected mouse brain at day 6 post-infection. Brains from two rats or mice were analyzed for each time point

(Fu et al. 1993), cyclooxygenase-2 (COX-2) (MORIMOTO et al. 1996; ROEHRENBECK et al. 1998), and calcitonin gene-related peptide (CGRP) (ROEHRENBECK et al. 1998). In contrast, infection with RV results in a strong reduction in the expression of enkephalin (Fu et al. 1993), neuronal nitric oxide synthase (AKAIKE et al. 1995), the 5-hydroxytryptamine receptor (CECCALDI 1993), and the housekeeping gene G3PDH (Fu et al. 1993). These differences support the concept that the pathogenesis of Borna disease is immune-dependent while that of rabies is the result of a direct effect of the virus on infected neurons.

3 Virus-CNS Interaction in Rabies

There are several likely alternatives that may lead to a down-regulation of late host response gene expression and ultimately to cell death in rabies including: (1) depletion of metabolic pools by excessive replication of the virus, (2) induction of apoptosis, and (3) activation of double stranded (ds)-RNA-dependent protein kinase (PKR), which can lead to the inhibition of protein synthesis through the phosphorylation of eIF-2α (JACOBS and LANGLAND 1996). PKR mRNA expression is up-regulated in the CNS in both Borna disease and rabies (Fig. 2) which argues against a general role of PKR in the down-regulation of protein synthesis. Moreover mice with a targeted disruption in the gene encoding PKR succumb to RV infection indistinguishably from their normal counterparts (data not shown). We therefore consider it more likely that RV causes neuronal death through the depletion of metabolic pools and the induction of apoptosis, two mechanisms which are not mutually exclusive. While there is evidence that rabies can induce apoptosis in vivo and in vitro, recent studies have revealed a potential controversy regarding the contribution of apoptosis to rabies pathogenesis (JACKSON et al. 1997; MORIMOTO et al. 1998). The pathogenicity of RV strains is inversely proportional to their capacity to induce apoptosis in neuronal cell culture (MORIMOTO et al. 1998). Interestingly, the extent of apoptosis correlates with the level of rabies glycoprotein (G protein) expression in the infected neuron (THOULOUZE et al. 1997; MORIMOTO et al. 1998). The differential expression of G protein appears to be largely determined by post-translational mechanisms that effect its stability (MORIMOTO et al. 1998). Down-regulation of G protein expression in neuronal cells evidently contributes to rabies pathogenesis by preventing apoptosis, which may be one of the most important defense mechanisms against virus infection (TEODORO and BRANTON 1997; ITOH et al. 1998). Apoptosis is known to lead to depolymerization of actin filaments (KOTHAKOTA et al. 1997), which, in the case of rabies, would prevent the transport of N protein and, likely, neuronal spread of the virus (MORIMOTO et al. 1998). Preliminary results indicate that street RV strains (i.e. dog- and silver-haired bat-derived RV strains), which are considerably more pathogenic than tissue culture adapted strains, express very limited levels of G protein in neuronal cells and do not induce apoptosis until very late in the infection cycle. Thus the regulation of G protein expression is very likely to be relevant to the

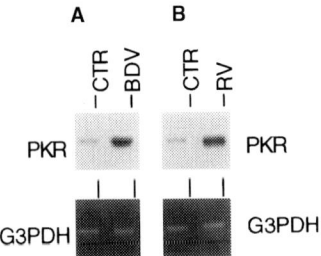

Fig. 2. Reverse transcriptase-polymerase chain reaction (RT-PCR) analysis of double stranded (ds)-RNA-dependent protein kinase (*PKR*) mRNA in Borna disease virus (*BDV*)-infected rat brain (**A**) and rabies virus-infected mouse brain (**B**). RNA was extracted from rat brain or mouse brain, and subjected to RT-PCR analysis for PKR. Amplification of G3PDH mRNA served as an internal control. *CTR*, RNA from normal uninfected rat or mouse brain; *BDV*, RNA from BDV-infected rat brain at day 22 post-infection; *RV*, RNA from rabies virus-infected mouse brain at day 6 post-infection. Brains from two rats or mice were analyzed for each time point

pathogenesis of street RV as it provides a means for more pathogenic variants to survive and spread within the nervous system without triggering a protective immune response. Nevertheless, we have evidence that neurons infected with highly pathogenic RV strains eventually may undergo apoptosis but at a time after the infection has already spread trans-synaptically.

4 Virus-CNS Interaction in Borna Disease

In contrast to rabies, immune cells invading the CNS in response to viral antigens are central to neuronal damage in Borna disease. Animals immunologically tolerant of BDV antigen or immunosuppressed show minimal overt signs of disease and limited CNS pathology despite a high virus load in the CNS (STITZ et al. 1995; MORIMOTO et al. 1996). In this case, a major element in the induction of CNS pathology is in the provision of a stimulatory signal from the infected cells in the CNS to the immune system in the periphery. We have evidence that the induction of certain factors in CNS resident cells by BDV infection may be responsible for establishing this connection. Using dexamethasone treatment to separate the responses of CNS resident cells to BDV infection from those of invading inflammatory cells, we have demonstrated that tumor necrosis factor (TNF)-α, macrophage inflammatory protein (MIP)-1β, interleukin (IL)-6, and the CXC chemokine *mob*-1 are all induced in CNS resident cells directly by BDV infection (Fig. 3). As these factors are all known to participate in the induction of inflammatory responses, we consider that these findings provide evidence of a link between innate immunity (i.e. the response of the cells to infection) and the adaptive immune response. Of note is the recent finding that genes associated with inflammatory mediators (i.e. COX-2, CGRP) are induced in CNS resident cells, both infected and non-infected, during Borna disease (ROEHRENBECK et al. 1998). This provides the first direct evidence of communication of a signal between infected and non-infected CNS resident cells and likely represents the first link in the chain of communication between the BDV-infected cells and the immune system. Neuronal damage in Borna disease is clearly the result of destructive elements of the adaptive immune response to BDV antigens with interferon (IFN)-γ-producing CD 4 cells and IFN-γ-dependent CD8 T cells as well as activated cells of the monocyte lineage all contributing to the pathogenesis. The importance of the latter cell populations in the pathogenesis of BDV is underscored by the tremendous up-regulation of monocyte-associated genes including the complement component C1q (DIETZSCHOLD et al. 1995), T-kininogen (MORIMOTO et al. 1996), and inducible nitric oxide synthase (iNOS) (AKAIKE et al. 1995). With regards to the role of iNOS in neuropathogenesis, it is unlikely that its product nitric oxide (NO) directly mediates neuronal damage. We have recently demonstrated that peroxynitrite (ONOO$^-$), the reaction product of NO and superoxide, is largely responsible for the immune-dependent neuropathogenesis in experimental allergic encephalomyelitis (EAE)

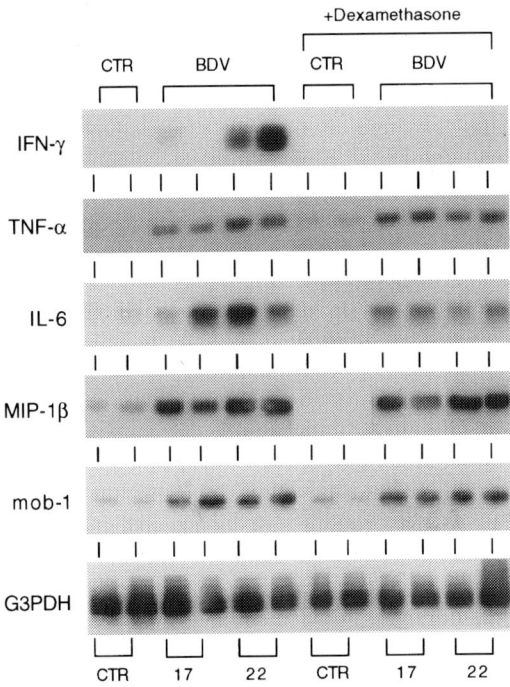

Fig. 3. Effect of dexamethasone on the expression of mRNAs for the cytokines interferon (IFN)-γ, tumor necrosis factor (TNF)-α, and interleukin (IL)-6 and chemokines macrophage inflammatory protein (MIP)-1β and mob-1 in Borna disease virus (BDV)-infected rat brains. Total RNA was extracted from rat brain and subjected to reverse transcriptase-polymerase chain reaction (RT-PCR) by using gene-specific primers, and the amplified DNA was analyzed by Southern blotting using ^{32}P-labeled gene-specific oligonucleotide probes. *CTR*, brain RNA from normal uninfected rats; *BDV*, brain RNA from BDV-infected rats at days 17 and 22 post-infection; + *Dexamethasone*: brain RNA from dexamethasone-treated rats. (From MORIMOTO et al. 1996)

(HOOPER et al. 1998a). Since uric acid, a potent scavenger of ONOO$^-$, is therapeutic in Borna disease (Hooper, unpublished), we believe that there is a significant pathogenic contribution from ONOO$^-$ to this disease as well.

5 Mechanisms of Virus Clearance from the CNS

For the most part, neurons are non-renewable; therefore classic anti-viral immune responses, which include the destruction of infected cells, can be detrimental if infection has spread into a significant number of neurons. This is the case for Borna disease, in which the immune response to the virus normally develops after the majority of the neurons in the brain have become infected. However, Borna disease can be prevented by the adoptive transfer of BDV-immune CD4 T cells prior to infection (RICHT et al. 1994). This causes the rapid onset of an extensive but transient inflammation in the brain and clears the virus (RICHT et al. 1994). Thus, BDV-specific cellular immune mechanisms can contribute to clearance of the virus if activated prior to the infection reaching a substantial number of neurons. The contribution of antibody to the clearance of BDV in this model is unknown but an accumulation of B cells in the brain was noted in the adoptive transfer experiments (RICHT et al. 1994).

The situation in rabies is somewhat different than in Borna disease, as RV G-protein-specific neutralizing antibodies play a predominant role in the clearance of RV from the CNS (DIETSCHOLD et al. 1992; DIETZSCHOLD 1993; HOOPER et al. 1998b). Treatment of RV-infected rats with monoclonal virus-neutralizing antibodies prevents lethal disease and promotes clearance of the virus from the CNS (DIETSCHOLD et al. 1992; DIETZSCHOLD 1993). Because of the existence of the blood–brain barrier, this clearance must be dependent upon innate immune mechanisms, induced by the infected neurons, which afford antibody access to the infected cells. To further investigate the concept that various humoral and cellular aspects of immunity collaborate in the clearance of RV from the brain, we studied the clearance of virus from the brains of knockout mice lacking either B and T cells, CD8+ cytotoxic T cells, B cells, IFN-α/β receptors, IFN-γ receptors, or C3 and C4 complement. Transient virus replication is seen in the brain after intranasal infection of normal adult mice with an attenuated RV, CVS-F3 (Fig. 4), and clearance of the virus is associated with the natural production of neutralizing antibodies (HOOPER et al. 1998b). Mice lacking either B and T cells or B cells alone fail to clear the virus from the CNS (Fig. 4), develop a progressive disease, and succumb to CVS-F3 infection (HOOPER et al. 1998b). On the other hand, mice lacking either CD8+ T cells, IFN receptors or the complement components C3 and C4 showed no significant differences in the development of clinical signs compared to intact counterparts having the same genetic background (HOOPER et al. 1998b). However, while infectious virus and viral RNA could be detected in normal control mice only until day 8 p.i., in all the gene knockout mice studied, except those lacking C3/C4, virus infection persisted through day 21 p.i. (Fig. 4). Analysis of RV-specific antibody production together with histological assessment of brain inflammation in infected animals revealed that clearance of CVS-F3 by 21 days p.i. correlated with both a strong inflammatory response in the CNS early in the infection (day 8 p.i.), and the rapid (day 10 p.i.) production of significant levels of virus-neutralizing antibody (VNA) (HOOPER et al. 1998b). These studies confirm that RV-neutralizing antibodies are an absolute requirement for clearance of an established RV infection. However, for the latter to occur in a timely fashion, collaboration between VNA and inflammatory mechanisms is necessary.

The precise mechanism through which RV-specific neutralizing antibodies clear the virus from the infected CNS is unknown. A major question concerns whether or not virus can be cleared without destroying the infected neuron. There is evidence that antibody can mediate the clearance of Sindbis virus from infected neurons by restoring the IFN-α response and the integrity of ion channels (DEPRES et al. 1995). Termination of the production of infectious Sindbis in antibody-treated neurons is preceded by a decrease in virus protein synthesis (LEVINE et al. 1991). This observation has led to the conclusion that antibody mediates virus clearance through the restriction of virus gene expression (LEVINE et al. 1991), which also appears to be the case for other neurotropic viruses including measles (SCHNEIDER-SCHAULIES et al. 1992) and rabies (DIETSCHOLD et al. 1992; DIETZSCHOLD 1993). For instance, treatment of RV-infected neuroblastoma cells with neutralizing monoclonal antibodies results in a reduction in rabies virus RNA

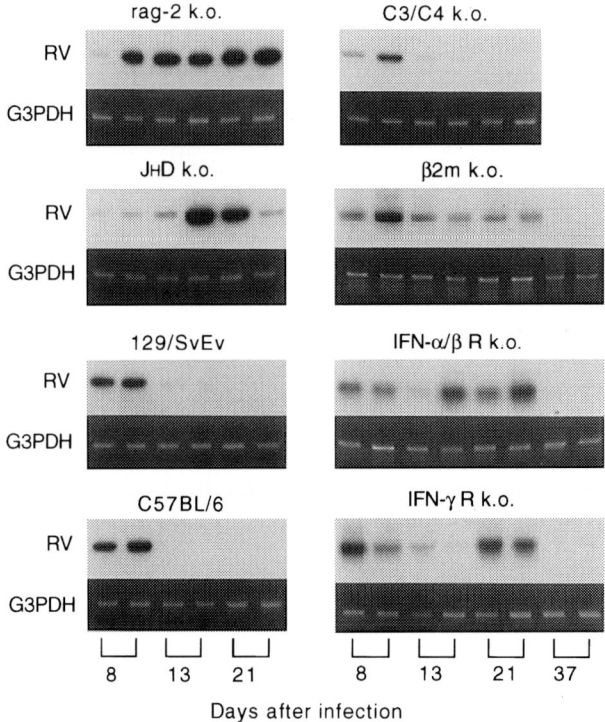

Fig. 4. Reverse transcriptase-polymerase chain reaction (RT-PCR) analysis of rabies virus (RV) genomic RNA in brain tissue from mice infected intranasally with the attenuated RV strain CVS-F3. Brains from two mice of each of the indicated strains were collected at 8, 13, 21, and 37 days after intranasal infection with CVS-F3, and RNA was extracted and subjected to RT-PCT analysis for rabies virus genomic RNA (*RV*). Amplification of G3PDH mRNA served as an internal control. Normal mice, C57BL/6J, 129/SvEv. Gene knockout mice: Rag 2 k.o (T and B cell deficient mice), C3/C4 k.o. (mice lacking the complement components C3 and C4), J$_H$D k.o (antibody-deficient mice), β2m k.o (mice deficient in CD4$^-$ CD8$^+$ T cells), IFN-α/β R k.o (mice lacking the interferon α/β receptor), IFN-γ R k.o (mice lacking the interferon γ receptor). (From HOOPER et al. 1998)

transcription (DIETZSCHOLD et al. 1992). Interestingly, the protective activity of a particular monoclonal antibody in vivo correlates with its ability to inhibit RV RNA transcription and its capacity to restrict cell to cell spread of the virus in vitro, rather than its virus-neutralizing activity (DIETZSCHOLD et al. 1992).

The mechanism through which antibodies affect virus transcription is again largely unresolved. It is possible that the virus-specific antibody may exert its inhibitory function after uptake, based on the observation that protective RV-specific monoclonal antibodies are internalized by the infected cell (DIETZSCHOLD et al. 1992). A more attractive hypothesis is that viral protein expressed on the infected cell surface may function as a signal-transducing receptor for an external signal mediated through antibody binding. For example, measles-virus-specific antibodies specifically induce an increase in inositol triphosphate which can act as a second messenger in the activation of protein kinase C in infected cells (WEINMANN-

DORSCH 1989). We have preliminary evidence that rabies G protein expressed on the surface of infected neurons can transduce a signal upon interaction with particular G-protein-specific monoclonal antibodies. In this case, the end result of the signal transduced by G protein appears to be apoptosis. The assumption that this mechanism is operative in the in vivo infection is based on the finding that RV-specific protective monoclonal antibodies given late in the infection exacerbate the disease.

6 Outlook

Is it possible to clear virus from neurons without massive cell damage which will inevitably result in severe neurological sequelae? The fact that, for several neurotropic viruses, the protective activity of particular antibodies has been associated with the ability to reduce expression of viral genes suggests the existence of a membrane signal transduction mechanism that may promote clearance of virus from infected cells. Delineation of the precise cell machinery involved in such a mechanism is necessary to utilize this system as a foundation for alternative strategies for anti-viral therapy. In the case of rabies, it appears that the signal induced by the interaction between antibody and G protein expressed on the cell surface blocks spread of the virus but likely causes cell death in at least some of the infected cells. Thus it appears that only early immunological intervention in rabies can clear the virus from the CNS without sequelae; but this is probably associated with a certain acceptable level of neuronal destruction. The assumption here is that RV must be cleared because of its cell toxicity, but it is conceivable that anti-viral immune mechanisms may be capable of reducing virus replication to a level consistent with the survival and continued function of the infected cell. BDV, at the other extreme, has minimal neurotoxicity compared to RV. The immune response to BDV supplies the pathogenic components of the disease and, in their absence, neuronal function is maintained despite the presence of high levels of virus replication. Because of the lack of direct neurotoxicity, persistent Borna infection is perhaps an excellent model to study the possibility of virus clearance without cell destruction.

References

Akaike T, Weihe E, Schaeffer MK-H, Fu ZF, Vogel WH, Schmidt HW, Koprowski H, Dietzschold B (1995) Effect of neurotropic virus infection on neuronal and inducible nitric oxide synthase activity in rat brain. J Neuro Virol 1:118 125

Baeuerle P (1991) The inducible transcription activator NF-kappa B: regulation by distinct protein subunits. Biochim Biophys Acta 1072:63–80

Ceccaldi PE, Fillion MP, Ermine A, Tsiang H, Fillion G (1993) Rabies virus selectively alters 5-TH 1 receptor subtypesin rat brain. Eur J Parmacol 245:129–138

Despres P, Griffin JW, Griffin DE (1995) Antiviral activity of alphainterferon in Sindbis virus-infected cells is restored by anti-E2 monoclonal antibody treatment. J Virol 69:7345–7348

Dietzschold B, Kao M, Zheng YM, Chen ZY, Maul G, Fu ZF, Rupprecht CE, Koprowski H (1992) Delineation of putative mechanisms involved in the antbody-mediated clearance of rabies virus from the central nervous system. Proc Natl Acad Sci USA 89:7252–7256

Dietzschold B (1993) Antibody-mediated clearance of viruses from the mamalien nervous system. Trends Microbiol 1:87–117

Dietzschold B, Schwaeble W, Schaeffer MK-H, Hooper DC, Zheng F, Petry H, Fink T, Loos M, Koprowski H, Weihe E (1995) Expression of C1q, a subcomponent of the rat complement system, is dramatically enhanced in brains of rats with either Borna disease or experimental allergic encephalomyelitis. J Neurol Sci 130:11–16

Fu ZF, Weihe E, Zheng YM, Schaeffer MK-H, Sheng H, Corisdeo S, Rauscher III FJ, Koprowski H, Dietzschold B (1993) Differential Effects of rabies and Borna disease on immediate-early- and late-response gene expression in brain tissue. J Virol 67:6674–6681

Gosztonyi G, Ludwig H (1995) Borna disease – neuropathology and pathogenesis. Curr Top Microbiol Immunol 190:39–69

Gourmelon P, Briet D, Court L, Tsiang H (1986) Electrophysiological and sleep alterations in experimental mouse rabies. Brain Res 396:128–140

Gourmelon P, Briet D, Clarencon D, Court L, Tsiang H (1991) Sleep alterations in experimental street rabies virus infection occur in the absence of major EEG abnormalities. Brain Res 554:159–165

Hooper DC, Spitsin S, Kean RB, Champion JM, Dickson GM, Chaudhry I, Koprowski H (1998a) Uric acid, a natural scavenger of peroxynitrite, in experimental allergic encephalitis and multiple sclerosis. Proc Natl Acad Sci 95:675–680

Hooper DC, Morimoto K, Bette M, Weihe E, Koprowski H, Dietzschold B (1998) Collaboration of antibody and inflammation in the clearance of rabies virus from the CNS. J Virol 72:3711–3719

Itoh M, Hotta H, Homma M (1998) Increased induction of apoptosis by a Sendai virus mutant is associated with attenuation of mouse pathogenicity. J Virol 72:2927–2934

Jackson AC, Rossiter JP (1997) Apoptosis plays an important role in experimental rabies virus infection. J Virol 71:5603–5607

Jacobs BL, Langland JO (1996) When two strands are better than one: the mediators and modulators of the cellular responses to double-stranded RNA. Virology 219:339–349

Jonat C, Rahmsdorf HJ, Park K-K, Cato ACB, Gebel S, Ponta H, Herrlich P (1990) Antitumor promotion and antiinflammation: down-modulation of AP-1 (Fos/Jun) activity by glucocorticoid hormone. Cell 62:1189–1204

Kothakota S, Azuma T, Reinhard C, Klippel A, Tang J, Chu K, McGarry TJ, Kirschner MW, Koths K, Kwiatkowski DJ, Williams LT (1997) Caspase-3-generated fragment of gelsolin: effector of morphological change in apoptosis. Science 278:294–297

Levine B, Hardwick JM, Trapp BD, Crawford TO, Bollinger RC, Griffin DE (1991) Antibody-mediated clearance of alpha-virus infection from neurons. Science 245:856–859

Morimoto K, Hooper DC, Bornhorst A, Corisdeo S, Bette M, Fu ZF, Schaeffer MK-H, Koprowski H, Weihe E, Dietzschold B (1996) Intrinsic responses to Borna disease virus infection of the central nervous sysem. Proc Natl Acad Sci USA 93:1345–1350

Morimoto K, Hooper DC, Spitsin S, Koprowski H, Dietzschold B (1999) Pathogenicity of different rabies virus variants inversely correlates with apoptosis and rabies virus glycoprotein expression in infected primary neuron cultures. J Virol 73:510–518

Richt J, Stitz L, Deschl U, Frese K, Rott R (1990) Borna disease virus-induced meningoencephalitis caused by a virus-specific CD4+ T cell-mediated immune reaction. J Gen Virol 71:2565–2573

Richt JA, Schmeel A, Freese K, Carbone KM, Narayan O, Rott R (1994) Borna disease virus-specific T cells protect against or cause immunopathological Borna disease. J Ex Med 170:1467–1473

Roehrenbeck A, Bette M, Hooper DC, Nyberg F, Eiden L, Dietzschold B, Weihe E (1998) Virus-induced encephalitis is characterized by upregulation of COX-2 and CGRP in cerebrocortical neurons and by induction of COX-2 in endothelial cells. Neurobiol Dis 6:15–34

Schneider-Schaulies S, Liebert UG, Segev Y, Rager-Zisman B, Wolfson M, ter Meulen V (1992) Antibody-dependent transcriptional regulation of measles virus in persistently infected neuronal cells. J Virol 66:5534–5541

Shankar V, Dietzschold B, Koprowski H (1991) Direct entry of rabies virus into the central nervous system without prior local replication. J Virol 65:2736–2738

Shankar V, Kao M, Hamir AN, Sheng H, Koprowski H, Dietzschold B (1992) Kinetics of virus spread and changes in levels of several cytokine mRNAs in the brain after intranasal infection of rats with Borna disease virus. J Virol 66:992–998

Stein B, Baldwin AS Jr (1993) Distinct mechanisms for regulation of the interleukin-8 gene involve synergism and cooperativity between C/EBP and NF-κB. Mol Cell Biol 13:7191–7198

Stitz L, Dietzschold B, Carbone KM (1995) Immunopathogenesis of Borna desease. Curr Top Microbiol Immunol 190:75–89

Teodoro JG, Branton PE (1997) Regulation of apoptosis by viral gene products. J Virol 71:1739–1746

Thoulouze M-I, Lafage M, Montano-Hirose JA, Lafon M (1997) Rabies virus infects mouse and human lymphocytes and induces apoptosis. J Virol 71:7372–7380

Weinmann-Dorsch C, Koschel K (1989) Coupling of viral membrane proteins to phosphatidylinosite signaling system. FEBS Letters 247:185–188

Borna Disease Virus Infection of Adult and Neonatal Rats: Models for Neuropsychiatric Disease

M. Hornig, M. Solbrig, N. Horscroft, H. Weissenböck, and W.I. Lipkin

1	Introduction: Biology, Epidemiology, Pathogenesis	157
2	Mechanisms for Neurotropism: Phosphorylation of BDV Phosphoprotein (P) by Protein Kinase C-ε	158
3	Mechanisms for BDV Persistence	159
4	Borna Disease Rat Models for Human CNS Disorders	160
4.1	Adult Infection: A Viral Model for Movement Disorders	161
4.2	Neonatal Infection: A Viral Model for Neurodevelopmental Disorders	164
5	Summary	173
References		174

1 Introduction: Biology, Epidemiology, Pathogenesis

Borna disease virus (BDV) is a newly classified RNA virus (Briese et al. 1994; Cubitt et al. 1994; Schneemann et al. 1995), worldwide in distribution, that infects the central nervous system (CNS) of warm-blooded animals to cause behavioral disturbances reminiscent of autism, schizophrenia, and mood disorders (Lipkin et al. 1995). It is not lytic in vitro or in vivo, replicates at lower levels than most known viruses and is dissimilar in nucleic acid and protein sequence to other infectious agents (de la Torre 1994; Schneemann et al. 1995). Thus, BDV eluded characterization until its nucleic acids were cloned by subtractive hybridization (Lipkin et al. 1990; VandeWoude et al. 1990). The molecular biology of BDV is unusual in many respects, including a nuclear localization for replication and transcription, overlap of open reading frames (ORFs) and transcription units, post-transcriptional modification of subgenomic RNAs, and marked conservation of coding sequence across a wide variety of animal species and tissue culture systems. These features led to its recent recognition by the International Committee on Taxonomy of Viruses as the prototype of a new family, Bornaviridae, within the nonsegmented negative-strand RNA viruses (de la Torre 1994; Schneemann

Laboratory for the Study of Emerging Diseases, 3101 Gillespie Neuroscience Research Facility, University of California, Irvine, CA 92697-4292, USA
e-mail: ilipkin@uci.edu

et al. 1995). Natural infection has been confirmed in horses, sheep, cattle, birds and cats. Primates can be infected experimentally (STITZ et al. 1980) and recent reports suggest an increased prevalence of BDV infection in mood disorders and schizophrenia (AMSTERDAM et al. 1985; BODE et al. 1988, 1992, 1993; FU et al. 1993; KISHI et al. 1995; WALTRIP II et al. 1995).

Most previous work in BDV pathogenesis has focused on adult immunocompetent rodents and ungulates, in which infection results in dramatic disturbances in behavior, limbic circuitry, and monoamine neurotransmitter systems. Although these models are intriguing, they are associated with marked CNS inflammation, loss of brain mass, and gliosis, and may be less relevant to neuropsychiatric diseases than those in neonatally-infected rats, in which BDV induces subtle disturbances of behavior without robust inflammatory cell infiltration. Rats infected with BDV during the neonatal period have dysgenesis of hippocampal and cerebellar structures (BAUTISTA et al. 1994, 1995; CARBONE et al. 1991), hyperactivity (BAUTISTA et al. 1994), and learning deficits (DITTRICH et al. 1989), paralleling the neurodevelopmental abnormalities of autism (RAPIN and KATZMAN 1998), schizophrenia (WEINBERGER 1987; STEVENS 1973), and bipolar mood disorder (VAN OS et al. 1997; YOLKEN and TORREY 1995; GUTIERREZ et al. 1998; BECKMANN and JAKOB 1991). The more modest immunopathology in the neonatal rat model suggests that damage is due to viral products that impact neuronal migration, connectivity, or viability. Although studies in adult infected animals have shown tropism for and pathology in monoaminergic circuits, there are at present no detailed studies of pharmacology, neurochemistry, or soluble immune or growth factors in the neonatal rat model. This chapter will first review mechanisms for neurotropism and viral persistence as well as the immunopathologic and neuropharmacologic disturbances noted in the more widely studied, immune-mediated adult rodent model of Borna disease, and then examine recent findings relevant to neuropathogenesis following neonatal BDV infection.

2 Mechanisms for Neurotropism: Phosphorylation of BDV Phosphoprotein (P) by Protein Kinase C-ε

The basis for the neurotropism of BDV is likely to be multifactorial. The integrity of the humoral immune response is critical to restriction of virus to neural compartments (STITZ et al. 1998); however, this alone cannot account for limbic tropism, as replication is still higher in limbic structures in animals with compromised humoral immunity. The presence of specific viral receptors on cell surfaces is a common mechanism for establishing patterns of tropism (KAUFFMAN and FIELDS 1985) and may play a role in BDV's limbic tropism. Although viral attachment proteins have been identified (gp18, G), the cellular receptors for BDV remain unknown. Another means by which preferential replication of BDV in limbic tissues might occur is through restricted distribution of the enzymatic machinery required for the virus' life cycle. Phosphorylation of BDV P, for example, is thought to be important for its functional activation and ability to serve as a transcription factor; the ε-isotype of protein kinase C (PKC-ε) is primarily responsible for the phosphorylation of BDV P (SCHWEMMLE et al. 1997). PKC-ε is concentrated heavily in limbic circuitry, including intense staining in hippocampus and olfactory tubercle, and moderate staining in anterior olfactory nuclei, olfactory bulb, cerebellum, nucleus accumbens, lateral septal nuclei, and caudate-putamen (SAITO et al. 1993). Given the extensive overlap in the regional distributions of BDV and PKC-ε in rat brain, the possibility exists that the localization of PKC-ε, through its phosphorylation effects, may influence the tropism of BDV for limbic circuitry.

3 Mechanisms for BDV Persistence

Unlike other CNS viral infections, in which the presence of an intact immune system results in either viral clearance or host mortality, BDV has the capacity to cause persistent infection, a feature that favors both virus and host survival. This does not reflect a failure to induce an immune response. Indeed, cellular immunity is essential to expression of classical disease and the presence of CD8+ T cells in the brains of adult-infected rats parallels the upregulation of MHC class I antigen expression in brain (CARBONE et al. 1991) and the appearance of signs of neurologic dysfunction (PLANZ et al. 1995). Rather, persistence appears to be due to induction of tolerance by mechanisms that remain poorly understood. Potential mechanisms include altered viral gene expression or modulation of the immune response. The first possibility is unlikely because viral titers in brain do not change substantively over the course of disease (NARAYAN et al. 1983a; CARBONE et al. 1987). Furthermore, brain levels of N and P, the only BDV proteins which can be readily quantitated, and of RNAs coding for other BDV proteins (HATALSKI 1996) are similar in early and late disease. There is, however, increasing evidence for modulation of the immune response over the course of Borna disease. Recent findings indicate the induction of BDV-specific Th_1 tolerance in chronic infection. Whereas lymphocytes isolated from brains of acutely infected rats have potent cytolytic activity, lymphocytes from brains of chronically infected rats do not lyse BDV-infected target cells (SOBBE et al. 1997). Induction of BDV-specific tolerance in chronic infection may reflect the time course for presentation of viral antigens in the thymus (RUBIN et al. 1995). Alternatively, Th_1 cells may become anergic or undergo apoptosis due to presentation of BDV antigens in brain without essential costimulatory signals (KARPAS et al. 1994; KHOURY et al. 1995; SCHWARTZ 1992). Support for this hypothesis is found in the observation that apoptosis of perivascular inflammatory cells is most apparent at 5–6 weeks post-infection, coincident with the onset of decline in encephalitis.

Our laboratory has used subtractive cloning technologies to address problems in neuropathogenesis since isolating the first BDV nucleic acid sequences in 1988

(LIPKIN et al. 1990). We were intrigued by the observation in adult infected Lewis rats that immune responses are curbed in late disease without changes in viral gene expression; thus, we pursued subtractive cloning (representational difference analysis, RDA) to enrich for messages present at higher levels in brains of chronically infected animals in which inflammation had resolved. The majority of clones isolated corresponded to mRNA for immunoglobulin. Based on these findings we established an RNase protection assay (RPA) for the purpose of measuring cytokine mRNAs linked to different aspects of the immune response. RPA studies of whole-brain RNA identified a $Th_1 \rightarrow Th_2$ shift, suggesting an intriguing mechanism for induction of tolerance in chronically infected animals. Consistent with such a shift was the observation of an isotype switch in peripheral blood from IgG to IgE (HATALSKI et al. 1998a,b). In addition, we observed that the humoral immune response may play an important role in limiting viral gene expression in the chronic phase of Borna disease in the adult-infected rat (HATALSKI et al. 1998a). Complementary work by HATALSKI et al. (1998b) in the chronic phase of Borna disease reported increases in intra-CNS production of IgG antibodies that parallel increases of antibodies with neutralizing activity against BDV in peripheral blood. Definitive evidence that humoral immunity contributes to BDV tropism emerged from recent work by STITZ et al. (1998), who found that passive transfer of neutralizing antibodies resulted in limitation of viral replication within the CNS. Similar results have been reported in other viral systems; for example, passive transfer of virus-specific antibodies limits viral replication in the CNS following infection with murine hepatitis virus type-4 (BUCHMEIER et al. 1984) or measles virus (LIEBERT et al. 1990) and induces clearance of virus following infection with rabies (DIETZSCHOLD et al. 1992) or Sindbis virus (LEVINE et al. 1991).

4 Borna Disease Rat Models for Human CNS Disorders

The mechanisms by which viral infections alter the architecture and function of limbic circuitry are not well-understood. In rats infected as adults with BDV, the cellular immune response appears critical to the development of the dramatic neurologic manifestations (PLANZ et al. 1995), whereas in neonatally-infected rats, the more subtle neurobehavioral syndrome has been reported to occur without invoking infiltrating inflammatory elements (STITZ et al. 1995). Thus, the functional and structural outcome of viral infections may depend on the differential timing of neurotoxic insults in relation to the specific phase(s) of neural and immune system maturation. When considered in the context of the wide variability in clinical presentations of human neuropsychiatric disorders such as autism, schizophrenia, and affective illnesses, this heterogeneity in neurobehavioral consequences of BDV infection underscores the relevance and utility of neurodevelopmental animal models for understanding the pathogenesis of neuropsychiatric diseases. Elucidation of mechanisms of viral–CNS interactions should permit more focused, informed investigations in humans and nonhuman primates regarding both the role of viral injury in neurodevelopmental disorders and the neuropathogenesis of specific neuropsychiatric disorders.

BDV pathogenesis has been extensively studied in rodent models. Two diseases have been defined: (1) an immune-mediated syndrome characterized by dramatic disturbances in movement and behavior (infection of adult immunocompetent rats) (LUDWIG et al. 1988; NARAYAN et al. 1983a; SOLBRIG et al. 1994, 1995, 1996a–c, 1998), and (2) a distinct syndrome characterized by cerebellar and hippocampal dysgenesis, hyperactivity, and learning disturbances (infection of neonatal rats) (BAUTISTA et al. 1994, 1995; CARBONE et al. 1996; DITTRICH et al. 1989).

4.1 Adult Infection: A Viral Model for Movement Disorders

As in autism (ANDERSON 1994; ERNST et al. 1997), schizophrenia (COOPER et al. 1991), and mood disorders (HAMNER and DIAMOND 1996; KELSOE et al. 1996; PARTONEN 1996), disorders of movement and behavior in adult Borna disease rats are linked to distinct changes in CNS dopamine (DA) systems (SOLBRIG et al. 1994, 1995, 1996a,b, 1998) and may be further linked to serotonin (5HT) abnormalities (SOLBRIG et al. 1995). The immune-mediated disorder in adult infected rats presents clinically as hyperactivity and exaggerated startle responses 10–14 days after intracerebral infection (NARAYAN et al. 1983a). The acute phase coincides with infiltration of monocytes into the brain, particularly in areas of high viral burden including the hippocampus, amygdala and other limbic structures (CARBONE et al. 1987). Two to three weeks later, rats show high-grade stereotyped motor behaviors (the continuous repetition of behavioral elements including sniffing, chewing, scratching, grooming, and self-biting), dyskinesias, dystonia, and flexed seated postures (SOLBRIG et al. 1994), in parallel with the widespread distribution of virus in limbic and prefrontal circuits. Five to ten percent of animals become obese, achieving body weights up to 300% of normal (LUDWIG et al. 1988).

Pharmacologic and lesion effects of adult infection are recognized for several neurotransmitter systems. Central DA systems of adult-infected Borna disease animals are more sensitive to DA agonists and antagonists than normal rats. Infected animals manifest increased locomotor and stereotypic behavior following administration of the mixed-acting DA agonist, dextroamphetamine (Fig. 1a,b; SOLBRIG et al. 1994). Similarly, enhanced locomotion and stereotypies are seen in adult infected rats in response to the DA re-uptake inhibitory effects of cocaine, indicating dose-dependent potentiation of DA neurotransmission (SOLBRIG et al. 1998). At low, presynaptic, autoreceptor doses of the direct DA agonist, apomorphine, hyperactivity is reduced, whereas higher doses increase locomotion (Fig. 1c). The movement and behavior disorder is improved following treatment with selective DA antagonists; whereas D2-selective antagonists (e.g., raclopride) do not affect locomotor responses in Borna disease rats, high doses of selective D1 antagonists (e.g., SCH23390) and atypical DA blocking agents with mixed D1 and

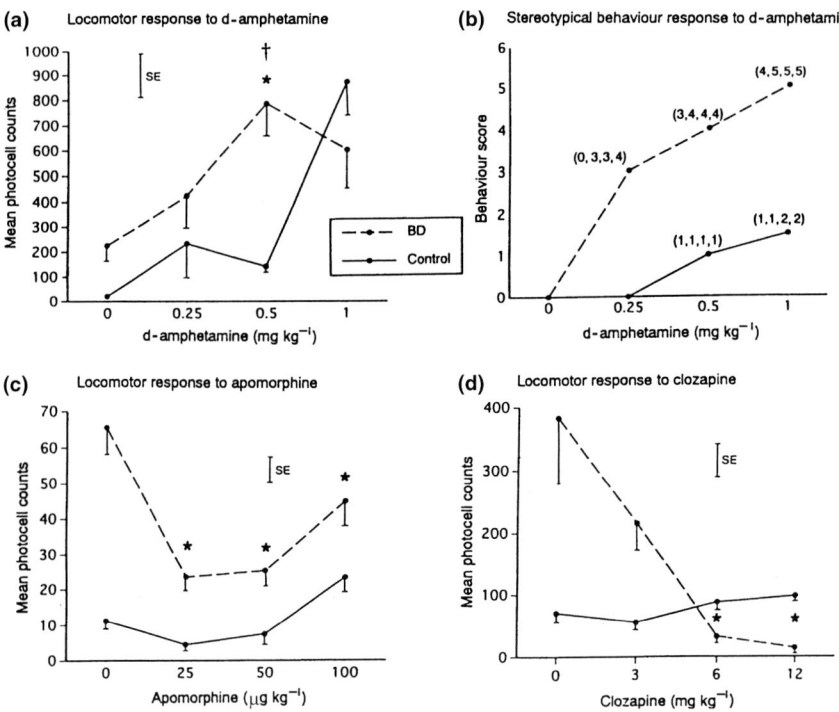

Fig. 1a–d. Locomotor and behavior responses to dopamine (*DA*) agonists and antagonists in adult Borna disease (*BD*) and noninfected (*NL*) rats. **a** Locomotor response to *d*-amphetamine in adult-infected BD (*Ad-BD*) and NL rats ($n = 4$ in each group). Values represent mean crossovers (the successive interruption of photocell beams) during a 180-min interval ± SEM; * indicates significant increase in locomotor activity in Ad-BD rats at this dose relative to saline injection (post-hoc Newman-Keuls test); † indicates significant increase in locomotor activity in Ad-BD group vs control group (main effect following significant effect on repeated measures ANOVA). **b** Stereotypical behavior response to *d*-amphetamine in Ad-BD and NL rats. Values represent median score using the MacLennan and Maier scale (MACLENNAN and MAIER 1983). Ad-BD rats typically had scores of 3 (0, 3, 3, 4) at the low dose, 4 (3, 4, 4, 4) at the 0.5mg/kg dose, and 5 (4, 5, 5, 5) at the highest dose. Stereotypical behavior ratings were significantly increased in Ad-BD rats at these doses. Information statistic (KULLBACK 1968) revealed a significant increase in stereotyped behavior ratings in the Ad-BD rats; subsequent examination at each dose revealed significant increases at doses 0.25, 0.50 and 1.00mg/kg. **c** Locomotor response to low-dose apomorphine in Ad-BD and NL rats ($n = 4$ each group). Values represent mean crossovers during a 15-min interval ± SEM; * indicates significant decrease in locomotor activity in Ad-BD rats at these doses relative to saline injection (post-hoc Newman-Keuls test). **d** Locomotor response to clozapine to Ad-BD and NL rats ($n = 4$ each group). Values represent mean crossovers during a 180-min interval ± SEM; * indicates significant decrease in locomotor activity in Ad-BD rats at these doses relative to pH-balanced water injection, (post-hoc Newman-Keuls test). (Reprinted with the permission of Academic Press from SOLBRIG et al. 1994)

D2 antagonist activity, such as clozapine, selectively reduce locomotor activity in Borna disease rats but not in controls (Fig. 1d; SOLBRIG et al. 1994).

The neuropathologic basis for these functional disturbances of DA neurocircuitry in the adult Borna disease system has begun to be defined. Greater decreases in DA than in dihydroxyphenylacetic acid (DOPAC, the major metabolite of DA) levels are noted by HPLC analysis of tissues from striatum, nucleus accumbens, and olfactory tubercle (SOLBRIG et al. 1994), whereas in prefrontal cortex the reverse pattern is seen, with a marked increase in DOPAC (SOLBRIG et al. 1996a). In addition, tyrosine hydroxylase-immunoreactive cells are depleted in substantia nigra and ventral tegmental area following adult BDV infection of Lewis rats (SOLBRIG et al. 1994). Taken together, these results suggest partial DA deafferentation with compensatory metabolic hyperactivity in nigrostriatal and mesolimbic DA systems. At the receptor level, both pre-and postsynaptic sites of the DA transmitter system appear to be damaged in striatum (caudate-putamen and nucleus accumbens), consistent with the neurochemical and neuropharmacologic data described above. DA uptake sites, as measured by binding of mazindol, are reduced in nucleus accumbens (SOLBRIG et al. 1996b) and caudate-putamen (SOLBRIG et al. 1998). D2 (but not D1) receptor binding is markedly reduced in caudate-putamen (Fig. 2); D2 and D3 receptor binding are reduced in nucleus accumbens (SOLBRIG et al. 1994, 1996a,b). In contrast, postsynaptic DA receptors (D1, D2, D3) remain intact in prefrontal cortex (SOLBRIG et al. 1996a).

Although the increased locomotor activity, stereotypic behaviors and dyskinesias of the adult Borna disease model are linked to distinct disturbances in DA pathways, additional neuromodulator abnormalities in adult Borna disease have

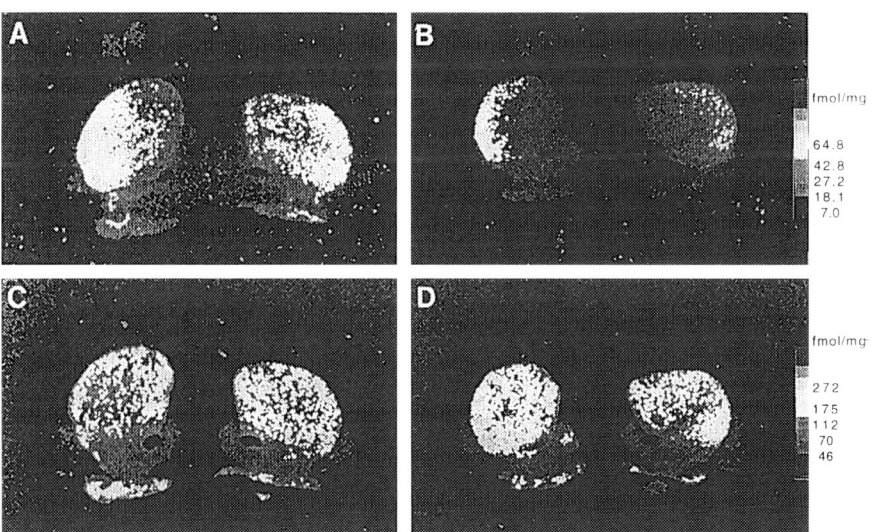

Fig. 2A–D. D1 and D2 receptors in caudate-putamen of adult-infected Borna disease (*BD*) and noninfected (*NL*) rats. Dopamine (DA) site labeling in coronal sections through caudate-putamen (*CP*) rostral to the decussation of the anterior commissure in normal (NL) (**A, C**) and adult-infected (*Ad-BD*) rats (**B, D**) 45 days post-intracerebral injection of either phosphate-buffered saline (PBS) or BDV, respectively. D2 site labeling by specific binding of 1nM [^{3}H]raclopride was significantly decreased in CP and each of its four subregions in Ad-BD (**B**) as compared to NL rats (**A**). D1 site labeling by specific binding of 2nM [^{3}H]SCH23390 in CP of Ad-BD rats (**D**) was unchanged from that seen in NL rats (**C**). (Reprinted with the permission of Academic Press from SOLBRIG et al. 1994)

also been noted. The expression of genes for neuromodulatory substances and their associated synthesizing enzymes, including somatostatin, cholecystokinin, and glutamic acid decarboxylase, is greatly reduced during the acute phase and recovers toward normal in the chronic phase of adult Borna disease (LIPKIN et al. 1988). The cholinergic system, a major participant in sensorimotor processing, learning, and memory, also appears to be affected in adult BDV infection. A decrease in the number of choline acetyltransferase-positive fibers has been observed to begin as early as day 6 post-infection (p.i.) and progresses to nearly complete loss of cholinergic fibers in hippocampus and neocortex by day 15 p.i. (GIES et al. 1998). Preliminary work on dysregulation of 5HT and norepinephrine (NE) systems suggests metabolic hyperactivity of 5HT (as evidenced by modest increase in the metabolite 5-hydroxyindoleacetic acid [5HIAA]) in striatum and of NE (as evidenced by a small increase in 3-methoxy-4-hydroxyphenethyleneglycol [MHPG]) in prefrontal and anterior cingulate cortex regions (SOLBRIG et al. 1995; M. Solbrig and W.I. Lipkin, personal communication). These changes may reflect compensatory upregulation or heterotypic sprouting following partial loss of DA afferents to these brain regions. Selective effects of BDV on 5HT and NE pre- or postsynaptic receptors have not yet been investigated. Pharmacologic and neurotransmitter-specific molecular probes have also been used to characterize endogenous opioid systems in the adult rat model. Infected animals respond abnormally to the opiate antagonist, naloxone, with hyperkinesis and seizures, and also demonstrate increases in striatal preproenkephalin mRNA at 14 and 21 days (FU et al. 1993b) and at 45 days after BDV infection (M. Solbrig and W.I. Lipkin, personal communication). However, the mechanisms by which these changes in endogenous opioid systems occur are unclear. The marked CNS inflammation in adult-infected rats makes it difficult to determine whether monoamine, cholinergic, and opiatergic dysfunction in Borna disease results from direct effects of the virus, virus effects on resident cells of the CNS, or a cellular immune response to viral gene products. More recently, our efforts have turned towards the neonatal model in an effort to identify the functional and structural consequences of BDV in a system that is linked to more direct interactions of the virus with CNS.

4.2 Neonatal Infection: A Viral Model for Neurodevelopmental Disorders

Neonatal rat infection may provide a more intriguing model for neuropsychiatric disorders. Indeed, the cerebellar and hippocampal dysgenesis that is observed in neonatally-infected animals is consistent with the more subtle neurodevelopmental abnormalities reported by some investigators in autism (KEMPER and BAUMAN 1993), schizophrenia (ALTSHULER et al. 1987; FISH et al. 1992), and affective disorders (SOARES and MANN 1997). Neonatally-infected animals also display a wide range of physiologic and neurobehavioral disturbances. They are smaller than uninfected littermates (CARBONE et al. 1991; BAUTISTA et al. 1994), without demonstrable alteration of glucose, growth hormone, or insulin-like growth factor-1 (BAUTISTA et al. 1994) or amount of food ingested (BAUTISTA et al. 1995);

they display a heightened taste preference for salt solutions; and they exhibit altered sleep-wake cycles (BAUTISTA et al. 1994). Neurobehavioral disturbances have been previously reported to be subtle in neonatally-infected animals. A study of behavioral and cognitive changes in Wistar rats infected in the neonatal period found spatial and aversive learning deficits, increased motor activity, and decreased anxiety responses (DITTRICH et al. 1989). Similar deficits in spatial learning and memory were found by Carbone and colleagues in neonatally-infected Lewis rats 23–73 days p.i. (CARBONE et al. 1996). More recently, play behavior has been reported to be abnormal in the neonatal model, with decreases in both initiation of nondominance-related play interactions and response to initiation of play by noninfected, age-matched control animals or by infected littermates (CARBONE et al. 1998). Thus, the neuropathologic, physiologic, and neurobehavioral features of BDV infection of neonates indicate that it not only provides a useful model for exploring the mechanisms by which viral and immune factors may damage developing neurocircuitry, but also has significant links to the range of biologic, neurostructural, locomotor, cognitive, and social deficits observed in human neuropsychiatric illnesses.

CNS dysfunction in neonatally infected animals has been proposed to be linked to direct viral effects on morphogenesis of the hippocampus and cerebellum, two structures in rodents that continue to develop after birth. CARBONE et al. (1996) found a quantitative relationship of limbic pathology to behavioral abnormalities in the neonatal infection model; the extent of neuronal loss in dentate gyrus appeared to be correlated with the severity of spatial learning and memory deficiencies in neonatally-infected Lewis rats. Although overt ambulatory or cerebellar dysfunction has not been reported (CARBONE et al. 1991), a preliminary investigation from our laboratory found impairments in balance and coordination during a dowel-walking task in neonatally-infected, but not sham-infected, animals (HATALSKI 1996). Because the cerebellum undergoes substantial postnatal development in many mammals, it is particularly vulnerable to injury from perinatal virus infection (MONJAN et al. 1971, 1973; OSTER-GRANITE and HERDON 1985). There is ample support for the role of the cerebellum in motor behavior and motor learning (for reviews see LLINAS and WELSH 1993; ROLAND 1993; THOMPSON and KIM 1996). Confirmation of subtle abnormalities in motor coordination in neonatally-infected rats would provide a functional correlate to anatomic alterations in cerebellum. However, further studies are needed to evaluate the mechanisms by which early postnatal exposure to BDV induces functional damage in either cerebellar or limbic circuitry.

Previous investigators reported an absence of cellular inflammatory response to BDV following neonatal infection (CARBONE et al. 1991; STITZ et al. 1995; GOSZTONYI and LUDWIG 1995), a phenomenon ascribed to the immaturity of the rat immune system in the postnatal period. Humoral immune response to BDV in neonatally-infected animals has also been reported to be restricted, with anti-BDV antibody titers remaining below 1:10 through 133 days p.i. (CARBONE et al. 1991). In contrast to the adult Borna disease model, MHC class I and II antigens do not appear to be upregulated (CARBONE et al. 1991). However, marked astrocytosis has been noted (CARBONE et al. 1991; GONZALEZ-DUNIA et al. 1996; BAUTISTA et al. 1995) in dentate gyrus and cerebellum, suggesting alternate, noninflammatory pathways for BDV to induce glial activation. Higher levels of message for tissue factor (TF) have been found in infected hippocampus. TF is a member of the class II cytokine receptor family, primarily produced by astrocytes, that plays important roles in cellular signal transduction, brain function, and neural development through its effects on coagulation protease cascades. Although this may be one mechanism by which BDV may alter CNS development (GONZALEZ-DUNIA et al. 1996), cerebellar changes cannot be explained by this mechanism, as TF upregulation is not observed in cerebellum, despite prominent astrocytosis. Furthermore, BDV infection of astrocytes appears to be required for TF upregulation (GONZALEZ-DUNIA et al. 1996), and cerebellar astrocytes are reported to be spared from BDV infection, at least through 30 days following neonatal infection (BAUTISTA et al. 1995).

Persistent tolerant BDV infection of neonatal rats has been linked to hippocampal and cerebellar disorganization (NARAYAN et al. 1983b; STITZ et al. 1995); however, cytoarchitectonic anomalies in other limbic regions have not been extensively explored. Dentate gyrus involution is evident along with the appearance of reactive glial cells (CARBONE et al. 1991), suggesting more direct pathways of viral cytopathic injury. Cerebellar size has been reported to be reduced, and there is evidence of reactive astrocytosis as demonstrated by glial fibrillary acidic protein (GFAP) reactivity as early as 3 days post-inoculation, preceding the identification of BDV proteins in the cerebellum. Furthermore, reactivity of cerebellar astrocytes and loss of cerebellar granule cells occurred without signs of BDV infection in those cell populations at all time points through to 30 days p.i. Curiously, Purkinje cells appeared to be the predominant cerebellar cell population demonstrating BDV proteins, although these cells did not appear to be selectively lost through day 30 p.i. (BAUTISTA et al. 1995). The mechanism by which astrocytes are activated in the absence of infection, be it directly by BDV or indirectly through elaboration of soluble factors by other cell types, is not known. Given the role of astrocytes in guiding migration of granule cells during cerebellar development, an assessment of the frequency of astrocyte reactivity without viral infection in conjunction with studies of apoptosis in limbic structures would aid our understanding of the relative contributions of migrational failure and programmed cell death in the evolution of BD neuropathogenesis.

Although previous work suggests subtle functional disturbances of limbic circuitry based upon analysis of complex learning behaviors, memory capacities, and emotional responses, the evolution of such disturbances and the mechanisms by which BDV induces their underlying neuropathology without invoking infiltrating inflammatory elements remains to be determined. In an effort to more fully define the nature and unfolding of the neurologic syndrome in neonatally-infected Lewis rats, and to begin to elucidate the mechanisms of neuropathogenesis in the neonatal model, we established neonatal infection in Lewis rats and serially assessed shifts in neuroanatomy, neurobehavior, and regional gene expression. This survey revealed more complex alterations of neurobehavioral and anatomic systems than had been noted in previously published work and identified shifts in levels of transcripts encoding soluble factors that correlate with neuropathologic changes.

Behavioral observations included measurement of locomotor activity, exploration of novel environments, and stereotypic movements. Locomotor activity and stereotypies were assessed in neonatally-infected animals 4, 6 and 12 weeks p.i. Following a 10-min adaptation period, behavior was monitored continuously for 90 min (three consecutive 30-min intervals) in $40 \times 25 \times 20$ cm cages equipped with two equally-spaced, horizontal, photocell beams across the short axis. Locomotor activity was quantified as the mean number of crossovers (the successive interruption of the two photobeams) within 30 min (averaged across all three observation periods). Photocell beam interruptions and crossovers were assessed by analysis of variance (ANOVA), with infected or noninfected groups forming the independent factor, at a significance level of 0.05. Stereotypic behavior was scored through direct observation (MACLENNAN and MAIER 1983). Four to eight animals were studied in each group.

Results of locomotor activity analyses indicated an overall significant increase in neonatally infected groups relative to controls at all timepoints tested (4, 6, and 12 weeks; ANOVA, $F = 25$, $p < 0.0001$; Fig. 3). Data extrapolated from our previous studies of locomotor activity in adult-infected animals (SOLBRIG et al. 1994; SOLBRIG et al. 1996b; SOLBRIG et al. 1998) are included in Fig. 3 to illustrate that the degree of heightened exploratory locomotor activity found at baseline in neonatally-infected animals at 6 and 12-weeks-p.i. is much greater than in

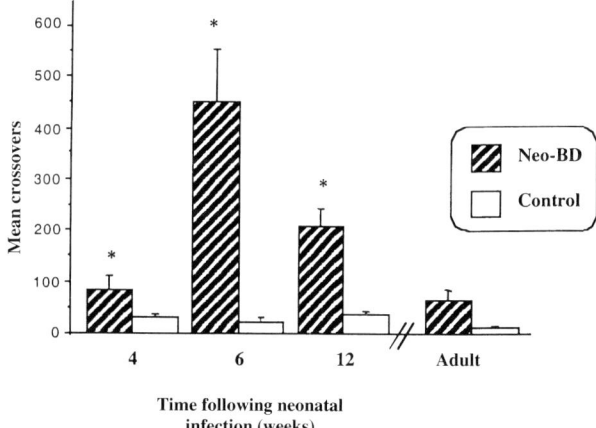

Fig. 3. Locomotor activity in rats neonatally infected with Borna disease virus (BDV). Locomotor activity at 4, 6, and 12 weeks following neonatal intracerebral inoculation with BDV (*Neo-BD, striped bars*) or phosphate-buffered saline (PBS; (*control, white bars*). Four weeks: $n = 4$ Neo-BD, $n = 8$ control; 6 weeks: $n = 8$ Neo-BD, $n = 8$ control; 12 weeks: $n = 9$ Neo-BD, $n = 3$ control. Comparison values for locomotor activity of adult-inoculated animals are derived from a meta analysis of previously published studies. Values represent mean crossovers (the successive interruption of photocell beams) during a 90-min interval ± SEM; * indicates significant overall difference in locomotor activity for Neo-BD as compared to control groups. (ANOVA, $F = 25$, $p < 0.0001$)

adult-infected animals. The differences between neonatal- and adult-infected rats cannot be explained by age or time since infection, as adult-infected animals were approximately 6 weeks p.i., or 11 weeks chronological age, at the time of behavioral analysis.

Locomotor activity across the 90-min observation period also differed for all neonatally-infected groups relative to noninfected controls. At 4 weeks following neonatal infection, animals exhibited prolonged behavioral inhibition upon introduction to the novel environment (first 30-min interval), consistent with greater anxiety in novel situations (Fig. 4). These data point toward an abnormal response to novel environments at 4 weeks following infection, suggesting dysfunction of the amygdala.

At the 60- and 90-min intervals, infected animals had greater mean activity measures than controls. Additionally, infected animals showed no attenuation in exploratory activity at 60 and 90 min, consistent with spatial memory deficits and hippocampal dysfunction (repeated measures ANOVA, $F=6$, $p=0.029$; Fig. 4).

Stereotypic behaviors were also increased in neonatally infected groups relative to noninfected controls, with a median stereotypy rating on the MacLennan and Maier scale of 4 at all time points (data not shown). In comparison, the median stereotypy rating for adult-infected rats in previous studies was less than 3 at

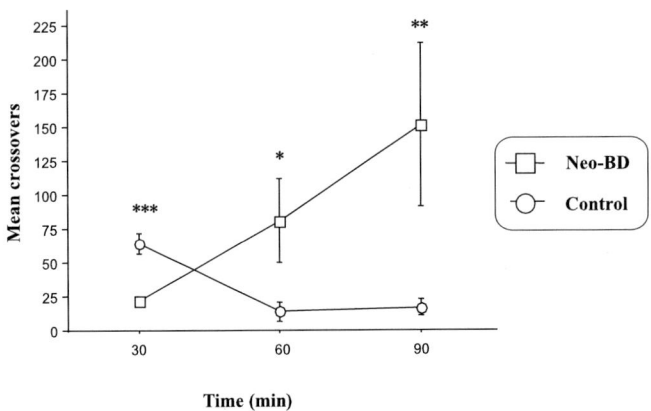

Fig. 4. Response to novel environment in neonatally infected rats at 4-weeks post-infection. Locomotor activity across three successive 30-min intervals was assessed in BDV (*Neo-BD*, *squares*) and control groups (*circles*) following neonatal inoculation with either BDV or phosphate-buffered saline (PBS), respectively. Values represent mean crossovers (successive beambreaks) during each 30-min interval. A significant overall group difference in locomotor activity (repeated measures ANOVA, main effect: $F=6$, $p=0.029$), and a significant difference in time course of locomotor activity over the total testing period (group × time interaction: $F=15$, $p<0.0001$) were found; *** indicates a significant difference between Neo-BD and control groups at the end of the first 30 min (unpaired t test, $p=0.003$); * indicates a significant difference between Neo-BD and control groups at 60 min (second 30-min interval; unpaired t test, $p=0.015$); ** indicates a significant difference between Neo-BD and control groups at 90 min (third 30-min interval; unpaired t test, $p=0.009$). Neo-BD rats exhibited prolonged behavioral inhibition upon introduction to the novel environment (first 30-min interval). Infected animals had greater mean activity measures than controls at 60 and 90 min and failed to show the expected attenuation in exploratory activity at these later timepoints

baseline (data not shown). Nonetheless, adult-infected animals were sensitive to pharmacologic challenges with the DA agonists dextroamphetamine and cocaine, indicating disruption of DA circuitry (SOLBRIG et al. 1994, 1998).

Serial analyses of differential gene expression of soluble factors (cytokines, neurotrophic factors, and apoptosis-related products) by RPA of samples from different brain regions were also pursued. One means by which a virus might disrupt neural function and development in the absence of inflammation is through the induction of neuronotrophic cytokines. These comprise a burgeoning set of immunoregulatory molecules, including the hematolymphopoietic factors, e.g., interleukins, tumor necrosis factor (TNF) family, interferons (IFNs), the TGF-β superfamily factors (including TGF-β1, 2, 3; glial-derived neurotrophic factor, or GDNF), and the classic neurotrophic factors, e.g., nerve growth factor (NGF), brain-derived neurotrophic factor (BDNF), neurotrophin-3 (NT3) and NT4. A large subset of the neuronotrophic, hematolymphopoietic cytokines may be roughly categorized according to their origin from one of two types of T-helper cells: Th_1 (cell-mediated immunity and stimulation of antigen-presenting cells) or Th_2 (humoral or B cell-mediated immunity). The potential mechanisms of cytokine-mediated damage in the context of the developing brain include: direct effects on neuronal elements; activation or suppression of second messenger/intracellular signaling pathways; induction of shifts in excitotoxic elements such as quinolinic acid or acute phase proteins such as neopterin or β-2-microglobulin; direct alterations of neuronal function (e.g., inhibition of long-term potentiation in hippocampus); activation or suppression of glial cells; or alteration of glial cell proliferation or differentiation (including expression of adhesion molecules such as the integrins) (BENVENISTE 1997; MEHLER et al. 1996). Given that the postnatal expression of neuronotrophic cytokine and cytokine receptor mRNAs in brain differs for each cytokine (BENVENISTE 1997), and that the sensitivity of neuronal populations to the trophic or apoptosis-inducing effects of cytokines changes during development, wide variation in the patterns of virus-induced, cytokine-related damage would be expected, depending on the relative maturity of the evolving nervous system at the time of infection. In addition, cell loss induced by either BDV or developmentally programmed changes may alter the capacity of resident CNS cells to both produce and respond to neuronotrophic cytokines.

One of the primary mechanisms of host defense following viral infection begins with the induction of IFN-γ and other cytokines, which in turn initiate a cascade of host responses in a wide variety of cell types. In the CNS, IFN-γ modulates oligodendrocyte, neuronal and glial cell functions and is important in activating glial cells to produce mediators of cell damage or death, including toxic intermediates of nitrogen and oxygen, and complement components (ST. PIERRE et al. 1996). Viral damage to neurodevelopmental circuitry may thus parallel the production of these downstream mediators following IFN-γ induction, and provide a means by which BDV might disrupt brain cell differentiation and function without inflammatory cell infiltration. There are only limited published data concerning cytokine expression in BDV-infected animals. Dietzschold and coworkers found elevated levels of mRNA encoding proinflammatory cytokines in brains of adult-infected rats with acute disease, including IFN-γ (SHANKAR et al. 1992). As noted above, we found changes in cytokine expression over the time course of adult infection, consistent with a shift from a predominantly Th_1-type pattern during the acute phase to a Th_2-type pattern in the chronic stages of disease (HATALSKI et al. 1998a). There are no published data concerning cytokine expression during neonatal infection. The experiments reported below suggest an important role for cytokines as mediators of BDV-related CNS injury in neonatally-infected rats.

Our results indicate that cytokine expression shifts over time in different brain regions, with maximal shift occurring at 4 weeks. Briefly, brains of infected and noninfected animals ($n = 5-8$ animals per group) were removed at serial time points following neonatal infection (2, 4, 6, 12, and 24 weeks) and selected brain regions were dissected out over ice using the "punch" approach, following stereotaxic coordinates of PAXINOS and WATSON (1986): hippocampus, amygdala, cerebellum, prefrontal cortex, nucleus accumbens, striatum, olfactory cortex, and hypothalamus. RNA samples from individual brain regions were subjected to RPA to quantitate level of transcripts encoding cytokines interleukin (IL)-1α, -1β, -2, -3, -4, -5, -6, -10; TNF-α, -β; IFN-γ; and TGF-β; and housekeeping genes L32 and GAPDH using the RiboQuant Multi-Probe RPA System (PharMingen).

Evidence was found for higher levels of mRNAs for cytokine products of CNS macrophages/microglia (IL-1α, -1β, -6; TNF-α) in hippocampus, amygdala, cerebellum, prefrontal cortex, and nucleus accumbens. Elevated levels of these proinflammatory cytokines were first apparent at 2 weeks, peaked at 4 weeks (see cytokine upregulation in prefrontal cortex, Fig. 5), then declined at 6 and 12 weeks. Alterations in other proinflammatory cytokines, including IL-2, IL-3, TNF-β, and IFN-γ, were not observed. The fact that cell populations other than macrophages or microglia – T cells, B cells, mast cells, bone marrow stromal cells – are the primary sources for the proinflammatory cytokines that remained static following neonatal infection suggests a selective effect of BDV on cells of microglial or macrophage lineage.

We also identified shifts in gene expression of neurotrophic factors following neonatal infection of Lewis rats. Serial RPA analysis of hippocampus, cerebellum, prefrontal cortex, nucleus accumbens and amygdala using a multi-probe RPA kit (RiboQuant Multi-probe RPA System, PharMingen) was pursued for the following panel of neurotrophic factors: nerve growth factor (NGF), brain-derived neurotrophic factor (BDNF), glial cell line-derived neurotrophic factor (GDNF), ciliary neurotrophic factor (CNTF), neurotrophin-3 (NT3), and neurotrophin-4 (NT4). In contrast to our findings of diffuse alterations in gene expression for cytokines described above, shifts in neurotrophic factor mRNAs were restricted to hippocampus. Decreased mRNA for BDNF and NT3 was prominent in hippocampus by 4 weeks p.i., but was still evident by 12 weeks p.i. Although decreased NT3 mRNA may reflect loss of the granule cell population in dentate gyrus, the role of BDNF in maintaining viability of cells suggests that its downregulation may be a more essential step in neonatal BDV pathogenesis.

Apoptosis, or programmed cell death, is a mechanism in which cells undergo chromosome condensation, DNA degradation, and morphologic change in the

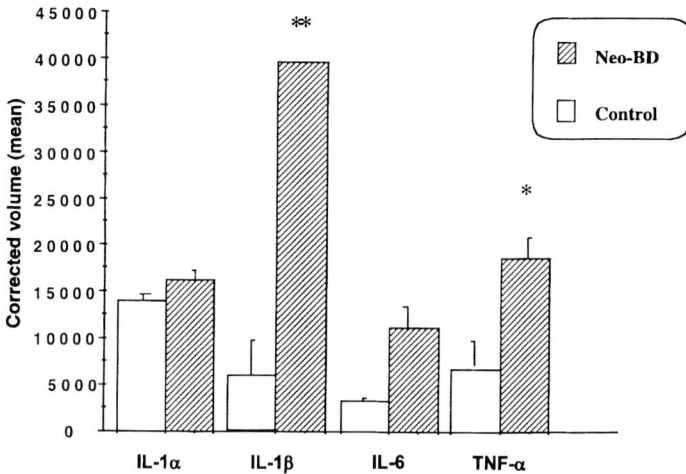

Fig. 5. Cytokine gene expression in prefrontal cortex of neonatally infected rats at 4 weeks post-infection. Levels of transcripts encoding cytokines interleukin (IL)-1α, -1β, -2, -3, -4, -5, -6, -10; tumor necrosis factor (TNF)-α, -β; interferon (IFN)-γ; and transforming growth factor (TGF)-β were quantitated by RNase protection assay (RPA) in prefrontal cortex from Lewis rats inoculated neonatally with BDV (*Neo-BD, striped bars*) or phosphate-buffered saline (PBS; *control, white bars*). Values were adjusted by correcting for the level of mRNA for housekeeping genes L32 and GAPDH in the same sample. No differences were noted in IL-2, -3, -4, -5, -10; TNF-β; IFN-γ; or TGF-β (data not shown); * indicates a significant difference between Neo-BD and control groups by unpaired t test at the $p = 0.042$ level; ** indicates a significant difference between Neo-BD and control groups at the $p = 0.0006$ level

nuclear membrane (WYLLIE 1995). Apoptosis plays an important role in CNS development and response to neuronal injury (BREDESEN 1995). It is conceivable that abnormal regulation of apoptosis, either failure of normal apoptotic sequences to proceed or excessive activity, may contribute to abnormal CNS architecture in neonatal infections with BDV or other neurotropic viruses. Furthermore, apoptosis of antigen-specific lymphocytes might provide an explanation for the immunotolerant state following neonatal BDV infections. Anergy or apoptosis of T cells may result if their stimulation by antigen-presenting cells resident to the CNS occurs in the absence of costimulatory signals required for immune activation, such as MHC class II antigens (MUNN et al. 1996).

Various stimuli, such as binding of TNF-α to its receptor, can trigger apoptosis; proteins such as the Bcl-2 and Bax proteins, NF-κB and caspase-1 (ICE)-related proteases have also been shown to play important roles in regulating apoptosis. TNF-α also stimulates apoptosis in a wide variety of cell types (BENVENISTE 1997). Furthermore, a host of excitants or neurotoxins including arachidonic acid, platelet-activating factor, free radicals (NO, O_2^-), glutamate, quinolinate, cysteine, cytokines (TNF-α, IL-1β, IL-6), amines, and as yet unidentified factors arising from stimulated macrophages and possibly reactive astrocytes may influence apoptosis by excessive activation of N-methyl-D-aspartate (NMDA) receptors (LIPTON 1996). Interestingly, GOSZTONYI and LUDWIG (1995) have proposed that the targeted pathology of BDV for two hippocampal cell layers, stratum oriens and stratum radiatum, may be due to their rich concentration of glutamate and aspartate receptors.

Lastly, complex alterations in mRNAs for apoptosis mediators were found. RPA multi-probe analysis was again pursued using serial analysis of mRNAs for Fas, Bcl-x, FasL, ICE, YAMA, ICH, Bax, and Bcl-2 (RiboQuant Multi-probe RPA System, PharMingen). Increased levels of mRNAs for Fas and ICE (caspase-1), two promoters of apoptosis, and decreased mRNA for Bcl-x, a factor that inhibits apoptosis, were identified in hippocampus (Fig. 6), amygdala, prefrontal cortex, nucleus accumbens, and cerebellum. These findings are consistent with promotion of apoptosis throughout the brains of rats neonatally infected with BDV by at least two strategies.

Consistent with previous reports (CARBONE et al. 1991; NARAYAN et al. 1983b), we found morphologic alterations in brains of rats infected as newborns, including loss of dentate gyrus granule cells and disorganization of cerebellar granule cell layer. Cerebellar abnormalities included decreased overall size and foliation of cerebellum, and thinning of cerebellar granule cell layers. Hippocampal changes in neonatally- and adult-infected animals included distinct loss of granule cells in dentate gyrus.

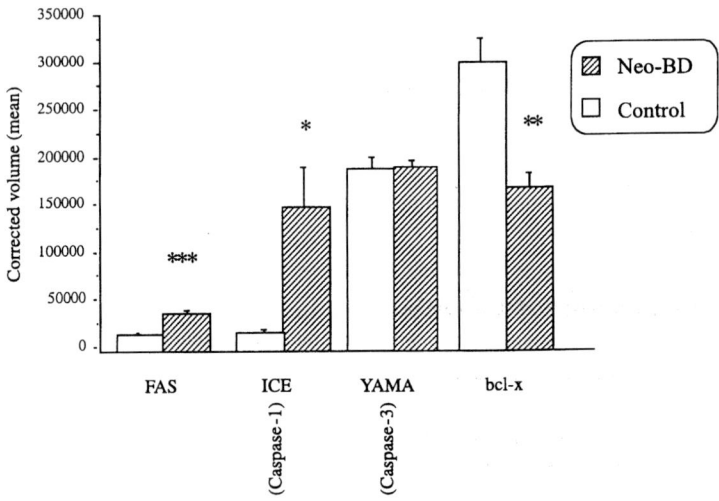

Fig. 6. Apoptosis-related gene expression in hippocampus of neonatally infected rats at 4 weeks post-infection. Levels of transcripts encoding apoptosis-related proteins (Fas, Bcl-x, FasL, ICE, YAMA, ICH, Bax, and Bcl-2) were quantitated by RNase protection assay (RPA) in hippocampus from Lewis rats inoculated neonatally with BDV (*Neo-BD, striped bars*) or phosphate-buffered saline (PBS; *control, white bars*). Values were adjusted by correcting for the level of mRNA for housekeeping genes L32 and GAPDH in the same sample. No differences were noted in FasL, ICH, Bax, or Bcl-2 (data not shown); * indicates a significant difference between Neo-BD and control groups by unpaired t test at the $p = 0.044$ level; ** indicates a significant difference between Neo-BD and control groups at the $p = 0.032$ level; and *** indicates a significant difference between Neo-BD and control groups at the $p = 0.017$ level

However, new findings in this study included: (1) transient inflammation restricted to meningeal and perivascular areas of motor, parietal and temporal cortex between 4 and 6 weeks (areas with the most distinct morphologic alterations, dentate gyrus and cerebellum, are intriguingly excluded from this brief inflammatory event), and (2) nearly complete loss of cerebellar Purkinje cells by week 6.

Consistent with our findings of multiple apoptosis-enhancing mechanisms in neonatal BDV-infected animals by RPA analysis, which peaks abruptly at 4 weeks p.i. and extends to at least 12 weeks p.i., we found evidence of apoptosis by terminal deoxynucleotidyl transferase dUTP-biotin nick end labeling (TUNEL) labeling in cerebral cortex and dentate gyrus peaking at 4 weeks p.i. and in granule cell layer of cerebellum of neonatally infected rats at weeks 4 and 6 p.i. Although apoptosis is described in normally developing rat hippocampus as late as day 7–10 of postnatal life, it is not found at later timepoints (Toth et al. 1998). Thus, two lines of evidence suggest that it is not the inflammatory cells themselves that are undergoing apoptosis: first, apoptosis in dentate gyrus and cerebellum continues to be detected as long as 3 weeks after resolution of inflammation; and second, apoptosis is most prominent in those areas that appear to be free of inflammatory infiltrates (namely, hippocampus and cerebellum). The anatomic location of the apoptotic cells suggests they represent dying neurons, but the possibility that the dying cells are an immune cell population cannot be excluded on the basis of these data.

Efforts are underway in this and other laboratories to ascertain the time course of the inflammatory response, characterize the participant immunocytes, determine whether the inflammatory response is specific for BDV, and identify the triggers for inflammatory cell recruitment.

5 Summary

Animal models provide unique opportunities to explore interactions between host and environment. Two models have been established based on Borna disease virus infection that provide new insights into mechanisms by which neurotropic agents and/or immune factors may impact developing or mature CNS circuitry to effect complex disturbances in movement and behavior.

Note in press:
Since this chapter was submitted, several manuscripts have been published that extend findings reported here and support the relevance of BDV infections of neonatal Lewis rats as models for investigating mechanisms of neurodevelopmental damage in autism. Behavioral abnormalities, including disturbed play behavior and chronic emotional overactivity, have been described by Pletnikov et al. (1999); inhibition of responses to novel stimuli were described by Hornig et al. (1999); loss

of Purkinje cells following neonatal BDV infection has been demonstrated by EISENMAN et al. (1999), HORNIG et al. (1999), and WEISSENBÖCK et al. (2000); and alterations in cytokine gene expression have been reported by HORNIG et al. (1999), PLATA-SALAMAN et al. (1999) and SAUDER et al. (1999).

References

Altshuler LL, Conrad A, Kovelman JA, Scheibel A (1987) Hippocampal pyramidal cell orientation in schizophrenia: a controlled neurohistologic study of the Yakovlev Collection. Arch Gen Psychiatry 44:1094–1098

Amsterdam J, Winokur A, Dyson W, Herzog S, Gonzalez F, Rott R, Koprowski H (1985) Borna disease virus: a possible etiologic factor in human affective disorders. Arch Gen Psychiatry 42:1093–1096

Anderson GM (1994) Studies on the neurochemistry of autism. In: Bauman ML, Kemper TL (eds) The neurobiology of autism. Johns Hopkins University Press, Baltimore, pp 227–242

Bautista JR, Rubin SA, Moran TH, Schwartz GJ, Carbone KM (1995) Developmental injury to the cerebellum following perinatal Borna disease virus infection. Dev Brain Res 90:45–53

Bautista JR, Schwartz GJ, de la Torre JC, Moran TH, Carbone KM (1994) Early and persistent abnormalities in rats with neonatally acquired Borna disease virus infection. Brain Res Bull 34:31–40

Beckmann H, Jakob H (1991) Prenatal disturbances of nerve cell migration in the entorhinal region: a common vulnerability factor in functional psychoses? J Neural Transm – Gen Sect 84:155–164

Benveniste EN (1997) Cytokine expression in the nervous system. In: Keane RW, Hickey WF (eds) Immunology of the nervous system. Oxford University Press, New York, pp 419–459

Bode L, Ferszt R, Czech G (1993) Borna disease virus infection and affective disorders in man. Arch Virol (Suppl) 7:159–167

Bode L, Riegel S, Lange W, Ludwig H (1992) Human infections with Borna disease virus: seroprevalence in patients with chronic diseases and healthy individuals. J Med Virol 36:309–315

Bode L, Riegel S, Ludwig H, Amsterdam J, Lange W, Koprowski H (1988) Borna disease virus-specific antibodies in patients with HIV infection and with mental disorders. Lancet ii:689

Bredesen D (1995) Neural apoptosis. Ann Neurol 38:839–851

Briese T, Schneemann A, Lewis AJ, Park YS, Kim S, Ludwig H, Lipkin WI (1994) Genomic organization of Borna disease virus. Proc Natl Acad Sci USA 91:4362–4366

Buchmeier MJ, Lewicki HA, Talbot PJ, Knobler RL (1984) Murine hepatitis virus-4 (strain JHM)-induced neurologic disease is modulated in vivo by monoclonal antibody. Virology 132:261–270

Carbone K, Park S, Rubin S, Waltrip R, Vogelsang G (1991) Borna disease: association with a maturation defect in the cellular immune response. J Virol 65:6154–6164

Carbone KM, Pletnikov M, Rubin SA (1998) Developmental brain injury associated with abnormal play behavior in neonatally Borna disease virus-infected Lewis rats: a model of autism. Second International Bornavirus Meeting, Freiburg, Germany, September 1998

Carbone KM, Duchala CS, Griffin JW, Kincaid AL, Narayan O (1987) Pathogenesis of Borna disease in rats: evidence that intra-axonal spread is the major route for virus dissemination and the determinant for disease incubation. J Virol 61:3431–3440

Carbone KM, Silvas PM, Rubin SA, Vogel M, Moran TH, Schwartz G (1996) Quantitative correlation of viral induced damage to the hippocampus and spatial learning and memory deficits. J Neurovirol 2:195

Cooper JR, Bloom FE, Roth RH (1991) The biochemical basis of neuropharmacology. Oxford University Press, New York, pp 332–334

Cubitt B, Oldstone C, de la Torre JC (1994) Sequence and genome organization of Borna disease virus. J Virol 68:1382–1396

de la Torre JC (1994) Molecular biology of Borna disease virus: prototype of a new group of animal viruses. J Virol 68:7669–7675

Dietzschold B, Kao M, Zheng YM, Chen ZY, Maul G, Fu ZF, Rupprecht CE, Koprowski H (1992) Delineation of putative mechanisms involved in antibody-mediated clearance of rabies virus from the central nervous system. Proc Natl Acad Sci USA 89:7252–7256

Dittrich W, Bode L, Ludwig H, Kao M, Schneider K (1989) Learning deficiencies in Borna disease virus-infected but clinically healthy rats. Biol Psychiatry 26:818–828

Ernst M, Zametkin AJ, Matochik JA, Pascualvaca D, Cohen RM (1997) Low medial prefrontal dopaminergic activity in autistic children. Lancet 350:638

Fish B, Marcus J, Hans SL, Auerbach JG, Perdue S (1992) Infants at risk for schizophrenia: sequelae of a genetic neurointegrative defect: a review and replication analysis of pandysmaturation in the Jerusalem Infant Development Study. Arch Gen Psychiatry 49:221–235

Fu ZF, Amsterdam JD, Kao M, Shankar V, Koprowski H, Dietzschold B (1993) Detection of Borna disease virus-antibodies from patients with affective disorders by western immunoblot technique. J Affect Disord 27:61–68

Fu ZF, Weihe E, Zheng YM, Schafer MKH, Sheng H, Corisdeo S, Rauscher FJ 3rd, Koprowski H, Dietzschold B (1993b) Differential effects of rabies and Borna Disease viruses on immediate-early- and late-gene expression in brain tissues. J Virol 67:6674–6681

Gies U, Bilzer T, Stitz L, Staiger JF (1998) Disturbance of the cortical cholinergic innervation in Borna disease prior to encephalitis. Brain Pathol 8:39–48

Gonzalez-Dunia D, Eddleston M, Mackman N, Carbone KM, de la Torre JC (1996) Expression of tissue factor is increased in astrocytes within the central nervous system during persistent infection with Borna disease virus. J Virol 70:5812–5820

Gosztonyi G, Ludwig H (1995) Borna disease: neuropathology and pathogenesis. In: Koprowski H, Lipkin WI (eds) Curr Top Microbiol Immunol 190:39–73

Gutierrez B, Van Os J, Valles V, Guillamat R, Campillo M, Fananas L (1998) Congenital dermatoglyphic malformations in severe bipolar disorder. Psychiatry Res 78:133–140

Hamner MB, Diamond BI (1996) Plasma dopamine and norepinephrine correlations with psychomotor retardation, anxiety, and depression in non-psychotic depressed patients: a pilot study. Psychiatry Res 64:209–211

Hatalski CG (1996) Alterations in the immune response within the central nervous system of rats infected with Borna disease virus: potential mechanisms for viral persistence. Ph.D. Thesis, University of California, Irvine

Hatalski CG, Hickey WF, Lipkin WI (1998a) Evolution of the immune response in the central nervous system following infection with Borna disease virus. J Neuroimmunol 90:137–142

Hatalski CG, Hickey WF, Lipkin WI (1998b) Humoral immunity in the central nervous system of Lewis rats infected with Borna disease virus. J Neuroimmunol 90:128–136.

Karpas WJ, Peterson JD, Miller SD (1994) Anergy in vivo: down regulation of antigen-specific CD4+ Th1 but not Th2 cytokine responses. Int Immunol 6:721–730

Kauffman RS, Fields BN (1985) Pathogenesis of viral infections. In: Fields BN (ed) Virology. Raven, New York, pp 153–167

Kelsoe JR, Sadovnick AD, Kristbjarnarson H, Bergesch P, Mroczkowski-Parker Z, Drennan M, Rapaport MH, Flodman P, Spence MA, Remick RA (1996) Possible locus for bipolar disorder near the dopamine transporter on chromosome 5. Am J Med Genet 67:533–540

Kemper TL, Bauman ML (1993) The contribution of neuropathologic studies to the understanding of autism. Neurol Clin North Am 11:175–187

Khoury SJ, Akalin E, Chandraker A, Turka LA, Linsley PS, Sayegh MH, Hancock WW (1995) CD28-B7 costimulatory blockade by CTLA4Ig prevents actively induced experimental autoimmune encephalomyelitis and inhibits Th1 but spares Th2 cytokines in the central nervous system. J Immunol 155:4521–4524

Kishi M, Nakaya T, Nakamura Y, Zhong Q, Iketa K, Sengo M, Kakinuma M, Kato S, Ikuta K (1995) Demonstration of human Borna disease virus RNA in human peripheral blood mononuclear cells. FEBS Lett 364:293–297

Kullback S (1968) Information theory and statistics. Dover, New York

Levine B, Hardwick JM, Trapp BD, Crawford TO, Bollinger RC, Griffin DE (1991) Antibody-mediated clearance of alphavirus infection from neurons. Science 254:856–860

Liebert UG, Schneider-Schaulies S, Baczko K, ter Meulen V (1990) Antibody-induced restriction of viral gene expression in measles encephalitis in rats. J Virol 64:706–713

Lipkin WI, Carbone KM, Wilson MC, Duchala CS, Narayan O, Oldstone MBA (1988) Neurotransmitter abnormalities in Borna disease. Brain Res 475:366–370

Lipkin WI, Schneemann A, Solbrig MV (1995) Borna disease virus: implications for human neuropsychiatric illness. Trends Microbiol 3:64–69

Lipkin WI, Travis G, Carbone K, Wilson M (1990) Isolation and characterization of Borna disease agent cDNA clones. Proc Natl Acad Sci USA 87:4184–4188

Lipton SA (1996) Similarity of neuronal cell injury and death in AIDS dementia and focal cerebral ischemia: potential treatment with NMDA open-channel blockers and nitric oxide-related species. Brain Pathol 6:507–517

Llinas R, Welsh JP (1993) On the cerebellum and motor learning. Curr Opin Neurobiol 3:958–965

Ludwig H, Bode L, Gosztonyi G (1988) Borna disease: a persistent disease of the central nervous system. Prog Med Virol 35:107–151

MacLennan AJ, Maier SF (1983) Coping and the stress-induced potentiation of stimulant stereotypy in the rat. Science 219:1091–1093

Mehler MF, Goldstein H, Kessler JA (1996) Effects of cytokines on CNS cells: neurons. In: Ransohoff RM, Benveniste EN (eds) Cytokines and the CNS. CRC Press, Boca Raton, pp 115–150

Monjan AA, Cole GA, Gilden DH, Nathanson N (1973) Pathogenesis of cerebellar hypoplasia produced by lymphocytic choriomeningitis virus infection of neonatal rats: evolution of disease following infection at 4 days of age. J Neuropathol Exp Neurol 32:110–124

Monjan AA, Gilden DH, Cole GA, Nathanson N (1971) Cerebellar hypoplasia in neonatal rats caused by lymphocytic choriomeningitis virus. Science 171:194–196

Munn DH, Pressey J, Beall AC, Hudes R, Alderson MR (1996) Selective activation-induced apoptosis of peripheral T cells imposed by macrophages: a potential mechanism of antigen-specific peripheral lymphocyte deletion. J Immunol 156:523–532

Narayan O, Herzog S, Frese K, Scheefers H, Rott R (1983a) Behavioral disease in rats caused by immunopathological responses to persistent Borna virus in the brain. Science 220:1401–1403

Narayan O, Herzog S, Frese K, Scheefers H, Rott R (1983b) Pathogenesis of Borna disease in rats: immune-mediated viral ophthalmoencephalopathy causing blindness and behavioral abnormalities. J Infect Dis 148:305–315

Oster-Granite ML, Herdon RM (1985) The pathogenesis of parvovirus-induced cerebellar hypoplasia in the Syrian hamster, *Mesociretus auratus*: fluorescent antibody, foliation, cytoarchitectonic, golgi and electron microscopic studies. J Comp Neurol 169:481–522

Partonen T (1996) Dopamine and circadian rhythms in seasonal affective disorder. Med Hyp 47:191–192

Paxinos G, Watson C (1986) The Rat brain in stereotaxic coordinates. Academic, San Diego

Planz O, Bilzer T, Stitz L (1995) Immunopathogenic role of T-cell subsets in Borna disease virus-induced progressive encephalitis. J Virol 69:896–903

Rapin I, Katzman R (1998) Neurobiology of autism. Ann Neurol 43:7–14

Roland PE (1993) Partition of the human cerebelllum in sensory-motor activities, learning and cognition. Can J Neurol Sci 20:S75–S77

Rubin SA, Sierra-Honigmann AM, Lederman HM, Waltrip RW II, Eiden JJ, Carbone KM (1995) Hematologic consequences of Borna disease virus infection of rat bone marrow and thymus stromal cells. Blood 85:2762–2769

Saito N, Itouji A, Totani Y, Osawa I, Koide H, Fujisawa N, Ogita K, Tanaka C (1993) Cellular and intracellular localization of epsilon-subspecies of protein kinase C in the rat brain; presynaptic localization of the epsilon-subspecies. Brain Res 607:241–248

Schneemann A, Schneider PA, Lamb RA, Lipkin WI (1995) The remarkable coding strategy of Borna disease virus: a new member of the nonsegmented negative strand RNA viruses. Virology 210:1–8

Schwartz RH (1992) Costimulation of T lymphocytes: the role of CD28, CTLA-4 and B7/BB1 in interleukin-2 production and immunotherapy. Cell 71:1065–1068

Schwemmle M, De B, Shi L, Banerjee A, Lipkin WI (1997) Borna Disease Virus P-protein is phosphorylated by protein kinase C-ε and casein kinase II. J Biol Chem 272:21818–21823

Shankar V, Kao M, Hamir AN, Sheng H, Koprowski H, Dietzschold B (1992) Kinetics of virus spread and changes in levels of several cytokine mRNAs in the brain after intranasal infection of rats with Borna disease virus. J Virol 66:992–998

Soares JC, Mann JJ (1997) The anatomy of mood disorders – review of structural neuroimaging studies. Biol Psychiatry 41:86–106

Sobbe M, Bilzer T, Gommel S, Noske K, Planz O, Stitz L (1997) Induction of degenerative brain lesions after adoptive transfer of brain lymphocytes from Borna disease virus-infected rats: presence of CD8+ T cells and perforin mRNA. J Virol 71:2400–2407

Solbrig MV, Fallon JH, Lipkin WI (1995) Behavioral disturbances and pharmacology of Borna disease virus. In: Koprowski H, Lipkin WI (eds) Curr Top Microbiol Immunol 190:93–101

Solbrig MV, Koob GF, Fallon JH, Lipkin WI (1994) Tardive dyskinetic syndrome in rats infected with Borna disease virus. Neurobiol Dis 1:111–119

Solbrig MV, Koob GF, Fallon JH, Reid S, Lipkin WI (1996a) Prefrontal cortex dysfunction in Borna disease virus (BDV)-infected rats. Biol Psychiatry 40:629–636

Solbrig MV, Koob GF, Joyce JN, Lipkin WI (1996b) A neural substrate of hyperactivity in Borna disease: changes in dopamine receptors. Virology 222:332–338

Solbrig MV, Koob GF, Lipkin WI (1996c) Naloxone-induced seizures in Borna disease virus-infected rats. Neurology 46:1170–1171

Solbrig MV, Koob GF, Lipkin WI (1998) Cocaine sensitivity in Borna disease virus infected rats. Pharmacol Biochem Behav 59:1047–1052

St Pierre BA, Merrill JE, Dopp JM (1996) Effects of cytokines on CNS cells: glia. In: Ransohoff RM, Benveniste EN (eds) Cytokines and the CNS. CRC Press, Boca Raton, pp 151–168

Stevens JR (1973) An anatomy of schizophrenia? Arch Gen Psychiatry 29:177–189

Stitz L, Dietzschold B, Carbone KM (1995) Immunopathogenesis of Borna disease. In: Koprowski H, Lipkin WI (eds) Curr Top Microbiol Immunol 190:75–92

Stitz L, Krey H, Ludwig H (1980) Borna disease in rhesus monkeys as a model for uveo-cerebral symptoms. J Med Virol 6:333–340

Stitz L, Nöske K, Planz O, Furrer E, Lipkin WI, Bilzer T (1998) A functional role for neutralizing antibodies in Borna disease: influence on virus tropism outside the central nervous system. J Virol 72:8884–8892

Stitz L, Thompson RF, Kim JJ (1996) Memory systems in the brain and localization of memory. Proc Natl Acad Sci USA 93:13438–13444

Toth Z, Yan XX, Haftoglou S, Ribak CE, Baram TZ (1998) Seizure-induced neuronal injury: vulnerability to febrile seizures in an immature rat model. J Neurosci 18:4285–4294

VandeWoude S, Richt J, Zink M, Rott R, Narayan O, Clements J (1990) A Borna virus cDNA encoding a protein recognized by antibodies in humans with behavioral diseases. Science 250:1276–1281

van Os J, Jones P, Lewis G, Wadsworth M, Murray R (1997) Developmental precursors of affective illness in a general population birth cohort. Arch Gen Psychiatry 54:625–631

Waltrip II RW, Buchanan RW, Summerfelt A, Breier A, Carpenter WT, Bryant NL, Rubin SA, Carbone K (1995) Borna disease virus and schizophrenia. Psychiatry Res 56:33–44

Weinberger DR (1987) Implications of normal brain devlopment for the pathogenesis of schizophrenia. Arch Gen Psychiatry 44:660–669

Wyllie AH (1995) The genetic regulation of apoptosis. Curr Opin Genet Dev 5:97–104

Yolken RH, Torrey EF (1995) Viruses, schizophrenia, and bipolar disorder. Clin Microbiol Rev 8: 131–145

The Mechanisms of Neuronal Damage in Retroviral Infections of the Nervous System

V.J. Sanders[1], C.A. Wiley[2,*], and R.L. Hamilton[2]

1	Introduction	180
1.1	Retroviridae	180
2	Murine Leukemia Viruses	181
2.1	Pathology and Symptomatology	181
2.2	Soluble Factors	182
2.3	Viral Proteins	183
3	Maedi/Visna Virus	184
3.1	Pathology and Symptomatology	184
3.2	Soluble Factors	184
4	Feline Immunodeficiency Virus	185
4.1	Pathology and Symptomatology	185
4.2	Soluble Factors	185
4.3	Viral Proteins	186
5	Simian Immunodeficiency Virus	186
5.1	Pathology and Symptomatology	186
5.2	Soluble Factors	187
5.2.1	Cytokines	187
5.2.2	Chemokines	188
5.2.3	Macrophage Activation Factors	188
6	Human T-cell Leukemia Virus	188
6.1	Pathology and Symptomatology	188
6.2	Soluble Factors	189
6.3	Viral Proteins	190
7	Human Immunodeficiency Virus	190
7.1	Pathology and Symptomatology	191
7.2	Soluble Factors	192
7.2.1	Cytokines	192
7.2.2	Chemokines	193
7.2.3	Macrophage Activation Factors	193
7.3	Viral Proteins	194
8	Conclusion	195
References		195

[1] Department of Neuroscience, University of California, San Diego, La Jolla, CA 92302, USA
[2] A 506 Presbyterian University Hospital, Department of Pathology, Division of Neuropathology, 200 Lothrop St., Pittsburgh, PA 15213, USA
* e-mail: wiley@np.awing.upmc.edu

1 Introduction

Retroviruses were first described as the causative agent in avian sarcomas: Rous sarcoma virus and avian leukosis virus. The past 80 years have seen numerous other retroviruses identified and characterized, while the past 15 years have seen an explosion of research on the most infamous member, human immunodeficiency virus (HIV). Neurological manifestations of retroviral infection such as peripheral neuropathy, encephalopathy, and neuronal degeneration have been reported; however, the pathogenetic mechanisms remain elusive. This review will describe the neuropathology of these viral infections and detail possible events leading to neurodegeneration, focusing on recent publications.

1.1 Retroviridae

The retroviridae share common characteristics in genome and function. Genomically, retroviruses are divided into two groups: simple and complex. The simple retroviruses contain only the Gag core proteins, Pol enzymatic proteins and Env envelope glycoproteins, while the complex retroviruses contain a number of regulatory proteins (i.e. Nef, Rev, and Tat proteins) (see WEISS 1996 for review). Functionally, retroviruses share a unique replication cycle involving reverse transcription from RNA to double-stranded pro-viral DNA. The pro-viral DNA is integrated into the host genome via the integrase enzyme. Integrated pro-virus may

Table 1. General classification of Retroviridae

Genomic classification	Subfamily	Genus	Subgenus	Example isolate(s)
Simple	Oncovirinae	Type B oncovirus		MMTV
		Type C oncovirus	Mammalian	MuLV, FeLV
			Reptilian	VRV
			Avian	SNV, REV
		Type D oncovirus		MPMV, SRV-1
		Avian leukemia virus-related		RSV, AMV
Complex		HTLV-BLV		HTLV, BLV, HTLV-2, STLV
	Spumavirinae			SFV
	Lentivirinae		Ovine/caprine	Visna
			Feline	FIV
			Equine	EIAV
			Bovine	BIV
			Primate	HIV, HIV-2, SIV

Abbreviations: AMV, avian myeloblastosis virus; BIV, bovine immunodeficiency virus; BLV, bovine leukemia virus; EIAV, equine infectious anemia virus; FeLV, feline leukemia virus; FIV, feline immunodeficiency virus; GALV, gibbon ape leukemia virus; HIV, human immunodeficiency virus; HTLV, human T-cell lymphotropic virus; MMTV, mouse mammary tumor virus; MuLV, murine leukemia virus; MPMV, Mason-Pfizer monkey virus; REV, reticuloendotheliosis virus; RSV, Rous sarcoma virus; SFV, simian foamy virus; SIV, simain immunodeficiency virus; SNV, spleen necrosis virus; SRV-1, simian retrovirus-1; STLV, simian T-cell lymphotropic virus; VRV, viper retrovirus remain latent and be passively replicated if the infected cell proliferates. Cellular transcriptional regulators (e.g. NFκB) control viral DNA transcription to mRNA. Eventual transcription to mRNA leads to translation of proteins and formation of infectious virions.

There are three subfamilies of the retroviruses: oncoviruses, spumaviruses, and lentiviruses (Table 1). Most retroviruses belong to the Oncovirinae family and may cause leukemias, sarcomas, or lymphomas, although not all oncoviruses are oncogenic. Within this group, murine leukemia virus (MuLV) and human T-cell leukemia virus (HTLV) may result in neuronal damage. Spumaviruses cause "foamy" changes in cells in vitro and have been recovered from the central nervous system (CNS), but, as yet, no disease has been linked to them. Lentiviruses, the "slow" viruses, cause relapsing-remitting disease (equine infectious anemia) or progressive disease (visna). The Lentivirinae family contains the greatest number of viruses that lead to specific CNS involvement: Visna, feline immunodeficiency virus (FIV), simian immunodeficiency virus (SIV) and HIV. The pathogenesis of CNS damage differs between each viral family member.

2 Murine Leukemia Viruses

Murine leukemia viruses have been studied as an animal model for AIDS due to similarities with the T-cell abnormalities seen in both disease states. MuLV infection results in systemic hypergammaglobulanemia, splenomegaly, T and B-cell dysfunction and eventually neurological disorder (LIANG et al. 1996). There are a number of different strains of murine retroviruses that cause neurodegenerative disease: variants of Friend murine leukemia virus, wild-type CasBrE, and temperature-sensitive mutants of Moloney MuLV. Neuroinvasiveness seems to be determined in part by the glycosylated Gag sequence of these viruses (FUJISAWA et al. 1998).

2.1 Pathology and Symptomatology

Pathological changes include primarily non-inflammatory spongiform degeneration and lesions in the motor nuclei, motor cortex and spinal cord (Fig. 1). For the Moloney MuLV and CasBrE, infected cells include microglia, pericytes, and endothelial cells (GRAVEL et al. 1993; LYNCH et al. 1991; CZUB et al 1994; Fig. 2). Numerous virions can be identified in the extracellular spaces with ultrastructural studies (Fig. 3). Multinucleated giant cells have sometimes been noted (KUSTOVA et al. 1996). The cell type infected and the location of lesions differs between strain of mouse and virus used (see WILEY and GARDNER 1993 for review). Neurological disease consists of tremor, hind limb paralysis, impaired spatial learning, and sometimes hyperexcitability (SEI et al. 1992).

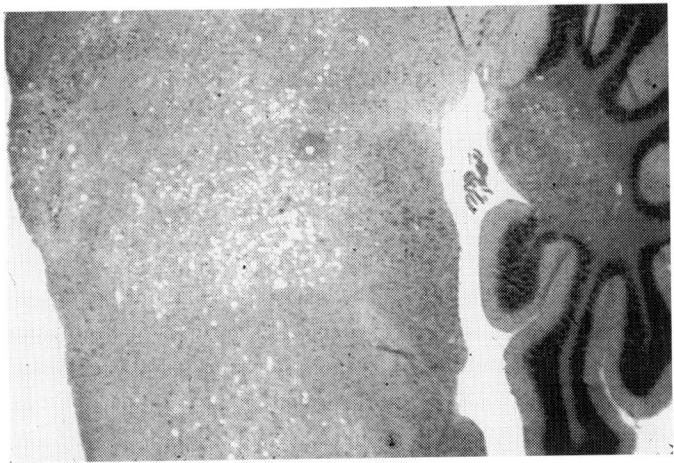

Fig. 1. Hematoxylin-and-eosin stained section from the brain of a mouse with murine leukemia virus (MuLV) encephalitis. At low magnification vacuoles are noted in the brainstem and deep gray matter of the cerebellum

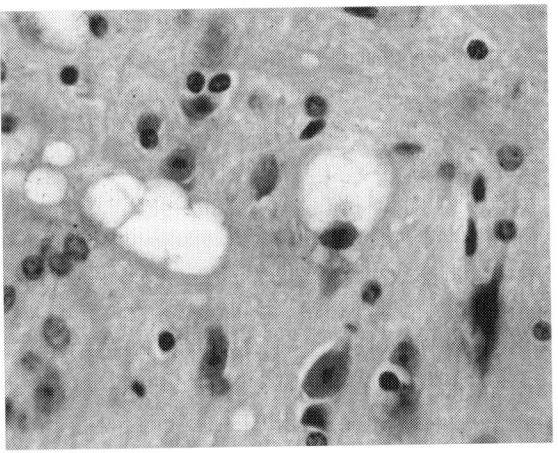

Fig. 2. Hematoxylin-and-eosin st-ained section from the brain of a mouse with murine leukemia virus (MuLV) encephalitis. At high magnification vacuoles in the cortical gray matter are seen to involve both neuropil and neuronal soma. Many of these regions contain abundant viral antigens

2.2 Soluble Factors

With all strains, lesions are associated with the presence of activated microglia, suggesting that damage is indirect and immune-mediated. Gliosis and neurochemical alterations precede morphological changes. Levels of substance-P and met-enkephalin are decreased in mice infected with the LP-BM5 strain of MuLV (SEI et al. 1998). Increased expression of tumor necrosis factor (TNF)-α and Fas, both proteins associated with programmed cell death, have been demonstrated in the CNS of mice infected with the *ts-1* mutant of Moloney MuLV. In this study, TNF-α and Fas were localized to astrocytes, while only Fas was seen in neurons.

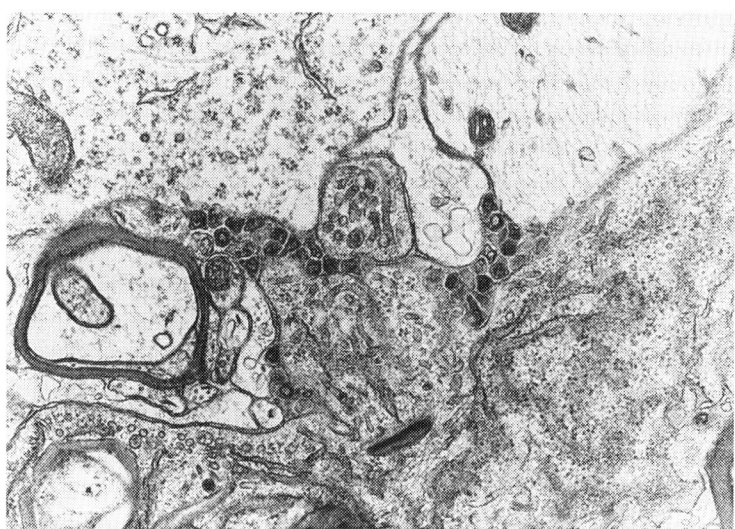

Fig. 3. Electron micrograph from the brain of a mouse with murine leukemia virus (MuLV) encephalitis. At high magnification numerous virions are observed in the extracellular space

Expression of these proteins is associated with spongiform degeneration suggesting Fas and TNF-α-mediated cell death.

Systemic and CNS levels of potential neurotoxins have been measured in mice infected with MuLV. One such factor, quinolinic acid (QUIN), is an excitotoxic NMDA agonist that has been implicated in a number of neuroimmune diseases. Increased QUIN levels in the blood, cerebrospinal fluid (CSF), and brain parenchyma have been noted. This increase is not seen in mice infected with the nonpathogenic ecotropic MuLV that invades the brain but does not cause disease (SEI et al. 1996). It is probable that QUIN production is not a result of productive retroviral infection alone, but may result as a response to other factors. Other potential neurotoxins include reactive oxygen intermediates. Constitutive nitric oxide synthase (cNOS) activity in neurons has been reported to be increased in LP-BM5-infected mice (LI et al. 1997). These changes were concomitant with microglial activation and preceded cognitive dysfunction. The authors hypothesized that increased nitric oxide synthase activity was a result of abnormal glutamatergic stimulation.

2.3 Viral Proteins

Few studies have examined the role of MuLV viral proteins in neurodegeneration. One study has sought to determine if the envelope protein could cause direct neurotoxicity. CasBrE MuLV-expressing cells were transplanted into mice resulting in spread of infection to microglia and spongiform neurodegeneration by 2 weeks. Spongiform changes directly correlated with envelope expression in microglia rather than expression from transplanted cells. These data suggest that the envelope protein of CasBrE is not itself neurotoxic, but that events beyond binding and fusion of the virus in microglia are necessary for the induction of disease (LYNCH et al. 1996).

3 Maedi/Visna Virus

Ovine Maedi/Visna virus (MVV) was the first lentivirus to be isolated and characterized. Unlike many of the other lentiviruses (FIV, SIV and HIV), MVV infects only macrophages and causes a pneumonia-like systemic disease, but may also result in neurological disease. It is not known what determines neurovirulence, although it has been reported that a Visna CNS isolate replicated more efficiently in choroid plexus cells than the Maedi lung isolate. These isolates showed differences in the envelope protein amino acid sequence (ANDRAESDAOTTIR et al. 1998).

3.1 Pathology and Symptomatology

Maedi/Visna virus most likely enters the CNS within infected monocytes that cross the blood–brain barrier during normal immune surveillance (PELUSO et al. 1985). Early lesions consist primarily of lymphocytic and monocytic infiltrates in the perivascular space. Later stages involve invasion of monocytes into the parenchyma. Different cell types within the CNS have been reported to contain viral antigen: microglia/macrophages, lymphocytes, endothelial cells, astrocytes, and oligodendrocytes (GEORGSSON et al. 1989), although macrophages and microglia are the most likely source of replicating virus. Multinucleated giant cells are rarely seen and only with in vitro derived clones. Demyelination in the presence of inflammation is common.

3.2 Soluble Factors

The pathogenesis of MVV encephalitis has not been clearly delineated. The pathologic hallmark of visna is a robust cell-mediated immune response. The lymphocytic component of the inflammatory infiltrates may be responsible for neurologic damage. Further support for this mechanism is the observation that inoculation with virus intramuscularly along with brain white matter in Freund's complete adjuvant (which causes an autoimmune reaction in the CNS similar to multiple sclerosis) results in CNS disease, whereas inoculation with virus alone results only in systemic disease. This study concluded that both activated CNS-specific T-cells and infected macrophages are essential for neuropathogenesis (CHEBLOUNE et al. 1998).

4 Feline Immunodeficiency Virus

This virus has become an important non-rodent model of AIDS because of its similarity with HIV. Like HIV, FIV causes immunodeficiency in its natural host. Furthermore, there is a progressive decline in CD4+ T-cells and inversion of the CD4/CD8 ratio (HARTMANN 1998; POLI et al. 1997; WILLETT et al. 1997). FIV infects primarily CD4+ T-cells but can infect a number of other CD4− cell types: B-cells, monocytes/macrophages, and neural cells (Dow et al. 1990). The receptor for naturally occurring FIV is unknown, although cell culture-adapted strains of FIV are able to use the chemokine receptor CXCR4 for cell fusion and infection (POESCHLA and LOONEY 1998; WILLETT et al. 1997).

4.1 Pathology and Symptomatology

In the early stages of FIV CNS infection, pathologic changes consist of moderate subcortical gliosis with glial nodules. Occasional white matter pallor and meningitis are also seen. At later stages, increased numbers of lesions and infected cells are noted along with perivascular infiltrates. Microglia and astrocytes appear to be infected (Dow et al. 1992). However, no multinucleated giant cells are seen and few cells contain actively replicating virus (BOCHE et al. 1996). One study has reported a case of FIV infection with bizarre cells, sometimes multinucleated (GUNN-MOORE et al. 1996). Neuropathology does not seem to correlate with clinical disease or neurophysiological alterations, and lesions may be seen in the absence of high levels of actively replicating virus (SILVOTTI et al. 1997).

Neuronal cell loss seems to be a progressive event. Loss of neuronal density (particularly large neurons) in the frontal cortex, parietal cortex and striatum has been correlated with a decrease of CD4/CD8 in FIV-infected cats asymptomatic for feline AIDS (MEEKER et al. 1997). Cortical cytoskeletal changes in pyramidal neurons and hippocampal axon reorganization have also been described (JACOBSON et al. 1997; MITCHELL et al. 1998). CSF protein measurements indicate that there is disruption of the blood–brain barrier (PODELL et al. 1997).

Neurological manifestations usually appear at the time of immunodeficiency and include hind limb paralysis, delayed reflexes, irritability, disorientation, and alterations in sleep (PHILLIPS et al. 1996). Neurological abnormalities occur in 20%–40% of naturally infected cats, while as many as 50% of experimentally infected and monitored cats show neurological abnormalities (Dow et al. 1990).

4.2 Soluble Factors

The mechanisms leading to CNS disease in FIV are not known. Like HIV, FIV does not infect neurons but does infect macrophages and microglia. This may lead to an increased immune activation state and release of potentially neurotoxic soluble factors. Excitotoxicity is a possible mechanism of neuronal loss in this model. Glutamate is an excitotoxin known to cause neuronal death. In one study, astrocytes infected with either of two molecular clones of FIV (FIV-34TF10 and FIV-PPR) showed decreased glutamate uptake ability (YU et al. 1998). Another study, using proton nuclear magnetic resonance, has reported neuronal loss associated with increased glutamate levels (POWER et al. 1997).

4.3 Viral Proteins

Feline immunodeficiency virus viral proteins may play a role in neurotoxicity. Systemic infection or intraventricular infusion of the FIV envelope protein results in sleep abnormalities (PROSPERO-GARCIA et al. 1994). Exposure of feline neurons to FIV Env protein alters intracellular calcium flux in response to exogenous glutamate that may lead to neuronal death (GRUOL et al. 1998). This observation parallels reports of HIV envelope protein-induced neurotoxicity.

5 Simian Immunodeficiency Virus

Of the lentiviruses, SIV is the most similar to HIV genotypically. Its natural host is the sooty mangabey, in which infection does not cause clinical disease. Experimental infection of rhesus macaques (*Macaca mulatta*) results in an immune disorder similar in progression to HIV: decrease in CD4+ T-cells, opportunistic infections, wasting disease, and often CNS complications. Molecular clones of SIV that are highly neurovirulent have been isolated and characterized. These clones share similar envelope sequences and are all macrophage tropic, macrophage tropism appears to be necessary but not sufficient to mediate encephalitis (MANKOWSKI et al. 1997).

5.1 Pathology and Symptomatology

Approximately 57% of macaques with SIV-related AIDS show evidence of neurologic involvement. Of these, 44% demonstrated signs of SIV encephalitis (WESTMORELAND et al. 1998a). SIV enters the brain through the same mechanisms as HIV, primarily within infected monocytes (LANE et al. 1996a). In addition, SIV has been reported to infect brain capillary endothelial cells (MANKOWSKI et al. 1994; STRELOW et al. 1998). As early as 1 week after infection, SIV may be seen in the brain, primarily in perivascular monocytes and in the leptomeninges. Later, virus is also seen in microglia and macrophages (see Fox et al. 1997 for review). Pathology resembles that seen in HIV encephalitis: perivascular macrophages

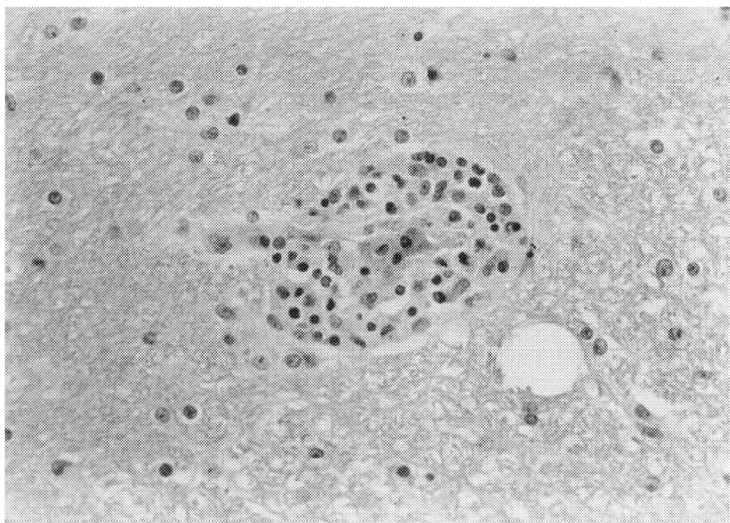

Fig. 4. Hematoxylin-and-eosin stained section from the brain of a rhesus macaque with simian immunodeficiency virus (SIV) encephalitis. Perivascular cuffs of macrophages are common and may be accompanied by multinucleated giant cells (not shown)

(Fig. 4), microglial nodules, multinucleated giant cells, neuronal damage in the form of loss of neuritic processes and synaptic connections, and loss of specific subpopulations of neurons (BERMAN et al. 1998; LACKNER et al. 1991; TRACEY et al. 1997). Abnormalities in the blood–brain barrier have been described as well (SMITH et al. 1994).

5.2 Soluble Factors

Since, like the other lentiviruses, SIV does not infect neurons, indirect mechanisms of neurodegeneration have been proposed.

5.2.1 Cytokines

Cytokines are a diverse group of secreted proteins that act as immune regulators (initiation, propagation, and suppression of immune activation). They generally act locally by binding to cell surface receptors on target cells. Within the brain, microglia/macrophages and astrocytes are the primary cytokine producers. The presence of macrophage- or astrocyte-derived cytokines interleukin (IL)-1β, TNF-α, and interferom (IFN)-γ, has been demonstrated by in situ hybridization in the brains of monkeys with SIV encephalitis (LANE et al. 1996b). The presence of inducible nitric oxide synthase (iNOS) was reported in this same report, suggesting that free radicals may play a role in neurodegeneration.

5.2.2 Chemokines

Chemokines are a subset of cytokines involved in the recruitment of inflammatory cells. Chemokines and their receptors have gained much attention after the discovery that specific receptors may act as co-receptors along with CD4 in HIV binding to target cells (DOMS and PEIPER 1997). Furthermore, chemokines and their receptors are expressed by CNS cells in response to injury or inflammation (RANSOHOFF et al. 1996).

Expression of chemokines and their receptors have been examined in SIV neurodisease. Brains from SIV-infected animals that show encephalitis have elevated expression of the C-C chemokines, macrophage inflammatory protein (MIP)-1α and MIP-1β, RANTES, monocyte chemotactic protein-3, and the C-X-C chemokine IFN-inducible protein-10 (SASSEVILLE et al. 1996). The chemokine receptors CCR3, CCR5, CXCR3, and CXCR4 have been noted in perivascular infiltrates in the CNS of SIV-infected monkeys. This study also reports CCR3, CCR5, and CXCR4 expression on subpopulations of neurons and glia (WESTMORELAND et al. 1998b). These findings suggest that chemokines and their receptors are involved in monocyte and lymphocyte recruitment to the brain in SIV encephalitis.

5.2.3 Macrophage Activation Factors

Other factors that may be involved in viral-induced neurodegeneration are macrophage activation factors. While no correlation between clinical deficits and inflammatory lesions have been noted, levels of QUIN in the CSF of monkeys with SIV infection correlate with degree of neurologic dysfunction (MURRAY et al. 1992), degree of perivascular infiltrates (LANE et al. 1996a), and progression of systemic disease (COE et al. 1997).

6 Human T-Cell Leukemia Virus

Only 1% of the HTLV-1 infected population develop a disease, either adult T-cell leukemia or chronic HTLV-associated myelopathy (HAM), also known as tropical spastic paraparesis (TSP). The neurologic disease incidence correlates with the geographical areas where HTLV-1 is endemic: southern Japan, the Caribbean, Africa, and some parts of the United States. HTLV-2 has not been tied definitively to any neurologic disease.

6.1 Pathology and Symptomatology

HAM/TSP is characterized by CNS infiltration of lymphocytes and macrophages with perivascular cuffing, inflammation, and demyelination with gliosis primarily in the spinal cord although white matter changes in the brain have been reported (IWASAKI 1993). Electron microscopy of lesions reveals disintegration of myelin sheaths, regular separation of the minor dense line of the myelin sheaths, and completely demyelinated axons. In long-standing lesions, there is degeneration of axons, replaced by glial scarring. HTLV-1-infected cells have been identified in the CNS of patients with HAM/TSP. In situ hybridization studies suggest that astrocytes may be infected (LEHKY et al. 1995); other studies detected only infected infiltrating lymphocytes (HARA et al. 1994). Patients with HAM/TSP exhibit signs of paraparesis and spasticity of the lower extremities. There is often bladder dysfunction with minimal sensory loss. Other symptoms are hand tremor, numbness of the lower legs, and lumbago. Interestingly, HTLV-1 has been associated with vascular dementia (KIRA et al. 1997).

6.2 Soluble Factors

Since direct infection of CNS cells has not been conclusively proven, it is likely that damage occurs secondary to an activated cellular and antibody-mediated immune response. Antibodies specific for HTLV are present in the serum and CSF of patients with HAM/TSP. There is also breakdown of the blood–brain barrier and evidence of intra-barrier IgG synthesis. HTLV-1-specific, cytotoxic CD8+ lymphocytes are found in the serum and CSF of patients with neurologic symptoms see (GESSAIN 1996 for reviews). A recent study has suggested molecular mimicry in the pathogenesis of HAM/TSP. Immunoglobulin from patients was immunoreactive with uninfected neurons. IgG from sero-positive individuals without neurologic disease did not show this reactivity with neurons (LEVIN et al. 1998).

In vivo, lymphocytes and macrophages are the likely source of soluble factors. Levels of the cytokines IL-1, IFN-γ, TNF-α, and granulocyte/macrophage-colony-stimulating factor (GM-CSF) are higher in HAM/TSP patients than in seropositive asymptomatic controls (WATANABE et al. 1995). TNF-α has been localized within the spinal cord of HAM/TSP patients by in situ hybridization; however, this expression did not co-localize with T cells, microglia or macrophages. Serum CD8+ T cells produce significantly elevated levels of IFN-γ, TNF-α, and IL-2 in HAM/TSP patients compared to controls (KUBOTA et al. 1998). Another study reported increased levels of the chemokines MIP-1α, MIP-1β, and matrix metalloproteinase-9 (MMP-9), a protein involved in cell migration and synaptic plasticity and produced by peripheral CD8+ cells which may systemically contribute to the pathology seen within the CNS (BIDDISON et al. 1997). MMP-2 and MMP-9 have also been localized to macrophages in acute lesions. This same study demonstrated breakdown of the blood–brain barrier as determined by collagen and decorin immunostaining (UMEHARA et al. 1998). MMP-9 has also been detected in the CSF of patients with HAM/TSP (GIRAUDON et al. 1998).

A number of soluble molecules are produced after infection with HTLV-1 in vitro. Cell lines that may be infected by HTLV-1 include astroglioma, neuroblastoma, and oligodendroglioma cells as well as primary human neuroglial cultures. Persistent infection results in production of GM-CSF, a cytokine involved in the maturation of leukocytes in areas of inflammatory immune response (Nishiura et al. 1994). Neuroprogenitor cell lines, transiently infected with HTLV, show increased production of MMP that is compounded by exposure to TNF-α and IL-1 (Giraudon et al. 1996).

6.3 Viral Proteins

Unlike HIV, HTLV-1 does not produce viral proteins known to be directly toxic to cells of the CNS. The Tax protein (which is similar to HIV Tat) has been shown to transactivate a number of cellular proteins that may promote an immune response and thus indirectly lead to damage within the spinal cord. These proteins include IL-2 and IL-2 receptor, GM-CSF, IL-3, TGF-β, and class I and class II MHC molecules (Dhib-Jalbut et al. 1994). In vitro, human neuronal cells produce TNF-α in response to endogenous exposure to Tax (Cowan et al. 1997). Astrocytic cell lines transfected with *tax* produce TNF-α. Furthermore, these Tax-expressing cells may act as targets for cytolytic lymphocytes, since Tax-specific CD8+ cells have been detected in HAM/TSP patients (Höger et al. 1997; Méndez et al. 1997).

7 Human Immunodeficiency Virus

Human immunodeficiency virus is certainly the best known of all the retroviruses. Since its identification in 1983 as the causative agent of AIDS, HIV has had an enormous impact on both the scientific and lay community. The virus infects monocytes and CD4+ T-cells. Infection and death of these cells leads to chronic immune suppression, susceptibility to opportunistic infection, and ultimately death. Great strides have been made in elucidating the pathogenesis of AIDS, culminating in the development and implementation of the triple-drug therapy regime, known as HAART (highly active anti-retroviral therapy). Despite considerable advancement in research concerning systemic disease, the mechanisms leading to HIV-associated neuronal damage are still unresolved.

More than two-thirds of adult AIDS autopsies have evidence of HIV infection in the CNS (Navia et al. 1986a; Price et al. 1991), and approximately one-quarter of AIDS patients develop neurologic symptoms attributable to HIV infection within the brain (Anders et al. 1986; Rhodes 1993). Since HIV-infected patients who survive longer are more likely to develop HIV encephalitis (Soontornniyomkij et al. 1998), it is possible that these numbers will increase with the advent of triple-drug therapy. Termed HIV encephalitis (HIVE), this infection covers a broad spectrum of neuropathologic changes of varying severity (Budka et al. 1987; Navia et al. 1986a; Sharer 1992). In adults, HIVE occurs in late stages of HIV infection, after prolonged periods of depressed immunity have led to unrestricted replication and spread of the virus within the CNS. Neuro-HIV infection in the pediatric population results in greater pathology, most likely due to interference with normal development (DICKSON et al. 1989; IANNETTI et al. 1989; OJUKWU and EPSTEIN 1998).

7.1 Pathology and Symptomatology

Human immunodeficiency virus enters the brain primarily within infected monocytes and spreads to the resident brain macrophages, the microglia. Neurons do not appear to be infected and evidence of infection of astrocytes has not been widely accepted. The pathological hallmarks of HIV encephalitis are microglial nodules (Fig. 5) and multinucleated giant cells (Fig. 6). Nodules and giant cells are usually observed in proximity to blood vessels, supporting the notion that virus is expressed soon after crossing the blood–brain barrier. There is, however, evidence of direct infection of

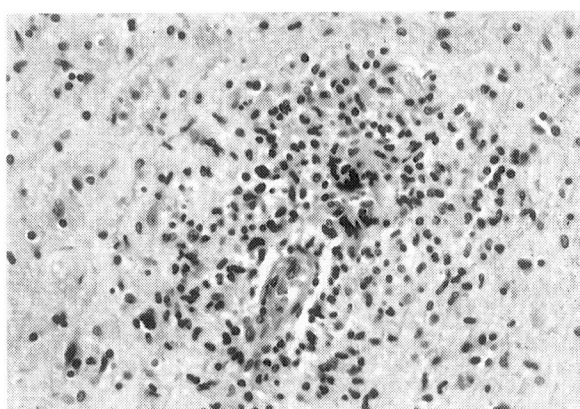

Fig. 5. Hematoxylin-and-eosin stained section from the brain of a patient with human immunodeficiency virus (HIV) encephalitis. Chronic perivascular inflammation blends with a parenchymal microglial nodule. Numerous microglia infiltrate beyond the nodule into the adjacent brain tissue

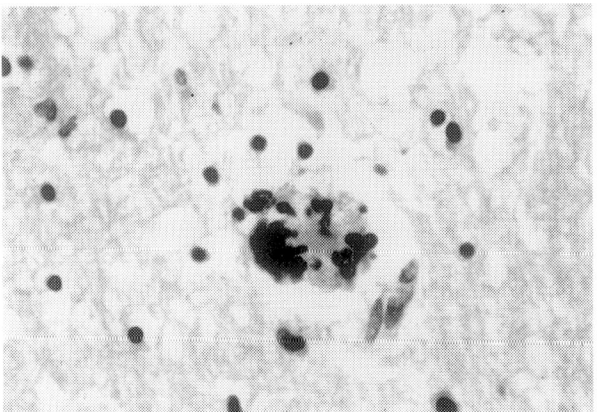

Fig. 6. Hematoxylin-and-eosin stained section from the brain of a patient with human immunodeficiency virus (HIV) encephalitis. At high magnification multinucleated giant cells are observed in the cortical white matter. Many of these macrophage elements are infected with HIV while adjacent neuroglial elements are not infected

endothelial cells, transcytosis of the virus by endothelial cells, and entry via the choroid plexus (BANKS et al. 1998; FALANGOLA et al. 1995; MOSES et al. 1993).

Neuronal loss has been observed within cortical gray matter (EVERALL et al. 1991; WEIS et al. 1993), and apoptotic neurons have been noted in pediatric and adult HIVE (ADLE-BIASSETTE et al. 1995; GELBARD et al. 1995; PETITO and ROBERTS 1995). Golgi impregnations have shown dendritic spine damage, while laser confocal immunomicroscopy studies have shown quantitative loss of pre- and post-synaptic elements in the cortex and basal ganglia, suggesting even more widespread neuronal damage without actual cell loss (MASLIAH et al. 1992; MASLIAH et al. 1997). There appears to be regional differences in neuropathology and viral load. Progressive volume loss, as seen by magnetic resonance imaging (MRI), specifically in the caudate, is associated with disease stage and decline in CD4+ cells (STOUT et al. 1998). Different neuronal populations seem to be selectively vulnerable, resulting in apoptosis. These include the large pyramidal neurons and interneurons of the cortex, spiny neurons of the putamen, and CA3 neurons in the hippocampus (EVERALL et al. 1996). Subcortical structures such as the basal ganglia and hippocampus, have been reported to contain the highest HIV mRNA levels (WILEY et al. 1998). How HIV infection of macrophages leads to neuronal and glial damage is still unknown.

The clinical presentation of HIV infection is one of a subcortical dementia. This dementia subtype is characterized by motor disturbances, attention deficits, and apathy. Loss of concentration, imbalance and weakness are primary symptoms in early stages of the disease. Severe dementia, ataxia, and motor weakness are primarily late-stage manifestations (NAVIA et al. 1986b). One of the more intriguing aspects of neuro-HIV disease is the lack of a straightforward association among viral load, pathology, and behavioral abnormalities. It is well-known that patients may present significant signs of HIV encephalitis at autopsy without ever demonstrating dementia or, conversely, may be demented but show little of the post-mortem hallmarks of HIVE. Because of these discrepancies, it is likely that there are multiple mechanisms of disease pathogenesis, as described below.

7.2 Soluble Factors

One mechanism is the production of soluble factors by infected microglia and macrophages within the brain. There have been a number of excellent recent reviews concerning HIV-induced neurodegeneration (EPSTEIN 1998; GRIFFIN 1997; LIPTON 1998). The following discussion will concentrate on more recent studies.

7.2.1 Cytokines

Since cytokines are known to be up-regulated in response to injury and in CNS disease (i.e. multiple sclerosis), much attention has been given to the profile seen in HIV infection. A number of cytokines have been reported to be up-regulated in brain tissue from AIDS patients: IL-1, IL-6, TNF-α, TGF-β, and IFN-γ (TYOR et al. 1992; WAHL et al. 1991; WESSELINGH et al. 1997). In vitro, macrophages cocultured with astrocytes and infected with HIV up-regulate production of IL-1β and TNF-α (GENIS et al. 1992).

TNF-α is a potent immune stimulator and has myelotoxic and neurotoxic effects in vitro. It has been localized to microglia and astrocytes in brain tissue from HIV patients, with greater abundance in white matter than in gray matter (ACHIM et al. 1993; WESSELINGH 1997). TNF-α may induce apoptosis in neurons in vitro (TALLEY et al. 1995; WESTMORELAND et al. 1996) and has recently been shown to decrease the spontaneous firing rate of neurons (KATAFUCHI et al. 1997), suggesting a more subtle role in damaging neurons. The HIV protein Tat has been reported to up-regulate TNF-α expression and induce neuronal damage via this cascade (see below). In addition to its direct actions on neurons, TNF-α may induce astrogliosis and may cause a decrease in glutamate uptake in astrocytes, contributing indirectly to excitotoxic neuronal cell death (FINE et al. 1996). It may also increase levels of other neurotoxins such as nitric acid, leukotrienes, and platelet activating factor (PAF). TNF-α might also play a role in paracellular trafficking of HIV through the blood–brain barrier (FIALA et al. 1997).

TGF-β can be either a suppressor or stimulator of cell functions and acts on both astrocytes and microglia. It may also act as a chemo-attractant for monocytes (WAHL et al. 1987). TGF-β is associated with AIDS brain tissue and is localized to astrocytes and microglial nodules in vivo (WAHL et al. 1991). In vitro, it is expressed by ameboid microglia after infection with HIV (DA CUNHA et al. 1997).

The presence of IL-6 and IL-1 have also been investigated. Both of these proteins have been localized to astrocytes, macrophages, and microglia (TYOR et al. 1992). IL-1 can stimulate the production of other cytokines as well as enhance HIV replication. IL-6 may be induced by gp120 or TNF-α and, in turn, modulate the activity of other cytokines. These molecules may promote gliosis and thus indirectly contribute to HIV-induced neuropathology.

7.2.2 Chemokines

Chemokines and their receptors have been demonstrated in HIVE (LAVI et al. 1997; ROTTMAN et al. 1997; SANDERS et al. 1998; VALLAT et al. 1998). Both are localized primarily to microglial nodules and perivascular infiltrates. Receptor expression has also been demonstrated on neurons in normal and diseased brain. The function of these receptors on neurons is unknown. In normal development they are hypothesized to play a role in neuronal migration (ZOU et al. 1998). Receptors on neurons in the proximity of HIV-up-regulated chemokines may be involved in neurodegeneration. CXCR4 is expressed by neurons and has been shown to mediate apoptosis after binding its natural ligand, SDF-1, or the viral envelope protein gp120 (HESSELGESSER et al. 1998).

7.2.3 Macrophage Activation Factors

Human immunodeficiency virus activation of macrophages may lead to synthesis and release of toxic factors, i.e. QUIN, prostaglandins, arachidonic acid, and PAF

(GELMAN et al. 1997; GRIFFIN et al. 1994; PEMBERTON et al. 1997; TYOR et al. 1995). QUIN, a metabolite of tryptophan, has been shown to be neurotoxic to neurons in vitro (KERR et al. 1998), and CSF levels correlate with neuropsychological impairment (HEYES et al. 1991). QUIN levels have been shown to be increased in the brain parenchyma of patients with HIV compared to controls (HEYES et al. 1998). PAF is produced by HIV-infected monocytes and is present in brains of AIDS patients with dementia. Conditioned media from infected monocytes induces neuronal apoptosis, which may be inhibited by PAF acetylhydrolase, the enzyme that catabolizes PAF (PERRY et al. 1998).

7.3 Viral Proteins

There is substantial evidence supporting the role of HIV gene products to induce and/or perpetuate an inflammatory reaction as well as directly mediate neuronal damage. The two strongest candidates are Tat and the envelope protein gp120. HIV membrane protein gp120 binds to CD4 and the appropriate co-receptor (CXCR4 or CCR5 among others) and thus mediates cell tropism and cell fusion. HIV Tat protein interacts with the RNA target transactivating region and serves to increase HIV transcription.

The envelope protein gp120 is neurotoxic to neurons in vitro and acts via an NMDA receptor (LIPTON 1993; SAVIO and LEVI 1993). Injection of recombinant gp120 into rat neocortex results in increased expression of cyclooxygenase-2, an enzyme involved in the catabolism of arachidonic acid into prostaglandin. This increased expression was associated with neuronal apoptosis (BAGETTA et al. 1998). It has also been shown recently to induce cell death via binding to the chemokine receptor CXCR4 (HESSELGESSER et al. 1998). The envelope protein may contribute to neuropathogenesis indirectly as well. Monocytes increase production of nitric oxide after exposure to gp120 (DAWSON et al. 1996). The protein has been shown to up-regulate adhesion molecules on endothelial cells and act synergistically with morphine or anandamide to enhance monocyte adhesion (STEFANO et al. 1998). Intraventricular infusion of gp120 in rats results in increased IL-1β and TNF-α levels (ILYIN and PLATASALAMAN 1997), suggesting a role in an immune activation cascade. Exposure of glioblastoma or neuroblastoma cell lines to gp41 peptide results in increased MMP-2 activity, suggesting an indirect role for the viral protein in neuropathology (CHONG et al. 1998).

Similarly, Tat protein may act directly on neurons or indirectly by modulating cytokine expression. Tat has been shown to induce cell death via a non-NMDA receptor on a variety of cell types in vitro (MAGNUSON et al. 1995; NATH et al. 1996; NEW et al. 1997). Intraventricular injection of Tat into the rat CNS results in inflammation, ventricular enlargement, apoptosis and alteration of normal transmitter levels in the hippocampus (JONES et al. 1998). In response to Tat, in vitro transcription of TGF-β and TNF-α in macrophages and astrocytes has been shown to be increased (SAWAYA et al. 1998), and that of TNF-β decreased (YANG et al. 1997). Astrocytes exposed to Tat demonstrate increased production of the

Table 2. Neurotropic retroviruses and proposed pathogenesis

Virus	Cells infected	Neuropathology	Mechanism
MuLV	Endothelia, microglia, astrocytes, neurons	Spongiform degeneration	Direct and indirect microglial activation?
FIV	Astrocytes, macrophages/microglia	Gliosis, myelin pallor, neuronal loss	Indirect glutamate toxicity?
Visna virus	Astrocytes, macrophages/microglia, lymphocytes	Perivascular inflammation, demyelination	Indirect cell-mediated immune response?
SIV	Endothelia, macrophages/microglia	Perivascular inflammation, demyelination, multinucleated giant cells, neuronal loss	Indirect macrophage/microglial activation
HTLV	Lymphocytes, astrocytes?	Perivascular infiltrates, demyelination	Indirect cellular and humoral immune response
HIV	Endothelia, macrophages/microglia, astrocytes, lymphocytes, choroid plexus	Perivascular inflammation, demyelination, multinucleated giant cells, neuronal loss	Indirect macrophage/microglial and/or astrocytic activation Viral protein toxicity

MuLV, murine leukemia virus; FIV, feline immunodeficiency virus; SIV, simain immunodeficiency virus; HTLV, human T cell leukemia virus; HIV, human immunodeficiency virus.

chemokine MCP-1. This can initiate further recruitment of monocytes into the brain (CONANT et al. 1998). Tat may also perpetuate an immune cascade within the CNS by acting as a chemotactic molecule for monocytes (ALBINI et al. 1998; LAFRENIE et al. 1996).

8 Conclusion

Retroviruses cause CNS damage by a variety of mechanisms (Table 2). In the case of the important human pathogen HIV, direct association of clinical symptoms and neuropathological changes with specific "neurotoxic" factors has been difficult. Investigations into the human CNS disease have primarily used autopsy material and thus only present a "snapshot" of late stage disease. Given the variety of factors shown to be altered in these cases, it has not been possible to construct an elegant model to account for the neurodegenerative changes. The relative contribution of macrophage-associated pro-inflammatory molecules and viral proteins to specific pathological changes is unknown. Currently, prevention of CNS damage appears to inhibit the ingress of virally infected monocytes into the CNS, a goal that may have been achieved by HAART therapy.

References

Achim CL, Heyes MP, Wiley CA (1993) Quantitation of human immunodeficiency virus, immune activation factors, and quinolinic acid in AIDS brains. J Clin Invest 91:2769–2775

Adle-Biassette H, Levy Y, Colombel M, Poron F, Natchev S, Keohane C, Gray F (1995) Neuronal apoptosis in HIV infection in adults. Neuropathol. Appl Neurobiol 21:218–227

Albini A, Benelli R, Giunciuglio D, Cai T, Mariani G, Ferrini S, Noonan DM (1998) Identification of a novel domain of HIV tat involved in monocyte chemotaxis. J Biol Chem 273:15895–15900

Anders KH, Guerra WF, Tomiyasu U, Verity MA, Vinters HV (1986) The neuropathology of AIDS. UCLA experience and review. Am J Pathol 124:537–558

Andraesdaottir V, Tang X, Agnarsdaottir G, Andraesson OS, Georgsson G, Skraban R, Torsteinsdaottir S, Rafnar B, Benediktsdaottir E, Matthaiasdaottir S, Arnadaottir S, Heognadaottir S, Paalsson PA, Paetursson G (1998) Biological and genetic differences between lung- and brain-derived isolates of maedi-visna virus. Virus Genes 16:281–293

Bagetta G, Corasaniti MT, Paoletti AM, Berliocchi L, Nistico R, Giammarioli A, Malorni W, Finazziagro A (1998) HIV-1 gp120-induced apoptosis in the rat neocortex involves enhanced expression of cyclo-oxygenase type 2 (COX-2). Biochem Biophys Res Commun 244:819–824

Banks WA, Akerstrom V, Kastin AJ (1998) Adsorptive endocytosis mediates the passage of HIV-1 across the blood–brain barrier: evidence for a post-internalization coreceptor. J Cell Sci 111:533–540

Berman NEJ, Raymond LA, Warren KA, Raghavan R, Joag SV, Narayan O, Cheney PD (1998) Fractionator analysis shows loss of neurons in the lateral geniculate nucleus of macaques infected with neurovirulent simian immunodeficiency virus. Neuropath Appl Neurobiol 24:44–52

Biddison WE, Kubota R, Kawanishi T, Taub DD, Cruikshank WW, Center DM, Connor EW, Utz U, Jacobson S (1997) Human T cell leukemia virus type I (HTLV-I)-specific CD8+ CTL clones from patients with HTLV-I-associated neurologic disease secrete proinflammatory cytokines, chemokines, and matrix metalloproteinase. J Immunol 159:2018–2025

Boche D, Hurtrel M, Gray F, Claessensmaire MA, Ganiere JP, Montagnier L, Hurtrel B (1996) Virus load and neuropathology in the FIV model. J Neurovirol 2:377–387

Budka H, Costanzi G, Cristina S, Lechi A, Parravicini C, Trabattoni R, Vago L (1987) Brain pathology induced by infection with the HIV. Acta Neuropathol 75:185–198

Chebloune Y, Karr BM, Raghavan R, Singh DK, Leung K, Sheffer D, Pinson D, Foresman L, Narayan O (1998) Neuroinvasion by ovine lentivirus in infected sheep mediated by inflammatory cells associated with experimental allergic encephalomyelitis. J Neurovirol 4:38–48

Chong YH, Seoh JY, Park HK (1998) Increased activity of matrix metalloproteinase-2 in human glial and neuronal cell lines treated with HIV-1 gp41 peptides. J Mol Neurosci 102:129–141

Coe CL, Reyes TM, Pauza CD, Reinhard JF Jr (1997) Quinolinic acid and lymphocyte subsets in the intrathecal compartment as biomarkers of SIV infection and simian AIDS. Aids Res Hum Retroviruses 13:891–897

Conant K, Garzino-Demo A, Nath A, McArthur JC, Halliday W, Power C, Gallo RC, Major EO (1998) Induction of monocyte chemoattractant protein-1 in HIV-1 Tat-stimulated astrocytes and elevation in AIDS dementia. Proc Natl Acad Sci USA 95:3117–3121

Cowan EP, Alexander RK, Daniel S, Kashanchi F, Brady JN (1997) Induction of tumor necrosis factor alpha in human neuronal cells by extracellular human T cell lymphotropic virus type 1 Tax. J Virol 71:6982–6989

Czub S, Lynch WP, Czub M, Portis JL (1994) Kinetic analysis of spongiform neurodegenerative disease induced by a highly virulent murine retrovirus. Lab Invest 70:711–723

da Cunha A, Jefferson JJ, Tyor WR, Glass JD, Jannotta FS, Cottrell JR, Resau JH (1997) Transforming growth factor-beta1 in adult human microglia and its stimulated production by interleukin-1. J Interferon Cytokine Res 17:655–664

Dawson VL, Dawson TM, Uhl GR, Snyder SH (1996) HIV-1 coat protein neurotoxicity mediated by NO in primary cortical neurons. Proc Natl Acad Sci USA 90:3256–3259

Dhib-Jalbut S, Hoffman PM, Yamabe T, Sun D, Xia J, Eisenberg H, Bergey G (1994) Extracellular human T cell lymphotropic virus type 1 Tax protein induces cytokine production in adult human microglial cells. Ann Neurol 36:787–790

Dickson DW, Belman AL, Park YD, Wiley C, Horoupian DS, Llena J, Kure K, Lyman WD, Morecki R, Mitsudo S (1989) Central nervous system pathology in pediatric AIDS: an autopsy study. APMIS Suppl 8:40–57

Doms RW, Peiper SC (1997) Unwelcomed guests with master keys: how HIV uses chemokine receptors for cellular entry. Virology 235:179–190

Dow SW, Drietz MJ, Hoover EA (1992) Feline immunodeficiency virus neurotropism: evidence that astrocytes and microglia are the primary target cells. Vet Immun Immunopathol 35:23–35

Dow SW, Poss ML, Hoover EA (1990) Feline immunodeficiency virus: a neurotropic lentivirus. J AIDS 3:658–668

Epstein LG (1998) HIV neuropathogenesis and therapeutic strategies. Acta Paediatr Jap 40:107–111
Everall IP, Gray F, Masliah E (1996) Neuronal injury and apoptosis. In: Gendelman H, Lipton S, Epstein L, Swindells S (eds) The Neurology of AIDS. Chapman and Hall, New York, pp 261-274
Everall IP, Luthert PJ, Lantos PL (1991) Neuronal loss in the frontal cortex in HIV infection. Lancet 337:1119–1121
Falangola MF, Hanly A, Galvao CB, Petito CK (1995) HIV infection of human choroid plexus: a possible mechanism of viral entry into the CNS. J Neuropathol Exp Neurol 54:497–503
Fiala M, Looney DJ, Stins M, Way DD, Zhang L, Gan X, Chiappelli F, Schweitzer ES, Shapshak P, Weinand M, Graves MC, Witte M, Kim KS (1997) TNF-alpha opens a paracellular route for HIV-1 invasion across the blood–brain barrier. Mol Med 3:553–564
Fine SM, Angel RA, Perry SW, Epstein LG, Rothstein JD, Dewhurst S, Gelbard HA (1996) Tumor necrosis factor alpha inhibits glutamate uptake by primary human astrocytes. Implication for pathogenesis of HIV-1 dementia. J Biol Chem 27:15303–15306
Fox HS, Gold LH, Henriksen SJ, Bloom FE (1997) Simian immunodeficiency virus: a model for NeuroAIDS. Neurobiol Dis 4:265–274
Fujisawa R, McAtee FJ, Wehrly K, Portis JL (1998) The neuroinvasiveness of a murine retrovirus is influenced by a dileucine-containing sequence in the cytoplasmic tail of glycosylated Gag. J Virol 72:5619–5625
Gelbard HA, James HJ, Sharer LR, Perry SW, Saito Y, Kazee AM, Blumberg BM, Epstein LG (1995) Apoptotic neurons in brains from pediatric-patients with HIV-1 encephalitis and progressive encephalopathy. Neuropath Appl Neurobiol 21:208–217
Gelman BB, Wolf DA, Rodriguez-Wolf M, West AB, Haque AK, Cloyd M (1997) Mononuclear phagocyte hydrolytic enzyme activity associated with cerebral HIV-1 infection. Am J Pathol 151:1437–1446
Genis P, Jett M, Bernton EW, Boyle T, Gelbard HA, Dzenko K, Keane RW, Resnick L, Mizrachi Y, Volsky DJ, et al. (1992) Cytokines and arachidonic metabolites produced during human immunodeficiency virus (HIV)-infected macrophage-astroglia interactions: implications for the neuropathogenesis of HIV disease. J Exp Med 176:1703–1718
Georgsson G, Houwers DJ, Palsson PA, Petursson G (1989) Expression of viral antigens in the central nervous system of visna-infected sheep: an immunohistochemical study on experimental visna induced by virus strains of increased neurovirulence. Acta Neuropathol 77:299–306
Gessain A (1996) Virological aspects of tropical spastic paraparesis/HTLV-I associated myelopathy and HTLV-I infection [see comments]. J Neurovirol 2:299–306
Giraudon P, Buart S, Bernard A, Thomasset N, Belin MF (1996) Extracellular matrix-remodeling metalloproteinases and infection of the central nervous system with retrovirus human T-lymphotrophic virus type I (HTLV-I). Prog Neurobiol 49:169–184
Giraudon P, Vernant JC, Confavreux C, Belin MF, Desgranges C (1998) Matrix metalloproteinase 9 (gelatinase B) in cerebrospinal fluid of HTLV-1 infected patients with tropical spastic paraparesis. Neurology 50:1920
Gravel C, Kay DG, Jolicoeur P (1993) Identification of the infected target cell type in spongiform myeloencephalopathy induced by the neurotropic Cas-Br-E murine leukemia-virus. J Virol 67:6648–6658
Griffin DE (1997) Cytokines in the brain during viral infection: clues to HIV-associated dementia. J Clin Invest 100:2948–2951
Griffin DE, Wesselingh SL, McArthur JC (1994) Elevated central nervous system prostaglandins in human immunodeficiency virus-associated dementia. Ann Neurol 35:592–597
Gruol DL, Yu N, Parsons KL, Billaud JN, Elder JH, Phillips TR (1998) Neurotoxic effects of feline immunodeficiency virus, FIV-PPR. J Neurovirol 4:415–425
Gunn-Moore DA, Pearson GR, Harbour DA, Whiting CV (1996) Encephalitis associated with giant cells in a cat with naturally occurring feline immunodeficiency virus infection demonstrated by in situ hybridization. Vet Pathol 33:699–703
Hara H, Morita M, Iwaki T, Hatae T, Itoyama Y, Kitamoto T, Akizuki S, Goto I, Watanabe T (1994) Detection of human T lymphotrophic virus type I (HTLV-I) proviral DNA and analysis of T cell receptor V beta CDR3 sequences in spinal cord lesions of HTLV-I-associated myelopathy/tropical spastic paraparesis. J Exp Med 180:831–839
Hartmann K (1998) Feline immunodeficiency virus infection – an overview. Vet J 155:123–137
Hesselgesser J, Taub D, Baskar P, Greenberg M, Hoxie J, Kolson DL, Horuk R (1998) Neuronal apoptosis induced by HIV-1 gp120 and the chemokine SDF-1 alpha is mediated by the chemokine receptor CXCR4. Curr Biol 8:595–598

Heyes MP, Brew BJ, Martin A, Price RW, Salazar AM, Sidtis JJ, Yergey JA, Mouradian MM, Sadler AE, Keilp J, et al. (1991) Quinolinic acid in cerebrospinal fluid and serum in HIV-1 infection: relationship to clinical and neurological status. Ann Neurol 29:202–209

Heyes MP, Saito K, Lackner A, Wiley CA, Achim CL, Markey SP (1998) Sources of the neurotoxin quinolinic acid in the brain of HIV-1-infected patients and retrovirus-infected macaques. FASEB J 12:881–896

Höger TA, Jacobson S, Kawanishi T, Kato T, Nishioka K, Yamamoto K (1997) Accumulation of human T lymphotropic virus (HTLV)-I-specific T cell clones in HTLV-I-associated myelopathy/tropical spastic paraparesis patients. J Immunol 159:2042–2048

Iannetti P, Falconieri P, Imperato C (1989) Acquired immune deficiency syndrome in childhood. Neurological aspects. Child's Nervous Syst 5:281–287

Ilyin SE, Platasalaman CR (1997) HIV-1 envelope glycoprotein 120 regulates IL-1beta system and TNF-alpha mRNAs in vivo. Brain Res Bull 44:67–73

Iwasaki Y (1993) Human T cell leukemia virus type 1 infection and chronic myelopathy. Brain Pathol 3:1–10

Jacobson S, Henriksen SJ, Prospero-Garcia O, Phillips TR, Elder JH, Young WG, Bloom FE, Fox HS (1997) Cortical neuronal cytoskeletal changes associated with FIV infection. J Neurovirol 3:283–289

Jones M, Olafson K, Del Bigio MR, Peeling J, Nath A (1998) Intraventricular injection of human immunodeficiency virus type 1 (HIV-1) tat protein causes inflammation, gliosis, apoptosis, and ventricular enlargement. J Neuropathol Exp Neurol 57:563–570

Katafuchi T, Motomura K, Baba S, Ota K, Hori T (1997) Differential effects of tumor necrosis factor-alpha and -beta on rat ventromedial hypothalamic neurons in vitro. Am J Physiol 272:R1966–1971

Kerr SJ, Armati PJ, Guillemin GJ, Brew BJ (1998) Chronic exposure of human neurons to quinolinic acid results in neuronal changes consistent with AIDS dementia complex. AIDS 12:355–363

Kira J, Hamada T, Kawano Y, Okayama M, Yamasaki K (1997) An association of human T cell lymphotropic virus type 1 infection with vascular dementia. Acta Neurol Scand 96:305–309

Kubota R, Kawanishi T, Matsubara H, Manns A, Jacobson S (1998) Demonstration of human T lymphotropic virus type I (HTLV-I) tax-specific CD8+ lymphocytes directly in peripheral blood of HTLV-I-associated myelopathy/tropical spastic paraparesis patients by intracellular cytokine detection. J Immunol 161:482–488

Kustova Y, Sei Y, Goping G, Basile AS (1996) Gliosis in the LP-BM5 murine leukemia virus – infected mouse: an animal model of retrovirus-induced dementia. Brain Res 742:271–282

Lackner AA, Smith MO, Munn RJ, Martfeld DJ, Gardner MB, Marx PA, Dandekar S (1991) Localization of simian immunodeficiency virus in the central nervous system of rhesus monkeys. Am J Pathol 139:609–621

Lafrenie RM, Wahl LM, Epstein JS, Hewlett IK, Yamada KM, Dhawan S (1996) HIV-1-Tat protein promotes chemotaxis and invasive behavior by monocytes. J Immunol 157:974–977

Lane JH, Sasseville VG, Smith MO, Vogel P, Pauley DR, Heyes MP, Lackner AA (1996a) Neuroinvasion by simian immunodeficiency virus coincides with increased numbers of perivascular macrophages/microglia and intrathecal immune activation. J Neurovirol 2:423–432

Lane TE, Buchmeier MJ, Watry DD, Fox HS (1996b) Expression of inflammatory cytokines and inducible nitric oxide synthase in brains of SIV-infected rhesus monkeys: applications to HIV-induced central nervous system disease. Mol Med 2:27–37

Lavi E, Strizki JM, Ulrich AM, Zhang W, Fu L, Wang Q, O'Connor M, Hoxie JA, González-Scarano F (1997) CXCR-4 (Fusin), a co-receptor for the type 1 human immunodeficiency virus (HIV-1), is expressed in the human brain in a variety of cell types, including microglia and neurons. Am J Pathol 151:1035–1042

Lehky TJ, Fox CH, Koenig S, Levin MC, Flerlage N, Izumo S, Sato E, Raine C, Osame M, Jacobson S (1995) Detection of human T-lymphotropic virus Type 1 (HTLV-1) tax RNA in the central nervous system of HTLV-1-associated myelopathy/tropical spastic paraparesis patients by in situ hybridization. Ann Neurol 37:167–175

Levin MC, Krichavsky M, Berk J, Foley S, Rosenfeld M, Dalmau J, Chang G, Posner JB, Jacobson S (1998) Neuronal molecular mimicry in immune-mediated neurologic disease. Ann Neurol 44:87–98

Li Y, Kustova Y, Sei Y, Basile AS (1997) Regional changes in constitutive, but not inducible nos expression in the brains of mice infected with the LP-BM5 leukemia virus. Brain Res 752:107–116

Liang B, Wang JY, Watson RR (1996) Murine AIDS, a key to understanding retrovirus-induced immunodeficiency. Vir. Immunol. 9:225–239

Lipton SA (1993) Human immunodeficiency virus-infected macrophages, gp120, and N-methyl-D-aspartate receptor-mediated neurotoxicity. Ann Neurol 33:227–228

Lipton SA (1998) Neuronal injury associated with HIV-1: approaches to treatment. Ann Rev Pharmacol Toxicol 38:159–177

Lynch WP, Czub S, McAtee FJ, Hayes SF, Portis JL (1991) Murine retrovirus-induced spongiform encephalopathy: productive infection of microglia and cerebellar neurons in accelerated cns disease. Neuron 7:365–379

Lynch WP, Snyder EY, Qualtiere L, Portis JL, Sharpe AH (1996) Late virus replication events in microglia are required for neurovirulent retrovirus-induced spongiform neurodegeneration: evidence from neural progenitor-derived chimeric mouse brains. J Virol 70:8896–8907

Magnuson DSK, Knudsen BE, Geiger JD, Brownstone RM, Nath A (1995) Human immunodeficiency virus type 1 Tat activates non-N-methyl-D-aspartate excitatory amino acid receptors and causes neurotoxicity. Ann Neurol 37:373–380

Mankowski JL, Flaherty MT, Spelman JP, Hauer DA, Didier PJ, Amedee AM, Murphy-Corb M, Kirstein LM, Munoz A, Clements JE, Zink MC (1997) Pathogenesis of simian immunodeficiency virus encephalitis: viral determinants of neurovirulence. J Virol 71:6055–6060

Mankowski JL, Spelman JP, Ressetar HG, Strandberg JD, Laterra J, Carter DL, Clements JE, Zink MC (1994) Neurovirulent simian immunodeficiency virus replicates productively in endothelial cells of the central nervous system in vivo and in vitro. J Virol 68:8202–8208

Masliah E, Ge N, Morey M, Deteresa R, Terry RD, Wiley CA (1992) Cortical dendritic pathology in human immunodeficiency virus encephalitis. Lab Invest 66:285–291

Masliah E, Heaton RK, Marcotte TD, Ellis RJ, Wiley CA, Mallory M, Achim CL, Mccutchan JA, Nelson JA, Atkinson JH, Grant I (1997) Dendritic injury is a pathological substrate for human immunodficiency virus-related cognitive disorders. Ann Neurol 42:963–972

Meeker RB, Thiede BA, Hall C, English R, Tompkins M (1997) Cortical cell loss in asymptomatic cats experimentally infected with feline immunodeficiency virus. AIDS Res Hum Retroviruses 13:1131–1140

Mendez E, Kawanishi T, Clemens K, Siomi H, Soldan SS, Calabresi P, Brady J, Jacobson S (1997) Astrocyte-specific expression of human T cell lmyphotropic virus type 1 (HTLZ-1) Tax: induction of tumor necrosis factor alpha and susceptibility to lysis by CD8+ HTLV-1 specific cytotoxic T cells. J Virol 71:9143–9149

Mitchell TW, Buckmaster PS, Hoover EA, Whalen LR, Dudek FE (1998) Axonal sprouting in hippocampus of cats infected with feline immunodeficiency virus (FIV). J Acquir Immune Defic Syndr Hum Retrovirol 17:1–8

Moses AV, Bloom FE, Pauza CD, Nelson JA (1993) Human immunodeficiency virus infection of human brain capillary endothelial cells occurs via a CD4/galactosylceramide-independent mechanism. Proc Natl Acad Sci USA 90:10474–10478

Murray EA, Rausch D, Lendvay J, Sharer L, Eiden L (1992) Cognitive and motor impairments associated with SIV infection in rhesus monkeys. Science 255:1246–1249

Nath A, Psooy K, Martin C, Knudsen B, Magnuson DS, Haughey N, Geiger JD (1996) Identification of a human immunodeficiency virus type 1 Tat epitope that is neuroexcitatory and neurotoxic. J Virol 70:1475–1480

Navia BA, Cho E-S, Petito CK, Price RW (1986a) The AIDS dementia complex: II Neuropathology. Ann Neurol 19:525–535

Navia BA, Jordan BD, Price RW (1986b) The AIDS dementia complex: I. Clinical features. Ann Neurol 19:517–524

New DR, Ma MH, Epstein LG, Nath A, Gelbard HA (1997) Human immunodeficiency virus type 1 Tat protein induces death by apoptosis in primary human neuron cultures. J Neurovirol 3:168–173

Nishiura Y, Nakamura T, Takino H, Ichinose K, Nagasato K, Ohishi K, Tsujihata M, Nagataki S (1994) Production of granulocyte-macrophage colony stimulating factor by human T-lymphotropic virus type I-infected human glioma cells. J Neurol Sci 121:208–214

Ojukwu IC, Epstein LG (1998) Neurologic manifestations of infection with HIV. Pediatr Infect Dis J 17:343–344

Peluso R, Haase A, Stowring L, Edwards M, Ventura P (1985) A Trojan Horse mechanism for the spread of visna virus in monocytes. Virology 147:231–236

Pemberton LA, Kerr SJ, Smythe G, Brew BJ (1997) Quinolinic acid production by macrophages stimulated with IFN-gamma, TNF-alpha, and IFN-alpha. J Interferon Cytokine Res 17:589–595

Perry SW, Hamilton JA, Tjoelker LW, Dbaibo G, Dzenko KA, Epstein LG, Hannun Y, Whittaker JS, Dewhurst S, Gelbard HA (1998) Platelet-activating factor receptor activation. An initiator step in HIV-1 neuropathogenesis. J Biol Chem 273:17660–17664

Petito CK, Roberts B (1995) Evidence of apoptotic cell death in HIV encephalitis. Am J Pathol 146: 1121–1130
Phillips TR, Prosperogarcia O, Wheeler DW, Wagaman PC, Lerner DL, Fox HS, Whalen LR, Bloom FE, Elder JH, Henriksen SJ (1996) Neurologic dysfunction caused by a molecular clone of feline immunodeficiency virus, FIV-PPR. J Neurovirol 2:388–396
Podell M, Hayes K, Oglesbee M, Mathes L (1997) Progressive encephalopathy associated with CD4/CD8 inversion in adult FIV-infected cats. J AIDS Hum Retrovirol 15:332–340
Poeschla EM, Looney DJ (1998) CXCR4 is required by a nonprimate lentivirus: heterologous expression of feline immunodeficiency virus in human, rodent, and feline cells. J Virol 72:6858–6866
Poli G, Abramo F, Diiorio C, Cantile C, Carli MA, Pollera C, Vargo L, Tosoni A, Costanzi G (1997) Neuropathology in cats experimentally infected with feline immunodeficiency virus- a morphological, immunocytochemical and morphometric study. J Neurovirol 3:361–368
Power C, Moench T, Peeling J, Kong PA, Langelier T (1997) Feline immunodeficiency virus causes increased glutamate levels and neuronal loss in brain. Neuroscience 77:1175–1185
Price RW, Sidtis JJ, Brew BJ (1991) AIDS dementia complex and HIV-1 infection: a view from the clinic. Brain Pathol 1:155–162
Prospero-Garcia O, Herold N, Waters AK, Phillips TR, Elder JH, Henriksen SJ (1994) Intraventricular administration of a FIV-envelope protein induces sleep architecture changes in rats. Brain Res 659:254–258
Ransohoff RM, Glabinski A, Tani M (1996) Chemokines in immune-mediated inflammation of the central nervous system. Cytokine Growth Factor Rev 7:35–46
Rhodes RH (1993) Histopathologic features in the central nervous system of 400 acquired immunodeficiency syndrome cases: implications of rates of occurrence. Human Pathol 24:1189–1198
Rottman JB, Ganley KP, Williams K, Wu L, Mackay CR, Ringler DJ (1997) Cellular localization of the chemokine receptor CCR5. Correlation to cellular targets of HIV-1 infection. Am J Pathol 151: 1341–1351
Sanders VJ, Pittman CA, White MC, Wang GJ, Wiley CA, Achim CL (1998) Chemokines and receptors in HIV encephalitis. AIDS 12:1021–1026
Sasseville VG, Smith MM, Mackay CR, Pauley DR, Mansfield KG, Ringler DJ, Lackner AA (1996) Chemokine expression in simian immunodeficiency virus-induced AIDS encephalitis. Am J Pathol 149:1459–1467
Savio T, Levi G (1993) Neurotoxicity of HIV coat protein gp120, NMDA receptors, and protein kinase C: a study with rat cerebellar granule cell cultures. J Neuroscie Res 34:265–272
Sawaya BE, Thatikunta P, Denisova L, Brady J, Khalili K, Amini S (1998) Regulation of TNFalpha and TGFbeta-1 gene transcription by HIV-1 Tat in CNS cells. J Neuroimmunol 87:33–42
Sei Y, Kustova Y, Li Y, Morse HC, Skolnick P, Basile AS (1998) The encephalopathy associated with murine acquired immunodeficiency syndrome. Ann NY Acad Sci 840:822–834
Sei Y, Makino M, Vitkovic L, Chattopadhyay SK, Hartley JW, Arora PK (1992) Central nervous system infection in a murine retrovirus-induced immunodeficiency syndrome. J Neuroimmunol 37:131–140
Sei Y, Paul IA, Saito K, Layar R, Hartley JW, Morse HC, Skolnick P, Heyes MP (1996) Quinolinic levels in a murine retrovirus-induced immunodeficiency syndrome. J Neurochem 66:296–302
Sharer LR (1992) Pathology of HIV-1 infection of the central nervous system. A review. Neuropathol Appl Neurobiol 51:3–11
Silvotti L, Corradi A, Brandi G, Cabassi A, Bendinelli M, Magnan M, Piedimonte G (1997) FIV induced encephalopathy – early brain lesions in the absence of viral replication in monocyte/macrophages – a pathogenic model. Vet Immunol Immunopathol 55:263–271
Smith MO, Sutjipto S, Lackner AA (1994) Intrathecal synthesis of IgG in simian immunodeficiency virus (SIV)-infected rhesus macaques (Macaca mulatta). Aids Res Hum Retroviruses 10:81–89
Soontornniyomkij V, Nieto-Rodríguez JA, Martínez AJ, Kingsley LA, Achim CL, Wiley CA (1998) Brain HIV burden and length of survival after AIDS diagnosis. Clin Neuropathol 17:95–99
Stefano GB, Salzet M, Bilfinger TV (1998) Long-term exposure of human blood vessels to HIV gp120, morphine, and anandamide increases endothelial adhesion of monocytes: uncoupling of nitric oxide release. J Cardiovasc Pharmacol 31:862–868
Stout JC, Ellis RJ, Jernigan TL, Archibald SL, Abramson I, Wolfson T, McCutchan JA, Wallace MR, Atkinson JH, Grant I (1998) Progressive cerebral volume loss in human immunodeficiency virus infection: a longitudinal volumetric magnetic resonance imaging study. HIV Neurobehavioral Research Center Group. Arch Neurol 55:161–168

Strelow LI, Watry DD, Fox HS, Nelson JA (1998) Efficient infection of brain microvascular endothelial cells by an in vivo-selected neuroinvasive SIVmac variant. J Neurovirol 4:269–280

Talley AK, Dewhurst S, Perry SW, Dollard SC, Gummuluru S, Fine SM, New D, Epstein LG, Gendelman HE, Gelbard HA (1995) Tumor necrosis factor alpha-induced apoptosis in human neuronal cells: protection by the antioxidant N-acetylcysteine and the genes bcl-2 and crmA. Mol Cell Biol 15:2359–2366

Tracey I, Lane J, Chang I, Navia B, Lackner A, Gonzalez RG (1997) H-1 magnetic resonance spectroscopy reveals neuronal injury in a simian immunodeficiency virus macaque model. J AIDS Hum Retrovirol 15:21–27

Tyor WR, Glass JD, Griffin JW, Becker PS, McArthur JC, Bezman L, Griffin DE (1992) Cytokine expression in the brain during the acquired immunodeficiency syndrome. Ann Neurol 31:349–360

Tyor WR, Wesselingh SL, Griffin JW, McArthur JC, Griffin DE (1995) Unifying hypothesis for the pathogenesis of HIV-associated dementia complex, vacuolar myelopathy, and sensory neuropathy. [Review]. J Acquir Immune Def Syndr Hum Retrovirol 9:379–388

Umehara F, Okada Y, Fujimoto N, Abe M, Izumo S, Osame M (1998) Expression of matrix metalloproteinases and tissue inhibitors of metalloproteinases in HTLV-I-associated myelopathy. J Neuropathol Expl Neurol 57:839–849

Vallat A, Girolami UD, He J, Mhashilkar A, Marasco W, Shi B, Gray F, Bell J, Keohane C, Smith TW, Gabuzda D (1998) Localization of HIV-1 co-receptors CCR5 and CXCR4 in the brain of children with AIDS. Am J Pathol 152:167–178

Wahl SM, Allen JB, McCartney-Francis N, Morganti KM, Kossmann T, Ellingsworth L, Mai UE, Mergenhagen SE, Orenstein JM (1991) Macrophage- and astrocyte-derived transforming growth factor beta as a mediator of central nervous system dysfunction in acquired immune deficiency syndrome. J Exp Med 173:981–991

Wahl SM, Hunt DA, Wakefield LM, McCartney- Francis N, Wahl LM, Roberts AB, Sporn MB (1987) Transforming growth factor type beta induces monocyte chemotaxis and growth factor production. Proc Natl Acad Sci USA 84:5788–5792

Watanabe H, Nakamura T, Nagasato K, Shirabe S, Ohishi K, Ichinose K, Nishiura Y, Chiyoda S, Tsujihata M, Nagataki S (1995) Exaggerated messenger RNA expression of inflammatory cytokines in human T cell lymphotropic virus type 1-associated myelopathy. Arch Neurol 52:276–280

Weis S, Haug H, Budka H (1993) Neuronal damage in the cerebral cortex of AIDS brains: a morphometric study. Acta Neuropathol (Berl) 85:185–189

Weiss RA (1996) Retrovirus classification and cell interactions. J Antimicrob Chemo 37, Suppl. B:1–11

Wesselingh S, Takahashi K, Glass JD, McArthur JC, Griffin JW, Griffin DE (1997) Cellular localization of tumor necrosis factor mRNA in neurological tissue from HIV-infected patients by combined reverse transcriptase/polymerase chain reaction in situ hybridization and immunohistochemistry. J Neuroimmunol 74:1–8

Westmoreland SV, Halpern E, Lackner AA (1998a) Simian immunodeficiency virus encephalitis in rhesus macaques is associated with rapid disease progression. J Neurovirol 4:260–268

Westmoreland SV, Kolson D, González-Scarano F (1996) Toxicity of TNF alpha and platelet activating factor for human NT2N neurons: a tissue culture model for human immunodeficiency virus dementia. J Neurovirol 2:118–126

Westmoreland SV, Rottman JB, Williams KC, Lackner AA, Sasseville VG (1998b) Chemokine receptor expression on resident and inflammatory cells in the brain of macaques with simian immunodeficiency virus encephalitis. Am J Pathol 152:659–665

Wiley CA, Gardner M (1993) The pathogenesis of murine retroviral infection of the central nervous system. Brain Pathol 3:123–128

Wiley CA, Soontornniyomkij V, Radhakrishnan L, Masliah E, Mellors J, Hermann SA, Dailey P, Achim CL (1998) Distribution of brain HIV load in AIDS. Brain Pathol 8:277–284

Willett BJ, Flynn JN, Hosie MJ (1997) FIV infection of the domestic cat – an animal model for AIDS. Immunol Today 18:182–189

Yang LP, Morris GF, Wang ZD, Morris CB (1997) Repression of tumor necrosis factor-beta expression by the human immunodeficiency virus type-1 Tat protein in central nervous system-derived glial cells. Virus Res 50:195–203

Yu N, Billaud JN, Phillips TR (1998) Effects of feline immunodeficiency virus on astrocyte glutamate uptake: implications for lentivirus-induced central nervous system diseases. Proc Natl Acad Sci USA 95:2624–2629

Zou YR, Kottmann AH, Kuroda M, Taniuchi I, Littman DR (1998) Function of the chemokine receptor CXCR4 in haematopoiesis and in cerebellar development. Nature 393:595–599

Prion-Induced Neuronal Damage – The Mechanisms of Neuronal Destruction in the Subacute Spongiform Encephalopathies

A. GIESE* and H.A. KRETZSCHMAR

1 Prions and Prion Diseases	203
2 Pattern of Neuronal Cell Death in Prion Diseases and the Role of Apoptosis	205
3 Neurotoxicity of PrP^{Sc} – The "Gain of Function" Hypothesis	206
3.1 Neurotoxicity of PrP^{Sc} and PrP Fragments In Vitro	207
3.2 Neurotoxicity of PrP^{Sc} In Vivo	208
4 Function of PrP^{C} – The "Loss of Function" Hypothesis	210
4.1 Phenotype of PrP-Knockout Mice and the Function of PrP^{C}	210
4.2 PrP and Copper	211
4.3 PrP and the Response to Oxidative Stress	211
5 Summary	212
References	213

1 Prions and Prion Diseases

The transmissible spongiform encephalopathies, or prion diseases, constitute a group of transmissible, rapidly progressive, invariably fatal neurodegenerative diseases that can manifest as acquired, hereditary or idiopathic ("sporadic") disease. They include Creutzfeldt-Jakob disease in humans as well as scrapie and bovine spongiform encephalopathy (BSE) in animals, and are characterized by a long incubation period which may last up to decades after experimental or accidental transmission. The classic pathological features of prion diseases include spongiform change, gliosis and neuronal loss. In contrast to what is typically seen in infectious diseases caused by viruses, they lack a significant inflammatory response (PRUSINER 1993, 1998).

Prion diseases have received considerable scientific attention due to the unique properties of the transmissible agent, which has been termed prion (PRUSINER 1982). The infectious agent is very small and extremely resistant to treatments which destroy nucleic acids and inactivate conventional viruses (ALPER et al. 1966; BROWN et al. 1982; PRUSINER 1982), but it is susceptible to treatments which

---

Institute of Neuropathology, Ludwig-Maximilians-Universität, Marchioninistr. 17, 81377 München, Germany
*e-mail: Armin.Giese@inp.med.uni-muenchen.de

denature proteins. Attempts to purify the infectious agent yielded fractions highly enriched for a hitherto unknown protein which has been named prion protein (PrP) (BOLTON et al. 1982; PRUSINER et al. 1983, 1984; OESCH et al. 1985). No agent-specific nucleic acid has been found in these preparations (KELLINGS et al. 1992, 1993; RIESNER et al. 1993); rather, the prion protein is encoded in the host genome (OESCH et al. 1985; CHESEBRO et al. 1985; BASLER et al. 1986). In the brains of affected individuals, a pathognomonic accumulation of a specific disease-associated isoform of the prion protein, termed PrP^{Sc}, is found (Fig. 1). PrP^{Sc} is derived through an ill-defined post-translational process involving conformational changes from the normal cellular isoform of the prion protein (PrP^C) (PRUSINER 1998). PrP^C and PrP^{Sc} have the same amino acid sequence (STAHL et al. 1993), however, they differ in regard to conformation (PAN et al. 1993). PrP^{Sc} can be distinguished from PrP^C by its high content in β-sheet structure (PAN et al. 1993), its tendency to form large aggregates (PRUSINER et al. 1983), and its partial resistance to digestion with proteinase K (BOLTON et al. 1982, 1984; OESCH et al. 1985; MEYER et al. 1986).

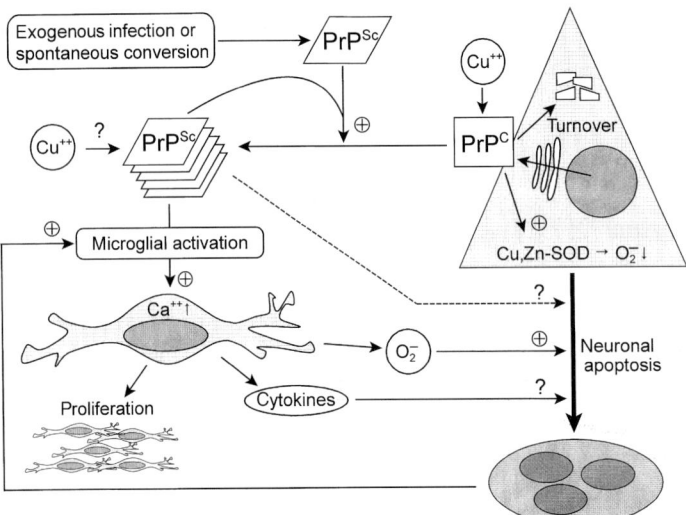

Fig. 1. Pathogenesis of neuronal cell death in prion disease. The normal cellular form of the prion protein (PrP^C) is a copper-binding glycoprotein located mainly at the cell surface of neurons, which has a rapid turnover rate and affects the activity of the antioxidant enzyme Cu,Zn-superoxide dismutase ($Cu,Zn\text{-}SOD$) (*upper right*). PrP^C is converted into the aggregated disease-specific isoform, termed PrP^{Sc}, in the presence of PrP^{Sc} by an autocatalytic mechanism involving conformational changes. This process may be initiated by exogenous infection or spontaneous generation of PrP^{Sc} (*upper left*). PrP^{Sc} activates microglial cells which in turn proliferate and release cytokines and reactive oxygen species such as superoxide (O_2^-) (*lower left*). Microglial cells and oxidative stress have been shown to be essential in mediating neuronal cell death in cell culture models of prion disease. Additionally, resistance to oxidative stress may be reduced due to impairment of PrP^C function. Only limited data are available on direct effects of PrP^{Sc} on neuronal cells. Potential effects include increased formation of an aberrant transmembrane form of PrP, termed ^{Ctm}PrP, and changes in plasma membrane properties (*dashed arrow*). Neuronal cell death occurs by apoptosis and contributes to microglial activation in a positive feedback loop (*lower right*)

PrP^{Sc} is the only molecule consistently present in infectious preparations and is believed to be the essential if not the only component of the prion (PRUSINER 1998). PrP^{Sc} appears to replicate by converting PrP^{C} synthesized by the infected cells into the PrP^{Sc} isoform in an autocatalytic fashion (PRUSINER 1982, 1998; KOCISKO et al. 1994, 1995).

2 Pattern of Neuronal Cell Death in Prion Diseases and the Role of Apoptosis

The clinical picture of prion diseases is characterized by severe and progressive neurological deficits (KRETZSCHMAR et al. 1996), which are most likely caused by a combination of synaptic pathology (HOGAN et al. 1987; JOHNSTON et al. 1997; JEFFREY et al. 1997; FERRER et al. 1999) and neuronal loss (MASTERS and RICHARDSON 1978; SCOTT and FRASER 1984). Neuronal loss is a salient feature in the pathology of prion diseases and can lead to gross atrophy of the brain. The pattern of neuronal loss is variable and depends on the species and genetic background of the host as well as on the strain of the agent (FRASER 1993; PARCHI et al. 1999). It is important to note that even in syngenic inbred mice different prion strains can be distinguished which are stable upon serial passage. These strains may vary in regard to incubation time, pattern of PrP^{Sc} deposition, lesion profile and pattern of neuronal loss (FRASER and DICKINSON 1973; BRUCE et al. 1976, 1994; BRUCE 1993; FRASER 1993). In terms of the "protein only" hypothesis of prion diseases, these strain properties have to be encoded by PrP^{Sc}. In fact, biochemical analysis of PrP^{Sc} associated with different prion strains has shown that there are differences in regard to the size of the protease-resistant fragment of PrP^{Sc} (BESSEN and MARSH 1994), which can most plausibly be attributed to differences in conformation, and differences in regard to glycosylation pattern (KASCSAK et al. 1986; SOMMERVILLE et al. 1997). Such differences have also been found in the analysis of different variants of human prion disease (PARCHI et al. 1996, 1997, 1999; COLLINGE et al. 1996). Further evidence that different prion strains are linked to conformational differences in PrP^{Sc} comes from recent studies based on infrared spectroscopy (CAUGHEY et al. 1998) and conformation-dependent antibody binding (SAFAR et al. 1998). Different strain-dependent lesion profiles in syngenic experimental animals suggest that the pattern of neuronal loss is not merely due to inherent differences in the susceptibility of different subsets of neurons to the disease process, but is also influenced by different strain-specific targeting of different neuronal populations.

Regarding the mechanisms of neuronal cell death in prion diseases, it has been shown that apoptosis plays a major role (GIESE et al. 1995; LUCASSEN et al. 1995). In contrast to necrosis, which typically results from severe and sudden injury and which is characterized by rapid cell lysis and a consecutive inflammatory response, apoptosis proceeds in an orderly manner following an active cellular suicide program (KERR et al. 1972; SEARLE et al. 1982; RAFFRAY and COHEN 1997). In general, a key feature of apoptosis is cleavage of nuclear DNA into oligonucleosome-length fragments by endonucleases (WYLLIE 1980; WYLLIE et al. 1984). Morphologically, apoptosis is characterized by condensation of nuclear chromatin, cellular and nuclear shrinkage and formation of apoptotic bodies. Usually, it is not accompanied by an inflammatory response (SEARLE et al. 1982).

Intrigued by the observation that apoptosis of neuronal cells can be induced in vitro by the addition of PrP^{Sc} (MÜLLER et al. 1993) or neurotoxic peptide fragments of PrP (FORLONI et al. 1993), we were interested in studying the potential role of apoptosis in prion diseases in vivo. Therefore, we performed a time course study on scrapie-infected mice using the in situ end-labeling technique (ISEL), which is also often referred to as the TUNEL assay. This technique is based on the incorporation of labeled nucleotides into fragmented DNA by enzymes such as terminal transferase and it can be used in histological sections (GAVRIELI et al. 1992; GOLD et al. 1993). In the brains and retinae of mice infected with the 79A strain of scrapie, labeled nuclei were detected in a well-defined temporal and spatial pattern during the second half of the incubation period and in clinically ill mice. In contrast, virtually no labeling was found in the first half of the incubation time and in mock-infected control animals at all time points (GIESE et al. 1995). ISEL-positive nuclei were also found in terminally ill mice infected with a number of other scrapie strains (LUCASSEN et al. 1995; GIESE et al. 1998). Subsequently, ISEL-positive neuronal nuclei have also been reported in human prion disease (DORANDEU et al. 1998; GRAY et al. 1999; FERRER 1999). It is well-known that the ISEL assay is not absolutely specific, especially when used in autopsy material, since random DNA fragmentation is also observed in necrosis or autolysis. Whereas the latter can be ruled out in adequately fixed tissues obtained from experimental animals (SCHALLOCK et al. 1997), further analysis has to be included to rule out necrosis. Circumstantial evidence for apoptotic cell death in prion disease includes the scattered distribution of labeled nuclei, nuclear condensation, and the well-known lack of inflammation (GIESE et al. 1995). In addition, the presence of cells showing all morphological features of apoptosis was demonstrated by electron microscopy in a number of studies (GIESE et al. 1995; WILLIAMS et al. 1997a). Furthermore, quantitative analysis of the number and percentage of ISEL-positive nuclei in the retina (GIESE et al. 1995) and in the hippocampus (WILLIAMS et al. 1997a) of scrapie-infected mice in relation to the time course of neuronal loss clearly shows that apoptosis is the principal mechanism of neuronal cell death in prion disease.

3 Neurotoxicity of PrP^{Sc} – The "Gain of Function" Hypothesis

Whereas it is now well-established that neuronal cell death in prion diseases is due to apoptosis, the mechanisms that lead to apoptosis are still a subject of debate. Basically, there are two main concepts, which are by no means mutually exclusive. On the one hand, the accumulation of PrP^{Sc} during the disease process may be directly or indirectly neurotoxic. This concept has also been termed the "gain of function" hypothesis. Alternatively, the conversion of the physiological cellular isoform of the prion protein (PrP^C) into PrP^{Sc} may reduce the amount of functional PrP^C available, thereby affecting cell survival – the "loss of function" hypothesis (KRETZSCHMAR et al. 1997).

3.1 Neurotoxicity of PrP^{Sc} and PrP Fragments In Vitro

PrP^{Sc} has been shown to be toxic to primary neuronal cell cultures in experiments using cultured rat cortical cells and purified PrP^{Sc} incorporated in liposomes (MÜLLER et al. 1993), and in experiments using primary mouse cerebellar cultures and direct addition of purified PrP^{Sc} (GIESE et al. 1998). Cell death in these experiments was shown to be due to apoptosis (MÜLLER et al. 1993) and was specific in that it was limited to cells derived from animals capable of expressing PrP (GIESE et al. 1998). In contrast, tumor cell lines such as N2a cells and PC12 cells persistently infected with prions can produce readily detectable amounts of PrP^{Sc} as well as infectivity without showing overt signs of cytotoxicity.

Most cell culture experiments addressing the neurotoxic mechanism of PrP^{Sc} have been performed with a synthetic peptide fragment corresponding to amino acids 106–126 of human PrP as a model system. The neurotoxicity of PrP106–126 was first described by Forloni and co-workers in a systematic analysis of different parts of the PrP peptide sequence in regard to neurotoxic properties (FORLONI et al. 1993). This specific peptide fragment has never been found to be formed in vivo. However, amino acids 106–126 (KTNMKHMAGAAAAGAVVGGLG) are contained in the protease-resistant core of PrP^{Sc}. In contrast, this sequence serves as a physiological cleavage site in the cellular processing of PrP^C. As part of the normal turnover of PrP^C, cleavage by an as yet unknown protease takes place between the positively charged N-terminal part and the hydrophobic C-terminal part of this sequence (PAN et al. 1992; HARRIS et al. 1993). Similar to PrP^{Sc}, PrP106–126 can form fibrillar aggregates rich in β-sheet-like structures (SELVAGGINI et al. 1993; DE GIOIA et al. 1994). However, PrP106–126 does not seem to be infectious, making it possible to separate infectivity and toxicity experimentally.

PrP106–126 has been shown to be toxic to cultured hippocampal neurons (FORLONI et al. 1993), cortical neurons (BROWN et al. 1994), and cerebellar neurons (BROWN et al. 1996) at a concentration of 80µM. This toxicity appears to be a specific effect, since PrP106–126 is not toxic to cultured neurons derived from PrP-knockout mice (BROWN et al. 1994, 1996). Additionally, a peptide of similar amino acid composition, however in a randomly scrambled sequence, is not toxic under these experimental conditions (FORLONI et al. 1993; BROWN et al. 1994, 1996). As a direct effect of PrP106–126 exposure, changes in plasma membrane microviscosity have been described (SALMONA et al. 1997), and it has been suggested that PrP106–126 may form large, non-selective ion-permeable channels (LIN et al. 1997). However, in cell culture experiments, it was shown that the toxic effect of PrP106–126 requires the presence of microglia (BROWN et al. 1996): primary neuronal cell cultures always contain a small percentage of contaminating astroglial and microglial cells. Pretreatment of these cell cultures with L-leucine methyl ester (LLME) effectively removes microglial cells from these cultures and, at the same time, blocks the toxicity of PrP106–126. Toxicity can be restored by adding microglial cells to a co-culture system. With this co-culture approach, the role of PrP^C-expression on neurons and microglia could be analyzed by using cell cultures prepared from either PrP-knockout mice or wild-type mice. PrP106–126 appears not to be toxic to neurons derived from PrP-knockout mice even after addition of wild-type microglia, whereas microglial cells derived from knockout mice can restore the toxicity of PrP106–126 in microglia-depleted wild-type cultures, albeit at a slightly lower level. Therefore, it can be concluded that the neurotoxic effect of PrP106–126 requires both neuronal expression of PrP^C and the presence of microglia (BROWN et al. 1996). This dual requirement has also been found in the analysis of the toxic effect of purified PrP^{Sc} on cerebellar cell cultures (GIESE et al. 1998).

In microglial cells, addition of PrP106–126 to the culture medium causes a rise in intracellular calcium concentration. This response can be found in microglia derived from wild-type and PrP-knockout mice, but it is somewhat more pronounced in wild-type cells (HERMS et al. 1997). In addition, tyrosine kinase-dependent inflammatory signal transduction cascades can be activated (COMBS et al. 1999) and the production of pro-inflammatory cytokines can be induced in microglia cells treated with PrP106–126 (PEYRIN et al. 1999). Microglial cells stimulated with PrP106–126 secrete increased amounts of reactive oxygen species (ROS) such as superoxide into the culture medium. Similar to the rise in intracellular calcium and the neurotoxic effect, this response is stronger in wild-type cells than in microglia derived from PrP-knockout mice (BROWN et al. 1996). The secretion of ROS appears to be essential for the mediation of neurotoxicity, since the toxic effect of PrP106–126 can be blocked by the addition of antioxidants to the culture medium (BROWN et al. 1996). Additionally, changes in the level of glutathione found after addition of PrP106–126 suggest an important role of oxidative stress (PEROVIC et al. 1997).

3.2 Neurotoxicity of PrP^{Sc} In Vivo

Deposition of PrP^{Sc} is the pathological hallmark of prion diseases. In the brains of terminally ill, experimentally infected animals, concentrations of several micrograms of PrP^{Sc} per gram of brain tissue are found (BEEKES et al. 1996). This corresponds well to the concentrations found to be toxic in cell culture experiments (GIESE et al. 1998). Regarding evidence of PrP^{Sc} neurotoxicity in vivo, it is well-established that, in general, the pattern and time course of PrP^{Sc} deposition corresponds to the pattern of neurodegeneration (DEARMOND et al. 1987; JENDROSKA et al. 1991). Since it appears that PrP^{Sc} toxicity in vitro depends on cellular expression of PrP^C and on the presence and activation of microglia cells, the question of whether this also applies to the in vivo situation has to be addressed.

It is well-known that PrP-knockout mice are resistant to scrapie and accumulate neither infectivity nor PrP^{Sc} after inoculation (BÜELER et al. 1993; SAILER et al. 1994). This is consistent with the "protein only" hypothesis of prion propagation, since no PrP^C is available for conversion into PrP^{Sc} in these animals. However, grafts of neuroectodermal tissue, derived from mice expressing PrP^C, that have been transplanted into the brains of PrP-knockout mice can be infected. In these grafts an accumulation of PrP^{Sc} and severe neurodegeneration can be found. PrP^{Sc} is also deposited in the surrounding brain tissue of the PrP-knockout host animals in this experimental setup. However, no signs of neurodegeneration and no neuronal loss were found in the host brains, suggesting that also in vivo PrP-knockout neurons remain healthy when exposed to a continuous source of PrP^{Sc} (BRANDTNER et al. 1996). Furthermore, it has been shown that comparatively low levels of PrP^{Sc} are found in transgenic mice overexpressing PrP^C when they become ill after infection with scrapie (FISCHER et al. 1996), whereas PrP^{Sc} seems to be less toxic for mice expressing reduced amounts of PrP^C (BÜELER et al. 1994). This is consistent with the requirement that PrP^C is expressed for the toxic effect of PrP^{Sc} and PrP fragments in vitro. However, the level of PrP^C expression does not seem to be the only factor affecting susceptibility to PrP^{Sc}-induced cell death, especially when comparing different types of neurons. For example, Purkinje cells in the cerebellum have been shown to strongly express PrP (KRETZSCHMAR et al. 1986), yet they are comparatively resistant to cell death in prion disease. In contrast, other GABAergic neuronal subpopulations such as parvalbumin-positive cortical neurons appear to be particularly affected in experimental scrapie (GUENTCHEV et al. 1998). The reason for these differences in susceptibility of different cell types has not been clarified yet.

A further unresolved issue related to the requirement of PrP^C expression for the disease process is the potential role of a transmembrane form of PrP, termed ^{Ctm}PrP. In addition to the fully translocated topological form of PrP^C, two different transmembrane forms which span the membrane with the central hydrophobic stretch of the protein have been described as forming in in-vitro-translation systems (YOST et al. 1990; LOPEZ et al. 1990; HEGDE et al. 1998). In wild-type animals, these aberrant topological forms of PrP^C could not be detected. However, mutations in the central part of PrP causing an increase in the percentage of ^{Ctm}PrP formed in vitro have been shown to cause neurodegeneration in transgenic mice in the presence of detectable levels of ^{Ctm}PrP (HEGDE et al. 1998). The diseases caused by these mutations do not seem to be prion diseases, since they are not transmissible and no PrP^{Sc} is found. However, it has been suggested that increased formation of ^{Ctm}PrP caused by the accumulation of PrP^{Sc} may contribute to the neuronal cell death observed in prion diseases (HEGDE et al. 1999).

As far as the role of microglial activation in vivo is concerned, it has been well-documented that there is an intense microglial reaction in mice terminally ill with scrapie, which corresponds to the pattern of neurodegeneration (WILLIAMS et al. 1994a). The question of whether microglial activation is secondary to neuronal cell death or whether microglial activation precedes neuronal cell death and is possibly involved in inducing it has been addressed in time course studies. Early activation of microglia preceding neuronal loss has been described in the hippocampus of VM mice infected with scrapie strain 310V (WILLIAMS et al. 1997a). In a large time course study using three different strains of scrapie in mice, it was shown that microglial activation was present early in the incubation period in all models studied. The pattern of microglial activation closely paralleled the pattern and time course of PrP^{Sc} accumulation. Microglial activation clearly preceded apoptotic neuronal cell death in all strains studied. Furthermore, quantitative analysis of microglial activation and cell death in the cerebellum showed that the time course and extent of neuronal cell death correlated with the time course of microglial activation (GIESE et al. 1998). Taken together with the data derived from cell culture studies, this indicates that microglial activation is also involved in mediating PrP^{Sc} toxicity in vivo. With regard to the role of oxidative stress in vivo, there are indications of increased oxidative stress in scrapie-infected animals such as changes in the level of both total ubiquinone and its reduced form in the brain (GUAN et al. 1996). Additionally, it has been suggested that secretion of cytokines by glial cells contributes to the disease process, since immunohistochemical and molecular genetic studies have shown an increased expression of some pro-inflammatory cytokines such as tumor necrosis factor (TNF)-α and interleukin (IL)-1β in scrapie-infected animals (WILLIAMS et al. 1994b, 1997b; CAMPBELL et al. 1994). However, the significance of these findings in regard to neuronal cell death remains to be demonstrated.

4 Function of PrP^C – The "Loss of Function" Hypothesis

4.1 Phenotype of PrP-Knockout Mice and the Function of PrP^C

While the role played by PrP^{Sc} in the pathogenesis of prion diseases has been analyzed in great detail, the normal cellular function of PrP^C still remains to be established. PrP^C is a cell surface glycoprotein which is attached to the outer leaflet of the plasma membrane by a glycosylphosphatidylinositol (GPI) anchor (STAHL et al. 1987, 1990; BALDWIN et al. 1990; CAUGHEY and RAYMOND 1991). PrP^C is encoded in the genome of all mammals investigated to date and shows a high degree of evolutionary conservation (SCHÄTZL et al. 1995; KRAKAUER et al. 1998), indicating an essential functional role. However, PrP-knockout mice, which lack expression of PrP^C and are resistant to infection with experimental scrapie, do not show an obvious phenotype attributable to lack of PrP^C (BÜELER et al. 1992; LIPP et al. 1998). Whether this is due to compensatory mechanisms in these mice remains to be established.

Electrophysiological differences regarding long-term potentiation in the hippocampus and GABAergic transmission (COLLINGE et al. 1994) as well as changes in circadian rhythm (TOBLER et al. 1996) have been described in some lines of PrP-knockout mice. However, expression of PrP^C, albeit particularly high in neurons, is not limited to the nervous system, making a purely synaptic function unlikely (KRETZSCHMAR et al. 1986; CASHMAN et al. 1990; BENDHEIM et al. 1992; MANSON et al. 1992).

4.2 PrP and Copper

The amino terminus of PrP contains the octapeptide sequence PHGGGWGQ, which is repeated four times and is among the best-preserved regions of mammalian PrP. Biophysical studies of synthetic peptides based on this octameric repeat region suggested that this region of the molecule may serve as a copper-binding domain (HORNSHAW et al. 1995). A highly specific, multivalent and cooperative binding of copper has been described in studies using recombinant PrP (BROWN DR et al. 1997a; STÖCKEL et al. 1998; VILES et al. 1999). A physiological function of this copper-binding property of PrP^C has been suggested based on experimental data showing a reduction in synaptosomal copper content and altered electrophysiological responses in the presence of excess copper in PrP-knockout mice (BROWN et al. 1997a), a phenotype which was also found in transgenic mice expressing mutant PrP lacking the copper-binding domain and which could be rescued by transgenic expression of full-length PrP (HERMS et al. 1999). Further evidence for a role of PrP^C in cerebral copper metabolism comes from the finding that the activity of the copper-dependent cytoplasmic enzyme Cu,Zn-superoxide dismutase (Cu,Zn-SOD) is significantly reduced in the brains of PrP-knockout mice (BROWN DR et al. 1997b).

A disturbance in cerebral copper homeostasis may well contribute to neurodegeneration. The activity of Cu,Zn-SOD depends on the level of available copper (MORGAN and O'DELL 1977; PROHASKA 1983). An inhibition of Cu, Zn-SOD has been shown to be able to induce apoptotic neuronal cell death (ROTHSTEIN et al. 1994). Other copper-dependent enzymes in the brain are involved in the synthesis of neurotransmitters such as norepinephrine (LINDER and HAZEGH-AZAM 1996). Consequently, copper deficiency impairs norepinephrine synthesis (MORGAN and O'DELL 1977; PROHASKA and SMITH 1982), and neurodegenerative changes have been described in animals fed a copper-deficient diet (CARLTON and KELLY 1969) and in animals treated with the copper chelator cuprizone (CARLTON 1969). Interestingly, the neuropathological changes in the latter have been suggested to be similar to the changes found in scrapie (PATTISON and JEBETT 1971); however, this opinion is not shared by us. At the moment, no comprehensive data on copper metabolism in scrapie-infected animals are available.

4.3 PrP and the Response to Oxidative Stress

In cell culture experiments, it has been shown that primary neuronal cell cultures derived from PrP-knockout mice show a somewhat reduced viability as compared to wild-type cells, which is most likely due to a reduced resistance to oxidative stress. This finding can be explained by the reduced activity of the antioxidant enzyme Cu,Zn-SOD in PrP-knockout mice (BROWN et al. 1997b). Recently, it has even been suggested that PrP^C itself may function as a superoxide dismutase (BROWN et al. 1999). Therefore, it is tempting to speculate that – in addition to inducing a microglia-mediated increase in the level of reactive oxygen species – the formation of PrP^{Sc} may reduce the cellular resistance towards oxidative stress by an impairment of PrP^C function. However, whereas a reduction in the activity of Cu,Zn-SOD in cell culture experiments using PrP106–126 has been described (BROWN et al. 1997b), it remains to be investigated if this also applies to the disease process in vivo.

5 Summary

Prion diseases are characterized by the accumulation of a specific disease-associated isoform of the prion protein (PrP), termed PrP^{Sc}, which is the main, if not the only, component of the infectious agent termed prion. PrP^{Sc} is derived by an autocatalytic post-translational process involving conformational changes from the normal host-encoded isoform of the prion protein, termed PrP^C. PrP^C is a copper-binding glycoprotein attached to the cell membrane of neurons and other cells by means of a GPI anchor. The pattern of neurodegeneration differs between variants of prion disease and is related to the pattern of PrP^{Sc} deposition and differences in susceptibility of different cell types to the disease process. The pattern of PrP^{Sc} deposition depends on the strain of the agent and the PrP genotype of the host. Strain properties of prions appear to be related to different pathological conformations of PrP^{Sc}.

Neuronal cell death is a salient feature in the pathology of prion diseases. Histological and electron microscopical studies have shown that cell death in prion disease occurs by apoptosis. Apoptosis of neuronal cells can also be induced in vitro by exposure to PrP^{Sc} or a neurotoxic peptide fragment corresponding to amino acids 106–126 of human prion protein (PrP106–126). Both in vitro and in vivo, the toxicity of PrP^{Sc} and PrP fragments appears to depend on neuronal expression of PrP^C and on microglial activation. Activated microglial cells release pro-inflammatory cytokines and reactive oxygen species. Cell culture experiments suggest an important role of microglia-mediated oxidative stress in the induction of neuronal cell death. Only limited data are available on direct effects of PrP^{Sc} on neuronal cells. Potential effects include increased formation of an aberrant transmembrane form of PrP, termed ^{Ctm}PrP, and changes in plasma membrane properties. In addition to direct and indirect toxic effects of PrP^{Sc}, a loss of function of PrP^C may contribute to neuronal cell death. Potential mechanisms include disturbances in cerebral copper metabolism and antioxidative defense mechanisms. A better understanding of the pathogenesis of neuronal cell death in prion diseases may also have important therapeutic implications in the future.

References

Alper T, Haig DA, Clarke MC (1966) The exceptionally small size of the scrapie agent. Biochem Biophys Res Commun 22:278–284

Baldwin MA, Stahl N, Reinders LG, Gibson BW, Prusiner SB, Burlingame AL (1990) Permethylation and tandem mass spectrometry of oligosaccharides having free hexosamine: analysis of the glycoinositol phospholipid anchor glycan from the scrapie prion protein. Anal Biochem 191:174–182

Basler K, Oesch B, Scott M, Westaway D, Walchli M, Groth DF, McKinley MP, Prusiner SB, Weissmann C (1986) Scrapie and cellular PrP isoforms are encoded by the same chromosomal gene. Cell 46:417–428

Beekes M, Baldauf E, Diringer H (1996) Sequential appearance and accumulation of pathognomonic markers in the central nervous system of hamsters orally infected with scrapie. J Gen Virol 77:1925–1934

Bendheim PE, Brown HR, Rudelli RD, Scala LJ, Goller NL, Wen GY, Kascsak RJ, Cashman NR, Bolton DC (1992) Nearly ubiquitous tissue distribution of the scrapie agent precursor protein. Neurology 42:149–156

Bessen RA, Marsh RF (1994) Distinct PrP properties suggest the molecular basis of strain variation in transmissible mink encephalopathy. J Virol 68:7859–7868

Bolton DC, McKinley MP, Prusiner SB (1982) Identification of a protein that purifies with the scrapie prion. Science 218:1309–1311

Bolton DC, McKinley MP, Prusiner SB (1984) Molecular characteristics of the major scrapie prion protein. Biochemistry 23:5898–5906

Brandtner S, Isenmann S, Raeber A, Fischer M, Sailer A, Kobayashi Y, Marino S, Weissmann C, Aguzzi A (1996) Normal host prion protein necessary for scrapie-induced neurotoxicity. Nature 379:339–343

Brown DR, Herms J, Kretzschmar HA (1994) Mouse cortical cells lacking cellular PrP survive in culture with a neurotoxic PrP fragment. Neuroreport 5:2057–2060

Brown DR, Schmidt B, Kretzschmar HA (1996) Role of microglia and host prion protein in neurotoxicity of a prion protein fragment. Nature 380:345–347

Brown DR, Quin K, Herms JW, Madlung A, Manson J, Strome R, Fraser PE, Kruck T, von Bohlen A, Schulz-Schaeffer W, Giese A, Westaway D, Kretzschmar HA (1997a) The cellular prion protein binds copper in vivo. Nature 390:684–687

Brown DR, Schulz-Schaeffer WJ, Schmidt B, Kretzschmar HA (1997b) Prion protein-deficient cells show altered response to oxidative stress due to decreased SOD-1 activity. Exp Neurol 146:104–112

Brown DR, Wong BS, Hafiz F, Clive C, Haswell SJ, Jones IM (1999) Normal prion protein has an activity like that of superoxide dismutase. Biochem J 344:1–5

Brown P, Gibbs CJ, Amyx HL, Kingsbury DT, Rohwer RG, Sulima MP, Gajdusek DC (1982) Chemical disinfection of Creutzfeldt-Jakob disease virus. N Engl J Med 306:1279–1282

Bruce ME (1993) Scrapie strain variation and mutation. Brit Med Bull 49:822–838

Bruce ME, Dickinson AG, Fraser H (1976) Cerebral amyloidosis in scrapie in the mouse: effect of agent strain and mouse genotype. Neuropathol Appl Neurobiol 2:471–478

Bruce ME, McBride PA, Jeffrey M, Scott JR (1994) PrP in pathology and pathogenesis in scrapie-infected mice. Mol Neurobiol 8:105–112

Büeler H, Fischer M, Lang Y, Bluethmann H, Lipp HP, DeArmond SJ, Prusiner SB, Aguet M, Weissmann C (1992) Normal development and behaviour of mice lacking the neuronal cell-surface PrP protein. Nature 356:577–582

Büeler H, Aguzzi A, Sailer A, Greiner RA, Autenried P, Aguet M, Weissmann C (1993) Mice devoid of PrP are resistant to scrapie. Cell 73:1339–1347

Büeler H, Raeber A, Sailer A, Fischer M, Aguzzi A, Weissmann C (1994) High prion and PrP^{Sc} levels but delayed onset of disease in scrapie-inoculated mice heterozygous for a disrupted PrP gene. Mol Med 1:19–30

Campbell IL, Eddleston M, Kemper P, Oldstone MB, Hobbs MV (1994) Activation of cerebral cytokine gene expression and its correlation with onset of reactive astrocyte and acute-phase response gene expression in scrapie. J Virol 68:2383–2387

Carlton WW (1969) Spongiform encephalopathy induced in rats and guinea pigs by cuprizone. Exp Mol Pathol 10:274–287

Carlton WW, Kelly WA (1969) Neural lesions in the offspring of female rats fed a copper-deficient diet. J Nutr 97:42–52

Cashman NR, Loertscher R, Nalbantoglu J, Shaw I, Kascsak RJ, Bolton DC, Bendheim PE (1990) Cellular isoform of the scrapie agent protein participates in lymphocyte activation. Cell 61:185–192

Caughey B, Raymond GJ (1991) The scrapie-associated form of PrP is made from a cell surface precursor that is both protease- and phospholipase-sensitive. J Biol Chem 266:18217–18223

Caughey B, Raymond GJ, Bessen RA (1998) Strain-dependent differences in beta-sheet conformations of abnormal prion protein. J Biol Chem 273:32230–32235

Chesebro B, Race R, Wehrly K, Nishio J, Bloom M, Lechner D, Bergstrom S, Robbins K, Mayer L, Keith JM, et al. (1985) Identification of scrapie prion protein-specific mRNA in scrapie-infected and uninfected brain. Nature 315:331–333

Collinge J, Whittington MA, Sidle KCL, Smith CJ, Palmer MS, Clarke AR, Jefferys JGR (1994) Prion protein is necessary for normal synaptic function. Nature 370:295–297

Collinge J, Sidle KCL, Meads J, Ironside J, Hill AF (1996) Molecular analysis of prion strain variation and the aetiology of "new variant" CJD. Nature 383:685–690

Combs CK, Johnson DE, Cannady SB, Lehman TM, Landreth GE (1999) Identification of microglial signal transduction pathways mediating a neurotoxic response to amyloidogenic fragments of beta-amyloid and prion proteins. J Neurosci 19:928–939

DeArmond SJ, Mobley WC, DeMott DL, Barry RA, Beckstead JH, Prusiner SB (1987) Changes in the localization of brain prion proteins during scrapie infection. Neurology 37:1271–1280

De Gioia L, Selvaggini C, Ghibaudi E, Diomede L, Bugiani O, Forloni G, Tagliavini F, Salmona F (1994) Conformational polymorphism of the amyloidogenic and neurotoxic peptide homologous to residues 106–126 of the prion protein. J Biol Chem 269:7859–7862

Dorandeu A, Wingertsmann L, Chretien F, Delisle MB, Vital C, Parchi P, Montagna P, Lugaresi E, Ironside JW, Budka H, Gambetti P, Gray F (1998) Neuronal apoptosis in fatal familial insomnia. Brain Pathol 8:531–537

Ferrer I (1999) Nuclear DNA fragmentation in Creutzfeldt-Jakob disease: does a mere positive in situ nuclear end labeling indicate apoptosis? Acta Neuropathol 97:5–12

Ferrer I, Rivera R, Blanco R, Marti E (1999) Expression of proteins linked to exocytosis and neurotransmission in patients with Creutzfeldt-Jakob disease. Neurobiol Dis 6:92–100

Fischer M, Rülicke T, Raeber A, Sailer A, Moser M, Oesch B, Brandtner S, Aguzzi A, Weissmann C (1996) Prion protein (PrP) with amino-terminal deletions restoring susceptibility of PrP knockout mice to scrapie. EMBO J 15:1255–1264

Forloni G, Angeretti N, Chiesa R, Monzani E, Salmona M, Bugiani O, Tagliavini F (1993) Neurotoxicity of a prion protein fragment. Nature 362:543–546

Fraser H (1993) Diversity in the neuropathology of scrapie-like diseases in animals. Brit Med Bull 49:792–809

Fraser H, Dickinson AG (1973) Scrapie in mice – agent-strain differences in the distribution and intensity of grey matter vacuolation. J Comp Pathol 83:29–40

Gavrieli Y, Sherman Y, Ben-Sasson SA (1992) Identification of programmed cell death in situ via specific labeling of nuclear DNA fragmentation. J Cell Biol 119:493–501

Giese A, Groschup MH, Hess B, Kretzschmar HA (1995) Neuronal cell death in scrapie-infected mice is due to apoptosis. Brain Pathol 5:213–221

Giese A, Brown DR, Groschup MH, Feldmann C, Haist I, Kretzschmar HA (1998) Role of microglia in neuronal cell death in prion disease. Brain Pathol 8:449–457

Gold R, Schmied M, Rothe G, Zischler H, Breitschopf H, Weckerle H, Lassmann H (1993) Detection of DNA fragmentation in apoptosis: application of in situ nick translation to cell culture systems and tissue sections. J Histochem Cytochem 41:1023–1030

Gray F, Chretien F, Adle-Biassette H, Dorandeu A, Ereau T, Delisle MB, Kopp N, Ironside JW, Vital C (1999) Neuronal apoptosis in Creutzfeldt-Jakob disease. J Neuropathol Exp Neurol 58:321–328

Guan Z, Soderberg M, Sindelar P, Prusiner SB, Kristensson K, Dallner G (1996) Lipid composition in scrapie-infected mouse brain: prion infection increases the levels of dolichyl phosphate and ubiquinone. J Neurochem 66:277–285

Guentchev M, Groschup MH, Kordek R, Liberski PP, Budka H (1998) Severe, early and selective loss of a subpopulation of GABAergic inhibitory neurons in experimental transmissible spongiform encephalopathies. Brain Pathol 8:615–623

Harris DA, Huber MT, van Dijken P, Shyng S-L, Chait BT, Wang R (1993) Processing of a cellular prion protein: identification of N- and C-terminal cleavage sites. Biochemistry 32:1009–1016

Hegde RS, Mastrianni JA, Scott MR, DeFea KA, Tremblay P, Torchia M, DeArmond SJ, Prusiner SB, Lingappa VR (1998) A transmembrane form of the prion protein in neurodegenerative disease. Science 279:827–834

Hegde RS, Tremblay P, Groth D, DeArmond SJ, Prusiner SB, Lingappa VR (1999) Transmissible and genetic prion diseases share a common pathway of neurodegeneration. Nature 402:822–826

Herms JW, Madlung A, Brown DR, Kretzschmar HA (1997) Increase in intracellular free Ca^{2+} in microglia activated by prion protein fragment. Glia 21:253–257

Herms J, Tings T, Gall S, Madlung A, Giese A, Siebert H, Schürmann P, Windl O, Brose N, Kretzschmar H (1999) Evidence of presynaptic location and function of the prion protein. J Neurosci 19:8866–8875

Hogan RN, Baringer JR, Prusiner SB (1987) Scrapie infection diminishes spines and increases varicosities of dendrites in hamsters: a quantitative Golgi analysis. J Neuropathol Exp Neurol 46:461–473

Hornshaw MP, McDermott JR, Candy JM (1995) Copper binding to the N-terminal tandem repeat regions of mammalian and avian prion protein. Biochem Biophys Res Commun 207:621–629

Jeffrey M, Goodsir CM, Bruce ME, McBride, Fraser JR (1997) In vivo toxicity of prion protein in murine scrapie: ultrastructural and immunogold studies. Neuropathol Appl Neurobiol 23:93–101

Jendroska K, Heinzel FP, Torchia M, Stowring L, Kretzschmar HA, Kon A, Stern A, Prusiner SB, DeArmond SJ (1991) Proteinase-resistant prion protein accumulation in Syrian hamster brain correlates with regional pathology and scrapie infectivity. Neurology 41:1482–1490

Johnston AR, Black C, Fraser J, MacLeod N (1997) Scrapie infection alters the membrane and synaptic properties of mouse hippocampal CA1 pyramidal neurons. J Physiol Lond 500:1–15

Kascsak RJ, Rubenstein R, Merz PA, Carp RI, Robakis NK, Wisniewski HM, Diringer H (1986) Immunological comparison of scrapie-associated fibrils isolated from animals infected with four different scrapie strains. J Virol 59:676–683

Kellings K, Meyer N, Mirenda C, Prusiner SB, Riesner D (1992) Further analysis of nucleic acids in purified scrapie prion preparations by improved return refocussing gel electrophoresis (RRGE). J Gen Virol 73:1025–1029

Kellings K, Meyer N, Mirenda C, Prusiner SB, Riesner D (1993) Analysis of nucleic acids in purified scrapie prion preparations. Arch Virol Suppl 7:215–225

Kerr JFR, Wyllie AH, Currie AR (1972) Apoptosis: A basic biological phenomenon with wide-ranging implications in tissue kinetics. Br J Cancer 26:239–257

Kocisko DA, Come JH, Priola SA, Chesebro B, Raymond GJ, Lansbury PT, Caughey B (1994) Cell-free formation of protease-resistant prion protein. Nature 370:471–474

Kocisko DA, Priola SA, Raymond GJ, Chesebro B, Lansbury PT, Caughey B (1995) Species specificity in the cell-free conversion of prion protein to protease-resistant forms: A model for the scrapie species barrier. Proc Natl Acad Sci USA 92:3923–3927

Krakauer DC, Zanotto PMA, Pagel M (1998) Prion's progress: Patterns and rates of molecular evolution in relation to spongiform disease. J Mol Evol 47:133–145

Kretzschmar HA, Prusiner SB, Stowring LE, DeArmond SJ (1986) Scrapie prion proteins are synthesized in neurons. Am J Pathol 122:1–5

Kretzschmar HA, Ironside JW, DeArmond SJ, Tateishi J (1996) Diagnostic criteria for sporadic Creutzfeldt-Jakob disease. Arch Neurol 53:913–920

Kretzschmar HA, Giese A, Brown DR, Herms J, Keller B, Schmidt B, Groschup M (1997) Cell death in prion disease. J Neural Transm (Suppl) 50:191–210

Lin MC, Mirzabekov T, Kagan BL (1997) Channel formation by a neurotoxic prion protein fragment. J Biol Chem 272:44–47

Linder MC, Hazegh-Azam M (1996) Copper biochemistry and molecular biology. Am J Clin Nutr 63:797S–811S

Lipp H-P, Stagliar-Bozicevic M, Fischer M, Wolfer DP (1998) A 2-year longitudinal study of swimming navigation in mice devoid of the prion protein: no evidence for neurological anomalies or spatial learning impairments. Behav Brain Res 95:47–54

Lopez CD, Yost CS, Prusiner SB, Myers RM, Lingappa VR (1990) Unusual topogenic sequence directs prion protein biogenesis. Science 248:226–229

Lucassen PJ, Williams A, Chung WCJ, Fraser H (1995) Detection of apoptosis in murine scrapie. Neurosci Lett 198:185–188

Manson J, West JD, Thomson V, McBride P, Kaufman MH, Hope J (1992) The prion protein gene: a role in mouse embryogenesis? Development 115:117–122

Masters CL, Richardson EP Jr (1978) Subacute spongiform encephalopathy (Creutzfeldt-Jakob disease). The nature and progression of spongiform change. Brain 101:333–344

Meyer RK, McKinley MP, Bowman KA, Braunfeld MB, Barry RA, Prusiner SB (1986) Separation and properties of cellular and scrapie prion proteins. Proc Natl Acad Sci USA 83:2310–2314

Morgan RF, O'Dell BL (1977) Effect of copper deficiency on the concentrations of catecholamines and related enzyme activities in the rat brain. J Neurochem 28:207–213

Müller WEG, Ushijima H, Schröder HC, Forrest JM, Schatton WF, Rytik PG, Heffner-Lauc M (1993) Cytoprotective effect of NMDA receptor antagonists on prion protein (Prion^{Sc})-induced toxicity in rat cortical cell cultures. Eur J Pharmacol 246:261–267

Oesch B, Westaway D, Wälchli M, McKinley MP, Kent SBH, Aebersold R, Barry RA, Tempst P, Teplow DB, Hood LE, et al. (1985) A cellular gene encodes scrapie PrP 27–30 protein. Cell 40: 735–746

Pan K-M, Stahl N, Prusiner SB (1992) Purification and properties of the cellular prion protein from Syrian hamster brain. Protein Sci 1:1343–1352

Pan K-M, Baldwin M, Nguyen J, Gasset M, Serban A, Groth D, Mehlhorn I, Huang Z, Fletterick RJ, Cohen FE, et al. (1993) Conversion of α-helices into β-sheets features in the formation of the scrapie prion proteins. Proc Natl Acad Sci USA 90:10962–10966

Parchi P, Castellani R, Capellari S, Ghetti B, Young K, Chen SG, Farlow M, Dickson DW, Sima AAF, Trojanowski JQ, et al. (1996) Molecular basis of phenotypic variability in sporadic Creutzfeldt-Jakob disease. Ann Neurol 39:767–778

Parchi P, Capellari S, Chen SG, Petersen RB, Gambetti P, Kopp N, Brown P, Kitamoto T, Tateishi J, Giese A, Kretzschmar H (1997) Typing prion isoforms. Nature 386:232–234

Parchi P, Giese A, Capellari S, Brown P, Schulz-Schaeffer W, Windl O, Zerr I, Budka H, Kopp N, Piccardo P, Poser S, Rojiani A, Streichemberger N, Julien J, Vital C, Ghetti B, Gambetti P, Kretzschmar H (1999) Classification of sporadic Creutzfeldt-Jakob disease based on molecular and phenotypic analysis of 300 subjects. Ann Neurol 46:224–233

Pattison IH, Jebbett JN (1971) Histopathological similarities between scrapie and cuprizone toxicity in mice. Nature 230:115–117

Perovic S, Schröder HC, Pergande G, Ushijima H, Müller WEG (1997) Effect of flupirtine on BCL-2 and glutathione level in neuronal cells treated in vitro with the prion protein fragment (PrP106–126). Exp Neurol 147:518–524

Peyrin JM, Lasmezas CI, Haik S, Tagliavini F, Salmona M, Williams A, Richie D, Deslys JP, Dormont D (1999) Microglial cells respond to amyloidogenic PrP peptide by the production of inflammatory cytokines. Neuroreport 10:723–729

Prohaska JR (1983) Changes in tissue growth, concentrations of copper, iron, cytochrome oxidase and superoxide dismutase subsequent to dietary or genetic copper deficiency in mice. J Nutr 113:2048–2058

Prohaska JR, Smith TL (1982) Effect of dietary or genetic copper deficiency on brain catecholamines, trace metals and enzymes in mice and rats. J Nutr 112:1706–1717

Prusiner SB (1982) Novel proteinaceous infectious particles cause scrapie. Science 216:136–144

Prusiner SB (1993) Genetic and infectious prion diseases. Arch Neurol 50:1129–1153

Prusiner SB (1998) Prions. Proc Natl Acad Sci USA 95:13363–13383

Prusiner SB, McKinley MP, Bowman KA, Bolton DC, Bendheim PE, Groth DF, Glenner GG (1983) Scrapie prions aggregate to form amyloid-like birefringent rods. Cell 35:349–358

Prusiner SB, Groth DF, Bolton DC, Kent SB, Hood LE (1984) Purification and structural studies of a major scrapie prion protein. Cell 38:127–134

Raffray M, Cohen GM (1997) Apoptosis and necrosis in toxicology: a continuum or distinct modes of cell death? Pharmacol Ther 75:153–177

Riesner D, Kellings K, Wiese U, Wulfert M, Mirenda C, Prusiner SB (1993) Prions and nucleic acids: search for "residual" nucleic acids and screening for mutations in the PrP-gene. Dev Biol Stand 80:173–181

Rothstein JD, Bristol LA, Hosler B, Brown RH, Kuncl RW (1994) Chronic inhibition of superoxide dismutase produces apoptotic death of spinal neurons. Proc Natl Acad Sci USA 91:4155–4159

Safar J, Wille H, Itri V, Groth D, Serban H, Torchia M, Cohen FE, Prusiner SB (1998) Eight prion strains have PrPSc molecules with different conformations. Nature Med 4:1157–1165

Sailer A, Büeler H, Fischer M, Aguzzi A, Weissmann C (1994) No propagation of prions in mice devoid of PrP. Cell 77:967–968

Salmona M, Forloni G, Diomede L, Algeri M, De Gioia L, Angeretti N, Giaccone G, Tagliavini F, Bugiani O (1997) A neurotoxic and gliotrophic fragment of the prion protein increases plasma membrane microviscosity. Neurobiol Dis 4:47–57

Schallock K, Schulz-Schaeffer WJ, Giese A, Kretzschmar HA (1997) Postmortem delay and temperature conditions affect the in situ end-labeling (ISEL) assay in brain tissue of mice. Clin Neuropathol 16:133–136

Schätzl HM, Da Costa M, Taylor L, Cohen F, Prusiner SB (1995) Prion protein gene variation among primates. J Mol Biol 245:362–374

Scott JR, Fraser H (1984) Degenerative hippocampal pathology in mice infected with scrapie. Acta Neuropathol 65:62–68

Searle J, Kerr JFR, Bishop CJ (1982) Necrosis and apoptosis: distinct modes of cell death with fundamentally different significance. Path Annu 17:229–259

Selvaggini C, DeGioia L, Cantu L, Ghibaudi E, Diomede L, Passerini F, Forloni G, Bugiani O, Tagliavini F, Salmona M (1993) Molecular characteristics of a protease-resistant, amyloidogenic and neurotoxic peptide homologous to residues 106–126 of the prion protein. Biochem Biophys Res Commun 194:1380–1386

Somerville RA, Chong A, Mulqueen OU, Birkett CR, Wood SCER, Hope J (1997) Biochemical typing of scrapie strains. Nature 386:564

Stahl N, Borchelt DR, Hsiao K, Prusiner SB (1987) Scrapie prion protein contains a phosphatidylinositol glycolipid. Cell 51:229–240

Stahl N, Borchelt DR, Prusiner SB (1990) Differential release of cellular and scrapie prion proteins from cellular membranes by phosphatidylinositol-specific phospholipase C. Biochemistry 29:5405–5412

Stahl N, Baldwin MA, Teplow DB, Hood L, Gibson BW, Burlingame AL, Prusiner SB (1993) Structural studies of the scrapie prion protein using mass spectrometry and amino acid sequencing. Biochemistry 32:1991–2002

Stöckel J, Safar J, Wallace AC, Cohen FE, Prusiner SB (1998) Prion protein selectively binds copper (II) ions. Biochemistry 37:7185–7193

Tobler I, Gaus SE, Deboer T, Achermann P, Fischer M, Rulicke T, Moser M, Oesch B, McBride PA, Manson JC (1996) Altered circadian activity rhythms and sleep in mice devoid of prion protein. Nature 380:639–642

Viles JH, Cohen FE, Prusiner SB, Goodin DB, Wright PE, Dyson HJ (1999) Copper binding to the prion protein: structural implications of four identical cooperative binding sites. Proc Natl Acad Sci USA 96:2042–2047

Williams AE, Lawson LJ, Perry VH, Fraser H (1994a) Characterization of the microglial response in murine scrapie. Neuropath Appl Neurobiol 20:47–55

Williams AE, van Dam AM, Man A Hing WK, Berkenbosch F, Eikelenboom P, Fraser H (1994b) Cytokines, prostaglandins and lipocortin-1 are present in the brains of scrapie-infected mice. Brain Res 654:200–206

Williams A, Lucassen PJ, Ritchie D, Bruce M (1997a) PrP deposition, microglial activation, and neuronal apoptosis in murine scrapie. Exp Neurol 144:433–438

Williams A, Van Dam AM, Ritchie D, Eikelenboom P, Fraser H (1997b) Immunocytochemical appearance of cytokines, prostaglandin E2 and lipocortin-1 in the CNS during the incubation period of murine scrapie correlates with progressive PrP accumulations. Brain Res 754:171–180

Wyllie AH (1980) Glucocorticoid-induced thymocyte apoptosis is associated with endogenous endonuclease activation. Nature 284:555–556

Wyllie AH, Morris RG, Smith AL, Dunlop D (1984) Chromatin cleavage in apoptosis: association with condensed chromatin morphology and dependence on macromolecular synthesis. J Pathol 142:67–77

Yost CS, Lopez CD, Prusiner SB, Myers RM, Lingappa VR (1990) Non-hydrophobic extracytoplasmic determinant of stop transfer in the prion protein. Nature 343:669–672

The Role of T-Cell-Mediated Mechanisms in Virus Infections of the Nervous System

R. DÖRRIES

1	Introduction	219
2	Expression of Immunoregulatory Molecules in CNS Tissue	221
2.1	Major Histocompatibility Antigens	221
2.1.1	Class I Major Histocompatibility Antigens	222
2.1.2	Class II Major Histocompatibility Antigens	223
2.2	B7-Molecules	224
2.3	Fas (CD95)/FasL	225
3	Afferent Events	225
3.1	Signaling from the CNS to Peripheral Lymphoid Tissue	225
3.2	Activation and Differentiation of T Cells	226
4	Efferent Events	227
4.1	Extravasation of T Cells	227
4.2	Homing of T Cells to Virus-Infected Sites	228
4.3	Profiles of Infiltration and Phenotypic Properties of T Lymphocytes	231
4.4	Effector Mechanisms	232
4.4.1	$CD4^+$ T Lymphocytes	232
4.4.2	$CD8^+$ T Lymphocytes	234
4.5	Downsizing of T Effector Lymphocytes	236
5	Consequences of T Cell Action	237
6	Summary	239
References		240

1 Introduction

For quite a long time, the paucity of immune system elements in the central nervous system (CNS) and its tight enclosure by the blood–brain barrier (BBB) have led to the assumption that this organ is an immunologically privileged site of the body that is rigorously excluded from immune surveillance. However, data accumulated in recent years suggest that this view has to be revised substantially, since the healthy as well as the injured CNS performs an intense and permanent cross-talk

Department of Virology, Institute of Medical Microbiology and Hygiene, University of Heidelberg, Theodor-Kutzer-Ufer 1-3, 68167 Mannheim, Germany
e-mail: Ruediger.Doerries@imh.ma.uni-heidelberg.de

with the immune system resulting usually in a well tuned and gradually adapted immune response in the brain (CSERR and KNOPF 1992; FABRY et al. 1994).

The healthy non-affected CNS is indeed an immunologically quiet organ with very little traffic of lymphoid cells and, dependent on the species, with no or very basal expression of major histocompatibility (MHC) antigens. Due to its vulnerability and poor capability of regeneration and in view of potentially tissue damaging properties of an immune response, it is not surprising that the CNS actively maintains this state of low immunoreactivity by creating a hostile environment for inflammation. This is mainly achieved by CNS-localized synthesis and maintenance of high levels of immunosuppressive substances for pro-inflammatory T lymphocytes, e.g. transforming growth factor (TGF)-β or nerve growth factor (NGF) (CSERR and KNOPF 1992; NEUMANN and WEKERLE 1998). Therefore, the BBB not only protects the CNS from challenges coming from the periphery of the body, but is also needed to keep the level of immunomodulatory substances high and prevent drainage of these important regulators into the periphery.

In case of injury or invasion of infectious agents, the state of immunological silence must be overcome very rapidly in order to efficiently defend the integrity of the organ. Virus infections of the CNS demonstrate very clearly the conflict which has to be solved by the immune system. The ultimate goal is a rapid and complete elimination of the virus. Nevertheless, in the attempts to achieve this goal, extreme care must be taken to keep tissue damage and functional disturbances on the level of neurons as low as possible. In this context, effector functions of $CD8^+$ T lymphocytes are of particular interest, since they can cause death of virus-infected cells by induction of the apoptotic pathway, either through delivery of enzymes through pores formed in the membrane of the target cell or by interaction with the Fas (CD95) molecule on the virus-infected cell (FROELICH et al. 1998). It is obvious that these potentially dangerous properties of $CD8^+$ T lymphocytes require strict control.

Cellular elements that exert regulator functions in the course of intracerebral immune reactions are provided by both the immune and the central nervous systems. It is the $CD4^+$ T lymphocyte which can overcome the immunosuppressive milieu of the CNS by secretion of pro-inflammatory tumor necrosis factor (TNF)-α and interferon (IFN)-γ. The latter substance is a potent inducer of MHC antigens on CNS resident cells like microglia, which thereby rapidly acquire antigen-presenting capabilities for both $CD8^+$ and $CD4^+$ T lymphocytes. Intimate contacts between $CD4^+$ T lymphocytes and antigen-presenting microglia are of major importance for attracting inflammatory cells from the blood, for homing of effector leukocytes to virus-infected areas and probably also for down-regulation of immune effector functions as well as for induction of tissue-damaging properties in microglial cells. This concert of regulation is further complicated by close inter-relationships between activated microglia and other cellular elements of the CNS, mainly astrocytes and to a certain extent nerve cells, which in turn might act back on the tissue-infiltrating T lymphocyte population.

This review will focus on some recent aspects of the role of the T lymphocyte compartment in the course and the outcome of viral CNS infection. Due to the extreme difficulties in obtaining clinical tissue specimens from the brains of patients suffering from acute viral CNS infection, almost all data about the immunological events and their dynamics in the brain parenchyma have been collected in natural and experimental animal models of virus-induced CNS disease. Although these models might not exactly describe the situation in humans, they have proven very valuable in our understanding of how the intracerebral immune system response is organized and how it determines clinical course and outcome of viral CNS infections to an extent which is much larger than we thought a few years ago.

2 Expression of Immunoregulatory Molecules in CNS Tissue

As pointed out in the Introduction, in the unaffected CNS constitutive expression of molecules which have regulatory functions in the immune system response is very rare. Nevertheless, coordinated action of T lymphocytes in the virus-infected brain tissue depends entirely on the presence and correct regulation of expression of these molecules. Therefore, in this section a brief discussion will focus on expression of the most important molecules which govern T cell activity.

2.1 Major Histocompatibility Antigens

Considering that antigen recognition of T cells depends entirely on presentation of antigenic peptides in the groove of MHC molecules, expression of these structures must be of critical importance for both $CD4^+$ and $CD8^+$ T cells in the virus-infected brain. In contrast to many other organs of the body, it was believed until recently that the CNS is part of a group of immunologically silent sites like the eye, the synovial spaces of the joints and the testis, which in a healthy state do not express MHC molecules. At least for the CNS, the absence of MHC antigens is not absolute. There are differences between species and even within one species there might be differential expression of these molecules in inbred strains of animals. For instance, Brown Norway rats express class II MHC antigens constitutively in healthy brain tissue on a level which is easily detectable by conventional immunohistochemistry. In contrast, Lewis rats show no or an extremely weak expression of these molecules in situ (SEDGWICK et al. 1993). These data indicate that under certain, so far unknown, circumstances, the unaffected brain is by no means immunologically secluded but is in a state of immunological alertness that allows immediate interaction with α/β T cell antigen receptor (TCR)-expressing T lymphocytes penetrating into the tissue.

Independently from the level of MHC antigen expression in the healthy brain, both class I and class II antigens are up-regulated strongly in the course of viral CNS infection. This poses a very important question, which to date is discussed very controversially, namely: What is(are) the cell type(s) which express and up-regulate MHC antigens in the brain? There is no doubt about the fact that infiltrating lymphoid cells are MHC-positive. This is true for all cells with respect to MHC class I and for many cells with respect to MHC class II. Inflammatory monocytes/macrophages, some T cell subsets and B lymphocytes do express class II antigens. However, whereas the latter expression serves mainly inter-lymphoid communication, it would be of major interest to know what type of brain-resident cell expresses MHC, because this will decide and influence in many ways the handling of T lymphocytes in the tissue.

2.1.1 Class I Major Histocompatibility Antigens

It is generally accepted that MHC class I expression can be up-regulated on microglia, astroglia and oligodendroglia by IFN-γ (WONG et al. 1984). This would allow interaction of $CD8^+$ T cells with these brain-resident cells in an antigen-dependent manner, either in order to destroy the contacted cell or to receive regulatory signals. In contrast there is considerable controversial discussion about the capability of nerve cells to express class I antigens. The answer to this question is of tremendous importance, because many viruses can infect nerve cells. In the absence of MHC class I expression, virus-infected nerve cells would be unrecognized by $CD8^+$ cytotoxic T cells, and up-regulation of these molecules would lead to a high risk of irreversible destruction of infected nerve cells resulting in a lethal threat to the host. Conflicting data were reported on this issue. Early data from JOLY et al. (1991) demonstrated very clearly that retroviral-transformed neurons can transcribe the genetic information of the MHC class I heavy chain, albeit on a level which is only 1%–3% of that detected in a fibroblast. In contrast, the other structural element of a functional MHC class I molecule, β2-microglobulin, is expressed at a level which is comparable to that found in fibroblasts of the same mouse strain. Thus, neurons are unable to express sufficient intact MHC class I molecules that would allow their killing after virus infection. In vivo data from the same laboratory confirmed this assumption (MUCKE and OLDSTONE 1992). During persistent infection of mice with lymphocytic choriomeningitis virus (LCMV), the virus is harbored in nerve cells, whereas during acute infection the virus is detected in cells of the leptomeninges. Transfer of $CD8^+$ T lymphocytes with specificity for LCMV causes very severe neurological disease in acute infection but not in persistent infection. Histopathological studies revealed that in the latter case no MHC class I expression could be observed, although neurons are heavily infected by the virus. In vitro, LCMV infection of neurons also fails to up-regulate MHC class I molecules, whereas stimulation of these cells with IFN-γ causes up-regulation of class I antigens and renders these cells susceptible to $CD8^+$ T cell-mediated killing.

These data are in sharp contrast to work from other investigators (BILZER and STITZ 1994), who reported strong up-regulation of MHC class I antigens on neurons of rats which were infected by Borna disease virus (BDV). Since neurons in the hippocampal area are the main target of this virus, one would expect that strong $CD8^+$ T cell infiltration of this area would cause extensive damage. Indeed it has been shown by BILZER and STITZ (1994) and others that $CD8^+$ T cells are an important pathogenic factor in BDV-induced CNS disease. A straightforward and simple explanation for these discrepancies could be the fact that virus-infected neurons are often wrapped by very fine and thin microglia processes. Up-regulation of MHC class I molecules on the microglia could suggest expression of these molecules on neurons if, as was done by Bilzer et al., the tissue is examined at the light microscopic level. Immune electron microscopy should solve this problem. Moreover, published work by NEUMANN et al. (1995) adds substantial support to the idea that expression of MHC class I on neurons is subject to very tight intracellular and extracellular control mechanisms. Analysis of neurons on a single-cell basis by patch-clamp techniques, confocal laser microscopy and reverse transcription (RT)-polymerase chain reaction (PCR) disclosed the remarkable fact that transcription of MHC class I heavy chain was rare in neurons with spontaneous electrical activity, whereas in electrically silent neurons both MHC class I heavy chain and β2-microglobulin were transcribed. However, β2-microglobulin transcription was more tightly controlled than that of MHC class I mRNA. Despite transcriptional activity no surface expression of MHC class I molecules was detectable in these "electrically defective" neurons. Treatment of these cells with IFN-γ resulted in detectable amounts of MHC class I at the cell membrane, making these cells susceptible to killing by virus-specific T cells. TNF-α, a potent stimulator of MHC antigens, acts differently on damaged neurons than does IFN-γ. Although strong up-regulation of MHC class I heavy chain transcription is detectable, a comparable effect on the β2-microglobulin message is absent, and consequently no expression of class I molecules is detected on the cell surface (NEUMANN et al. 1997).

A very careful conclusion from these data would be that nerve cells can transcribe the message for the heavy chain of MHC class I; expression of the molecule, however, is limited to severely injured neurons that are exposed to IFN-γ.

2.1.2 Class II Major Histocompatibility Antigens

In contrast to MHC class I antigens, expression of MHC class II antigens in the brain is less controversial. There is no doubt that brain endothelial cells, pericytes, microglial and astroglial cells can express these important regulatory molecules and that IFN-γ, and to a certain extent also TNF-α, are the important mediators of induction (FIERZ et al. 1985; FONTANA et al. 1984; FREI et al. 1987; MALE et al. 1987; PARDRIDGE et al. 1989; SEDGWICK et al. 1993; SEDGWICK et al. 1991b). In vivo, these mediators are very often supplied by infiltrating T lymphocytes – a fact that is reflected by observations that, in the course of virus-induced damage of the CNS, the dynamics of MHC class II expression on microglial cells follow the influx and disappearance of $CD4^+$ T lymphocytes in the tissue. Nevertheless, care has to be taken when looking at in vitro and in situ expression of these molecules in virus-infected cells. In astrocytes, MHC class II expression can be induced in vitro by IFN-γ (FIERZ et al. 1985). Very interestingly, viruses themselves, even if noninfectious, can up-regulate MHC class II on astrocytes in vitro in the presence of IFN-γ-neutralizing antibodies. This was shown by our laboratory earlier for the coronavirus JHM in primary rat astrocytes (MASSA et al. 1986). However, in situ, expression of MHC class II on astrocytes is difficult to observe even under conditions of severe inflammation (POPE et al. 1998; TONTSCH and ROTT 1993). Today it is known that MHC expression on astrocytes is tightly controlled by neurons (NEUMANN et al. 1996; TONTSCH and ROTT 1993). Hippocampal slice cultures from rats revealed only moderate class II up-regulation after treatment with IFN-γ and this up-regulation was focused on areas of severe neuronal degeneration. The same observation holds true for microglial cells; however, inducibility of class II antigens is less tightly controlled than in astrocytes. After poisoning neurons in the tissue slice with a sodium channel blocker and subsequent treatment with IFN-γ, class II molecules were up-regulated on both astrocytes and microglial cells, also in areas of intact neuronal architecture. Nevertheless the level of MHC class II expression on microglial cells was much higher than in astrocytes. It seems that intact neurons actively control the expression in their surroundings of MHC antigens that are important in immunoregulation, but in contrast to microglia, in astrocytes this control is much more stringent (NEUMANN et al. 1996). These data indicate that microglia are much more alert with respect to inducible antigen-presenting cell (APC) function, an interpretation which is supported by in vitro data showing easy induction of antigen-presentation in microglia, whereas astrocytes are more restrictive in providing these functions (SEDGWICK et al. 1991a). Possible mediators in this control function are neurotrophins released by intact neurons. NGF, brain-derived neurotrophic factor (BDNF) and neurotrophin-3 (NT-3) all actively suppress class II inducibility on microglial cells, whereas blockade of neuronal function by toxins or glutamate antagonists restores IFN-γ mediated inducibility of class II antigens (NEUMANN et al. 1998).

2.2 B7-Molecules

The fact that microglial cells (perivascular and ramified cells) can be induced by T lymphocyte-derived IFN-γ to up-regulate MHC class II expression has fed the speculation that priming and activation of T cells during virus infection of the brain might also occur in the CNS. In this case more than MHC molecules have to be expressed on microglial cells, namely, B7-molecules, which are necessary as a ligand for CD28 on T lymphocytes. Interaction of B7.2 with CD28 provides the second signal which is necessary, beside MHC/TCR engagement, for activation of unprimed T cells (JUNE et al. 1994).

Very recent work from the Theiler's virus model of demyelinating encephalomyelitis unraveled interesting aspects of differential expression of B7.1 and B7.2 molecules in the virus-infected brain (POPE et al. 1998). In contrast to inflammatory T cells, which express predominantly B7.2, nearly all MHC class II-positive macrophages and microglia were also positive for both B7.1 and B7.2. This is of particular interest because interaction of B7.1 with its counterpart CTLA-4 (CD 152) on the proliferating activated T lymphocyte can cause apoptosis (SCHEIPERS and REISER 1998). Therefore, the brain-resident microglial cell must be considered as a potent regulator of T lymphocyte activity.

2.3 Fas (CD95)/FasL

Expression of the CD95 molecule renders a cell susceptible to induction of the apoptotic pathway, induced by interaction with the appropriate ligand, FasL. This regulatory event is an entirely physiological mechanism which is used to control cell proliferation during ontogenesis and in the course of specific immune reactions. However, the signal transduced by CD95/FasL interaction does not necessarily result in apoptosis (LYNCH et al. 1995). Dependent on the state of differentiation, engagement of CD95 on prelytic $CD8^+$ T lymphocytes by $CD4^+$ T-lymphocyte-expressed FasL during MHC-mediated contact with APC can trigger differentiation of $CD8^+$ T cells to effector killing cells in an interleukin (IL)-2-independent manner. As such it may play a role in viral CNS infections during expansion of antigen-specific $CD8^+$ T lymphocytes in peripheral lymphoid tissue. Besides control of lymphoid expansion and elimination in immune system responses, killing of virus-infected target cells by class I-restricted $CD8^+$ T lymphocytes can also be mediated by this effector mechanism.

In search for CD95/FasL expression in the body, it turned out that some so-called immunologically sequestered organs, like joints, testis and eyes, express FasL at a high density (GRIFFITH and FERGUSON 1997). At present, this is interpreted as an immunosuppressive action of these organs against inflammatory T cells, which all express CD95 when activated. Although not expressed constitutively in the CNS, CD95 and FasL can be up-regulated on brain-resident as well as infiltrating lymphoid cells during inflammation of the CNS, as in autoimmune encephalomyelitis or virus infection (BECHER et al. 1998; CHOE et al. 1998).

3 Afferent Events

3.1 Signaling from the CNS to Peripheral Lymphoid Tissue

Due to the absence of a regular lymphoid drainage system and the lack of dendritic leukocytes in the brain, it is difficult to understand how and where T lymphocytes come into contact with antigens produced in the course of viral CNS infection. As has been shown very elegantly by Cserr and coworkers, priming of the immune system- vs brain-derived antigens occurs in peripheral lymphoid tissue. Radioactively labeled antigens that are applied to the ventricular cerebrospinal fluid (CSF) without damage to the BBB appear very rapidly in the superficial and deep cervical lymph nodes (CLNs), and a strong antigen-specific antibody response consisting mainly of IgM is noted within a few days (CSERR et al. 1992a). In contrast, the contribution of the spleen to this primary response is only marginal. Most likely, the applied antigen is drained with CSF which constantly leaks at the sealing points, which are formed by the meninges and the sensory brain nerves. One of the major places of antigen drainage is the cribriform plate, where the olfactory nerves penetrate the base of the skull. Antigen leaving the CNS at this site is transported by the oral mucosa to the lymphoid tissue of the neck area (CSERR et al. 1992b).

In the case of experimental viral CNS infection, this drainage system allows priming of the naive peripheral immune system. Natural viral infection of the CNS, however, occurs only under very rare circumstances directly via entry of virus particles into peripheral nerves and retrograde transport. Generally, such infections are the consequence of a concomitant or preceding infection in peripheral organs of the patient. Therefore, virus-specific priming of the T cell compartment occurs in secondary lymphoid tissues which are in close proximity to the viral entry site in the body, and contact of CNS-derived viral antigen in the cervical lymph nodes often represents a secondary encounter of already primed T cells with the antigen.

3.2 Activation and Differentiation of T Cells

Since CLNs are one of the major sites where T cells come into contact with brain-derived antigens, then activation and differentiation into effector cells should occur at this site. As we noted, antigen-specific, proliferative T cell responses are detectable in CLNs of experimentally infected animals a few days after intracerebral application of the virus (IMRICH et al. 1994). Interestingly, in rodent models of viral encephalitis, quality as well as quantity of the CLN-located T cell response differs between inbred strains. The intracerebral infection of Lewis and Brown Norway rats can serve as a vicarious example to demonstrate the role of the genetic background in the immunological handling of a viral CNS infection. Both rat strains are susceptible to intracerebral infection with the murine hepatitis virus (coronavirus) strain JHM (JHMV); however, whereas in Lewis rats a paralytic course of the infection may cause death in a third of all animals within a week, Brown Norway rats never show any clinical signs of infection (WATANABE et al. 1987). It is noteworthy that Lewis but not Brown Norway rats develop an autoimmune reaction against basic myelin protein (BMP) (WATANABE et al. 1983), which is a cell component of the viral target cell in the CNS, namely, the oligodendrocyte. The dynamics of the proliferative T cell response in the CLNs of these animals disclose remarkable differences. In 60%–70% of Lewis rats which survive the infection and recover from neurological disease, CLN-derived T cells exhibit ex vivo proliferation without further antigenic challenge in vitro (IMRICH et al. 1994). Proliferating activity of these T cells starts immediately after intracerebral infection and increases up to 12 days post-infection, when severity of the neurological disease peaks and the animals start to recover. Proliferation cannot be substantially enhanced by addition of viral antigen into the cultures. Obviously, intracerebral infection of this rat strain induces a strong and polyclonal activation of T cells in the CLNs, albeit with a considerable delay after onset of neurological disease. In sharp contrast, CLN-derived T cells from JHMV-infected Brown Norway rats show extremely low ex vivo proliferation for many weeks after infection, but a robust proliferative response to in vitro stimulation with viral antigens (IMRICH et al. 1994). Most remarkable, this virus-specific T cell response in the CLNs is biphasic, with a very early response within a few days after infection and a second wave approximately 2 weeks post-infection. In this rat strain, activation of the T cell compartment is rapid, restricted to a few clones and can occur multiple times in the course of the infection. Of course it is tentative to speculate that differences in the CLN-localized T cell reactivity of Lewis and Brown Norway rats mirrors in some way the completely different clinical and histopathological course of the infection. As will be discussed later in this review, it is indeed conceivable that quality and quantity of the peripheral T cell response in this animal model has a causative link to the way that these animals restrict viral spread in the brain and thereby minimize tissue destruction at the very beginning of the infection.

4 Efferent Events

4.1 Extravasation of T Cells

After clonal expansion and differentiation in the periphery, effector T cells must transmigrate through the BBB formed by the wall of brain blood vessels and adjacent foot processes of astrocytes. In this context, a whole series of questions arises: (1) What is the nature of the transmigrating T cell? (2) Is antigen specificity of T cells of importance for extravasation? (3) Do brain-resident cells influence the endothelial layer in a way that might help T cells to transmigrate through the BBB? (4) What signals are given to T cells that have migrated to the perivascular space and the brain parenchyma, and what are the consequences of this signaling? Only few, if any of these questions can be answered in detail for virus-specific T lymphocytes.

Both studies on lymphocyte traffic in sheep and an experimental autoimmune disease of the brain, experimental allergic encephalitis (EAE), in rats have paved the way for studies in this difficult field of CNS-localized immunoreactivity. Today, it is fairly well accepted that freshly activated and non-activated but primed memory T cells can enter any non-lymphoid tissue on a random basis independent of their antigenic specificity (WILLIAMS and HICKEY 1995). As pointed out above, in most cases of viral CNS infection the invading agent entered the host somewhere in the periphery, which then provoked a primary T cell response in the regional lymphatic tissues. Therefore, T lymphocytes with virus-specificity are already recirculating in the vast pool of peripheral T lymphocytes during initial replication of the virus in the CNS. What causes adhesion and extravasation of T cells at the site of infection?

Pathological events associated with viral replication in the CNS most likely cause initial up-regulation of adhesion molecules on the endothelial cell. Viruses are able to infect brain endothelial cells (BRANKIN et al. 1995; COSBY and BRANKIN 1995; CZUB et al. 1995; GAIRIN et al. 1991; GEORGSSON 1994; KRAKOWKA et al. 1987), and it is known that this event can not only up-regulate MHC antigens

(GAIRIN et al. 1991) but also molecules such as intercellular adhesion molecule (ICAM)-1 (BRANKIN et al. 1995), which are important for transendothelial migration of T lymphocytes (GREENWOOD et al. 1995). Besides increased stickiness of T cells at infected BBB endothelium, expression of MHC class I loaded with viral peptides probably causes $CD8^+$ T lymphocyte-mediated destruction of these targets – as has been suggested in LCMV infection of rodents (DOHERTY et al. 1990). This would result in serious damage of the BBB, creating further activation of the surrounding endothelial layer.

Moreover, although not understood in detail signals are transmitted from inside the CNS to the BBB, which cause luminal up-regulation of adhesion molecules on the endothelial cells. Infection of the brain by retroviruses such as human-, simian- and feline immunodeficiency virus (HIV, SIV, FIV) activates brain-resident microglia cells to secrete IL-1 and TNF-α (NOTTET and GENDELMAN 1995), two cytokines which are known to stimulate expression of endothelial adhesion molecules, e.g. vascular cellular adhesion molecule (VCAM)-1 (WEKERLE et al. 1991). VCAM-1 expression on brain endothelial cells seems to be important for adhesion of T lymphocytes in viral CNS infections (CHRISTENSEN et al. 1995; MARKER 1995; SOILU HANNINEN et al. 1997) and in CNS inflammatory conditions (WELLER et al. 1996).

Additional interesting observations come from model systems of tissue grafting to the brain (ISHIHARA et al. 1993). Iso- or autografting of skeletal muscle cells onto the surface of the brain (e.g. implantation into the fourth ventricle, between medulla oblongata and cerebellum) provides a focal opening of the BBB, which in other places is otherwise undisturbed. Whereas solutes enter the brain directly through the vessels of the graft, macrophages penetrate into the CNS through vessels that are adjacent to the graft rather than through vessels in the graft. The reason for this is the up-regulation of ICAM-1 on endothelial cells in the medulla close to the grafting site. This experiment demonstrates that a focal pathological event in the brain can result in activation of adjacent endothelial cells which will cause adhesion of activated mononuclear cells.

Once activated T lymphocytes have attached to an endothelial cell, they will enhance expression of adhesion molecules (ICAM-1) by secretion of IFN-γ and TNF-α (MARKER et al. 1995), thereby facilitating the adherence of more T cells (DEL POZO et al. 1996). Adherence of cells is then followed by transmigration into the perivascular space. Chemokines and adhesion molecules like T cell-expressed lymphocyte functional antigen (LFA)-1 and endothelial ICAM-1 are deeply involved in the cascade of events triggering this process (for a detailed discussion see SPRINGER 1994).

4.2 Homing of T Cells to Virus-Infected Sites

Once an activated T lymphocyte has passed the BBB, its fate depends on two aspects: (1) Is there antigen presentation in the perivascular space? (2) Is the TCR specific for any of the presented peptides? As has been shown by HICKEY et al.

(1991), the amount of activated T cells in the perivascular space drops rapidly to undetectable levels if no antigen-specific contact is possible with perivascular macrophages. T cells that are specific for a tissue-localized antigen are detectable much longer in the perivascular space. These cells interact with antigen-presenting macrophages, a process which leads to mutual stimulating events between incoming T cells and local APC. At the port of entry the APC are perivascular microglia (HICKEY and KIMURA 1988).

The regulating events that guide T cells through the tissue to the virus-infected area of the brain are completely unknown. Although purely speculative (LAMPSON 1998), it seems likely that, immediately after penetration of T cells into the brain tissue, the direction of movement is more random than targeted. Movement itself may be governed by control mechanisms which are known from other motile cells in the CNS (MERCER et al. 1994), and they probably follow tracks of "low tissue resistance" like perivascular spaces or white matter tracts when penetrating into densely packed neural tissue (EMMETT et al. 1991). Speed of movement, stop-and-go, and changes in direction are probably dictated by the extracellular matrix and the expression patterns of adhesion and cellular interaction molecules of both the migrating and the endogenous cells. If a browsing T cell arrives in close vicinity of a virus-infected site, movement becomes more specific and more directed, determined by a gradient of chemotactic factors. In the virus-infected brain, expression of many well known chemokines is up-regulated (ASENSIO and CAMPBELL 1997; MORIMOTO et al. 1996). It looks as if true homing to the virus-infected tissue is a rather late event in the course of T lymphocyte movement through the parenchyma.

How do T cells arrange in relation to virus-infected cells in the tissue? Very few studies on the dynamics and the spatial relationships between different lymphocyte subpopulations and virus-infected cells have been conducted so far. With the aid of computer-aided immunohistochemistry carried out at different times post-infection, we have analyzed these arrangements and their dynamic changes in the course of a viral CNS infection (DÖRRIES et al. 1991). Despite the fact that comparable amounts of $CD4^+$ and $CD8^+$ T lymphocytes can be recovered early after infection from the brain of coronavirus JHMV-infected rats, homing patterns of the two T cell subsets to virus infected areas differ. One week post-infection, with the onset of clinical symptoms, only a few $CD4^+$ T cells focus in these areas and they are outnumbered by far by accompanying $CD8^+$ T cells, which fill up the center of a virus-infected demyelinated plaque. A week later the picture changes completely. At the maximum of neurological disease expression, $CD8^+$ T cells are seen in intimate proximity to virus-infected cells, which form the border of the plaque. Numerous $CD4^+$ T lymphocytes have gathered in the area and the center of the plaque is filled with antibody-secreting plasma cells. Another week later, at the time of recovery, only few $CD8^+$ T lymphocytes are associated with the remainder of scattered virus-infected cells. The dominating population is clearly of the $CD4^+$ T cell subset. Over the entire observation period many macrophages/microglia concentrate in foci of T cell infiltration.

Comparable to the situation immediately after transmigration through the BBB, antigen presentation seems to play an important role in the spatial arrangement of lymphocytes in virus-infected areas because strong up-regulation of class I and II major histocompatibility antigens is seen in JHMV-infected foci. Besides cells of hematopoietic origin, class I antigens are also expressed on the viral target cell, which in case of JHMV in rats is the oligodendrocyte. Class II antigens are up-regulated strongly on the brain-resident microglia cell. This was shown unequivocally by flow cytometry, which allows differentiation of brain-resident microglia from infiltrating macrophages/monocytes (SEDGWICK et al. 1991b). This highlights the central role of MHC up-regulation in both the perivascular space and in the brain parenchyma for coordinated action and movement of T cells in the virus-infected brain. In this context the crucial question is what event causes up-regulation of MHC antigens in the virus-infected brain tissue? As pointed out above, IFN-γ is a strong up-regulating agent for MHC class I and class II antigens, and activated T lymphocytes, $CD8^+$ as well as $CD4^+$, are able to synthesize this cytokine abundantly. It is further known that expression of IFN-γ in the virus-infected brain is strongly dependent on a functional T cell compartment (PEARCE et al. 1994). However, the tentative conclusion that MHC expression in the brain and homing of T lymphocytes to virus-infected areas are entirely dependent on IFN-γ secretion would pose the serious problem of the chicken and the egg: If homing of T lymphocytes is dependent on appropriate MHC expression, and up-regulation of these molecules is T cell dependent, it is difficult to understand how MHC molecules are up-regulated very early after CNS infection to guide infiltrating T cells to their targets. There are multiple observations which suggest that initial up-regulation of MHC antigens in the virus-infected brain is mediated by IFN-α/β and is independent from T lymphocytes. PEARCE et al. (1994) noted that in nude mice which are intracerebrally infected by JHMV, MHC expression is as strong as in normal mice, even in the absence of detectable IFN-γ mRNA. NJENGA et al. (1997) reported that IFN-α/β-receptor knockout mice have deficiencies in up-regulate of MHC class I antigens in the CNS soon after infection with Theiler's virus. SANDBERG et al. (1994) observed that treatment of mice with anti-IFN-α/β antibodies reduced the inflammatory T cell reaction in the LCMV-infected brain. Finally, LCMV infection in IFN-γ knockout mice did not alter recruitment of T cells to the brain but had some consequences on the activation of macrophages (NANSEN et al. 1998), and up-regulation of class I molecules on measles virus-infected glial cells is strongly associated with IFN-β (DHIB JALBUT et al. 1995). However, in this context it is very important to note that up-regulation of MHC class I antigens by IFN-β does not operate in neurons (DHIB JALBUT et al. 1995), although an antiviral state can be induced in these cells by IFN-β treatment (WARD and MASSA 1995). Obviously, in neurons, strong IFN-γ signaling by inflammatory T cells and serious defects in the electrical properties of neurons are needed to allow up-regulation of class I molecules (see Sect. 2.1.1).

From these data it can be assumed that soon after infection, in the absence of infiltrating T cells, up-regulation of MHC antigens in the brain is strongly influenced by local production of IFN-α/β. Moreover, some viruses have been shown to induce MHC antigens (class I and class II) directly on brain-derived cells in the absence of IFN-γ (GOMBOLD and WEISS 1992; MASSA et al. 1986), or on brain endothelial cells in a cytokine-independent manner (GAIRIN et al. 1991). Once T cells have settled in these areas they strongly enhance MHC expression on brain-resident cells (including defect neurons) by secretion of IFN-γ and thereby potentiate the inflammatory reaction. The necessity of appropriate MHC/viral antigen presentation to induce an inflammatory response by T cells in the virus-infected brain was very elegantly shown by DOHERTY and ALLAN (1986). Using bone marrow chimeras, it was demonstrated that, for a maximal T lymphocyte-mediated inflammatory response in the brain, it is necessary that MHC-restricted virus-specific T cells expand in peripheral lymphoid tissue must find virus infected cells in the CNS expressing the same MHC as seen in the periphery.

4.3 Profiles of Infiltration and Phenotypic Properties of T Lymphocytes

In the course of coronavirus JHMV-induced encephalomyelitis, a first peak of T lymphocytes can be recovered from the brain parenchyma a few days after intracerebral infection. This is followed by a second peak approximately 3 days later (DÖRRIES et al. 1991; HEIN et al. 1995; IMRICH et al. 1994). Infiltrating T cells can phenotypically be characterized by expression of the CD45RC molecule. CD45RC is an isoform of the common leukocyte antigen, which is expressed in the rat on unprimed T lymphocytes. Freshly activated T cells lose expression of this molecule but can re-express CD45RC when entering the state of a memory cell (BELL et al. 1998; SARAWAR et al. 1993). Using flow cytometry and immunostaining for CD45RC and the TCR, infiltrating T cells can be subdivided into two categories. Whereas in the first peak a significant amount of $CD45RC^+$, TCR^+ cells accompanies the majority of TCR^+, $CD45RC^-$ cells, the second peak of infiltration contains more than 98% TCR^+, $CD45RC^-$ cells. Thus, an initial influx of activated T lymphocytes ($CD45RC^-$) accompanied by a small amount of naive or memory T cells ($CD45RC^+$) is followed by a second wave of infiltrating T lymphocytes, which is composed almost entirely of T cells with an activated phenotype ($CD45RC^-$).

In addition to a state of fresh activation, infiltrating T cells show another remarkable feature in JHMV-induced encephalomyelitis. Virtually none of them express a receptor for interleukin 2 (IL-2r), which implies that they are in a state of terminal differentiation, no longer able to respond to proliferation-stimulating signals. Any attempt to induce proliferation in these cells either with antigen or a mitogen fails over a period of 3 weeks post-infection (IMRICH et al. 1994). Comparable data were reported from the EAE model, in which BMP-specific effector T cells stop proliferation when entering the brain parenchyma (OHMORI et al. 1992). Moreover, results from the Sindbis virus model of encephalitis in mice also suggest that T cells in the brain tissue have lost expression of the IL-2r and thereby their capability to respond to proliferation signals (IRANI et al. 1997)

What is the origin of the proliferation switch-off signal? Amongst several possibilities an intriguing hypothesis to explain T lymphocyte switch-off in the CNS shall be discussed here: a strong and efficient antigen presentation on brain-resident professional APC-like microglia cells can send negative signals to tissue-infiltrating T cells which suppress the response to proliferative stimulation but which maintain effector functions of the T cell, as there is secretion of cytokines. There are multiple observations which lend support to this idea: (1) Application of BMP to the CNS without disturbing the BBB causes suppression of clinical EAE after transfer of BMP-specific T cell lines (HARLING BERG et al. 1991); (2) in contrast to peripheral macrophages, microglial cells are not able to induce proliferation and IL-2 secretion in $CD4^+$ T lymphocytes (FORD et al. 1995); and (3) although stimulating secretion of IFN-γ and TNF-α in $CD4^+$ T lymphocytes, MHC class II-positive microglial cells subsequently cause death in the contacting T cell population (FORD et al. 1996).

The nature of the negative signaling to T lymphocytes is not exactly defined. However, it is known that activation of tissue macrophages by TH1-cell-derived cytokines causes induction of the inducible nitric oxide synthetase (iNOS) which produces nitric oxide (NO) from L-arginine (for review see (KOLB and KOLB-BACHOFEN 1998)). Besides its static effects on infectious agents including viruses (KULKARNI et al. 1997; PERTILE et al. 1996), NO strongly suppresses the induction of proliferation in T lymphocytes (MILLS 1991) and may thereby be involved in the immunosuppressive effects of virus infections (BUTZ et al. 1994). In the brain, iNOS and NO are up-regulated after virus infection (AKAIKE et al. 1995; BI et al. 1995; HOOPER et al. 1995), and microglial cells have been shown to express iNOS in the presence of pro-inflammatory cytokines in HIV-infected patients (LANE et al. 1996) or during vesicular stomatitis virus (VSV) infection of the murine CNS (CHRISTIAN et al. 1996). Since up-regulation of iNOS occurs during acute viral CNS infection (OLESZAK et al. 1997), it seems possible that activated microglial cells contribute via NO to the inhibition of T cell proliferation in the CNS. If this assumption is correct, than microglial cells would fulfill a delicate regulatory role. On one hand, they act as stimulators of cytokine secretion in contacting T lymphocytes, which in turn enables them to assist in the elimination of infectious agents by enhanced phagocytosis and endosomal digestion. On the other hand, in a kind of feedback loop, they must prevent uncontrolled expansion of inflammatory T cells in order to avoid damage in the CNS.

4.4 Effector Mechanisms

4.4.1 $CD4^+$ T Lymphocytes

By virtue of the broad pattern of cytokines that can be secreted by $CD4^+$ T lymphocytes, these cells are central regulators of local immune effector functions. Dependent on the predominant cytokines secreted, $CD4^+$ regulator T cells are classified at least into four subpopulations, namely TH0–TH3. TH0 is an activated $CD4^+$ T cell which so far is not committed to differentiation along a certain pathway – a process that depends on external stimuli by cytokines. Under the influence of IL-12, which is secreted predominantly by APC, a TH0 cell may develop into a pro-inflammatory TH1 cell, characterized by strong IFN-γ and TNF-α secretion. In the absence of IL-12 and in the presence of Il-4, TH0 cells can differentiate into anti-inflammatory TH2 cells, which predominantly act via secretion of IL-10 and help B lymphocytes to differentiate into antibody-secreting plasma cells. And finally, a less well understood differentiation pathway is that of TH3 cells, which, due to their secretion of TGF-β, are characterized as suppressor cells (HAFLER et al. 1997).

Analysis of the cytokine milieu in the inflamed brain of virus-infected animals has revealed expression of IFN-γ and/or TNF-α (IRANI et al. 1997; LANE et al. 1996; MOKHTARIAN et al. 1996; MORRIS et al. 1997; PEARCE et al. 1994; WESSELINGH et al. 1994), a finding which implies a TH1-determined cytokine milieu. Antigen-presenting microglial cells can stimulate secretion of IFN-γ in $CD4^+$ T lymphocytes (FORD et al. 1996), and respond to IFN-γ by up-regulation of MHC I antigens (GRAU et al. 1997) and the CD40 molecule (NGUYEN et al. 1998), as well as by proliferation (SEDGWICK et al. 1998). Up-regulation of the CD40 molecule could allow interaction with activated T cells via the CD40 ligand, which on tissue resident macrophages usually results in the secretion of multiple pro-inflammatory substances like TNF-α, IL-12, macrophage inflammatory protein (MIP)-1α and NO (STOUT and SUTTLES 1996). Indeed, it has been shown that activated microglia can secrete these substances (ALOISI et al. 1997; FREI et al. 1987; HAYASHI et al. 1995) and, since they are better stimulators of TH1 cells than of TH2 cells (ALOISI et al. 1998), an increasing pro-inflammatory micro-environment will arise in areas where $CD4^+$ T cell/microglia interactions take place. All of these pro-inflammatory substances act as attractants for peripheral blood leukocytes, either directly as chemoattractants or indirectly by up-regulation of adhesion molecules such as ICAM-1 on adjacent endothelial cells. Additionally, an enhanced capability of antigen processing and presentation is achieved by the responding tissue macrophages, as shown by a strong up-regulation of MHC class II and class I molecules as well as of B7 molecules (MENENDEZ IGLESIAS et al. 1997).

All of these observations define a role for $CD4^+$ TH1 cells as potent activators that switch the immunologically calm CNS to a state of competent immunological reactivity. As a consequence of this regulatory circuit, more and more inflammatory cells including peripheral macrophages, mobilized microglial cells, $CD8^+$ T cells, and antibody-secreting plasma cells are trapped in virus-infected sites. With increasing activation of the microglia, however, negative regulatory effects on T lymphocytes may overcome positive signals of attraction and stimulation of effector functions. In particular, the release of NO and the overexpression of B7 molecules can result in T cell arrest or death (see Sect. 4.3), and chronic and strong stimulation of T lymphocytes by antigenic recognition up-regulates the CD95/FasL receptor system on these cells (LYNCH et al. 1995).

Very recently, published investigations support the concept that, besides regulating effects on the influx and homing of peripheral immune system cells, the presence of $CD4^+$ T effector lymphocytes is indispensable for $CD8^+$ T lymphocytes to exert their effector functions, namely, reducing the viral load of the brain by destruction of virus-infected cells and/or by inhibition of intracellular viral growth. As shown in coronavirus JHMV infected mice, clearance of virus from the CNS by $CD8^+$ T lymphocytes, but not their recruitment to the tissue, is dependent on the presence of a functional $CD4^+$ T lymphocyte compartment in these animals (STOHLMAN et al. 1998). Comparable data were described in BDV-infected rats. Here, it was demonstrated that transfer of primed $CD4^+$ T cells, which were not cytotoxic themselves, resulted in many more $CD8^+$ cytotoxic T cells in the brain of BDV-infected animals than in non-transferred animals (PLANZ et al. 1995).

Although the molecular basis of $CD4^+$ and $CD8^+$ T cell interactions is still unknown, studies on the role of CD95/FasL in $CD4^+$ and $CD8^+$ T lymphocyte cross-talk offers some interesting perspectives (LYNCH et al. 1995). It is known that CD95/FasL interaction can exert both stimulating and death-inducing effects on lymphocytes, dependent on the stage of differentiation of interacting cells. Freshly activated $CD4^+$ T cells rapidly up-regulate both CD95 and FasL. However, whereas CD95 expression is induced on both T cell subsets, TH1 and TH2, FasL is seen only on the TH1 T cell subset. Engagement of these receptors on freshly activated cells does not induce apoptosis but enhances proliferation and cytokine secretion (IFN-γ, IL-2 and TNF-α). TH1 cells expressing FasL can also interact with the CD95 molecule expressed on precytotoxic $CD8^+$ T lymphocytes when both cell subsets make contact with an APC cell via their antigen receptor. This FasL/CD95-mediated $CD4^+$/$CD8^+$ T lymphocyte cross-talk drives differentiation of the $CD8^+$ T lymphocyte into an activated, MHC class I-restricted killer cell in an IL-2 independent way. Although it is completely unknown if these mechanisms also operate in non-lymphoid tissue like the CNS, observations of STOHLMAN et al. (1998) support this idea. They reported that apoptosis of virus-specific $CD8^+$T lymphocytes in the CNS of JHMV-infected mice is considerably higher in animals that were depleted of $CD4^+$ T lymphocytes. It seems that $CD4^+$ T lymphocytes help $CD8^+$ T effector lymphocytes to survive and that they are important regulators of $CD8^+$ T-lymphocyte-mediated killing. This might explain why clearance of the virus from the brain by $CD8^+$ T lymphocytes is $CD4^+$ T-lymphocyte-dependent, or at least substantially enhanced by TH1 $CD4^+$ T lymphocytes.

4.4.2 $CD8^+$ T Lymphocytes

$CD8^+$ T lymphocytes can kill virus-infected cells by at least two mechanisms which, although acting on different signal pathways, both will end in the induction of the apoptotic killing machinery of the affected target cell. The perforin pathway acts via pore formation in the membrane of the target cell by perforins and transfer of granzyme B (GrB) through this pore into the cytoplasm of the target cell. Here, GrB activates and converts pro-caspases to active caspases, which are mediators of apoptosis. The FasL/CD95 pathway acts via engagement of the CD95 molecule on the target cell with the FasL expressed on the killing $CD8^+$ T cell. Upon engagement, a stimulating signal is transduced by the CD95 molecule via intracellular death domain proteins, which convert pro-caspases into death-inducing caspases (FROELICH et al. 1998) in the virus-infected cell.

There is no doubt that $CD8^+$ T cells extracted from the virus-infected CNS can kill virus-infected cells in vitro in a MHC class I-restricted manner (HEIN et al. 1995; PLANZ et al. 1993). This idea is also supported by in vivo data showing a disastrous clinical course of viral CNS infection in animals supplemented with virus-specific $CD8^+$ T cells at the time of virus infection. From this experience and the results from several in vivo manipulations of $CD8^+$ T lymphocytes, e.g. depletion of $CD8^+$ T cells (BILZER and STITZ 1994; SUBAK SHARPE et al. 1993), CNS infection in β2-microglobulin- or perforin-deficient mice (ROSSI et al. 1998), and constitutive transgenic expression of MHC class I molecules in neurons (RALL et al. 1995), it was concluded that $CD8^+$ T cells indeed can kill virus-infected brain cells in vivo. In the case of neurons, however, the cell must be seriously damaged by the virus infection before MHC class I expression is up-regulated and T cell-mediated killing can take place (see Section 2.1.1). If this happens, in vitro experiments from Rensing Ehl's group suggest that the perforin-dependent pathway is preferred by $CD8^+$ T effector cells (RENSING EHL et al. 1996).

Despite this strong but indirect experimental support of the idea that $CD8^+$ T lymphocytes kill cells in the brain parenchyma, there are a substantial data which suggest that this is the exception rather than the rule. In various animal models $CD8^+$ T lymphocytes extracted from the virus-infected brain cannot kill ex vivo; rather, they have to be restimulated with strong signals, sometimes including allo- or even xenogenic TCR engagement, before substantial killing activity can be detected on virus-infected syngeneic target cells (HEIN et al. 1995). Work from KÜNDIG et al. (1993) helps to explain the virus-cleansing property of T lymphocytes in the CNS by soluble factors rather than by killing. After immunization of mice with VSV, the animals were challenged with two different recombinant vaccinia viruses (vacc) one expressing the nucleoprotein of VSV (vacc-VSV-NP) and the other expressing the glycoprotein of LCMV (vacc-LCMV-GP). Surprisingly, animals challenged with the VSV-unrelated vacc-LCMV-GP could restrict growth of a vacc-LCMV-GP in the brain, but not in the testis or ovaries. This bystander effect in the CNS was partly inhibited by treatment with anti-IFN-γ antibodies. This effect was seen in mice with the haplotype $H-2^b$, which usually clear LCMV by $CD8^+$ T lymphocytes, as well as in animals with the haplotype $H-2^k$, which use $CD4^+$ T cells for LCMV clearance.

Non-cytolytic clearance of virus by MHC class I-restricted T cells seems to be the regular effector mechanism if viruses persist in neurons without causing lethal damage to their host cells. Work from Oldstone's laboratory described removal of virus from persistently infected neurons after adoptive transfer of virus-specific cytotoxic T cells (OLDSTONE et al. 1986; TISHON et al. 1993). Although lymphoid infiltration was noted in the brain parenchyma and viral load was reduced considerably, no neuronal loss occurred. Most remarkably, clearance of virus by non-lytic mechanisms is limited to the CNS and takes much longer than in peripheral organs, where tissue destruction and inflammation can be seen. The mechanism of this elimination process remains to be unraveled.

Cumulative data suggest that $CD8^+$ T lymphocytes can successfully combat virus-infections in the brain. The strategy, however, seems to differ from that followed in peripheral infections. Due to the lethal hazard that comes from irreversible destruction of neurons, $CD8^+$ T lymphocytes try to cure the cell instead of killing it, at least in the case of neurons which are functionally not impaired by the virus (concerning Ig-induced clearance of viruses from neurons, see chapter by Dietzschold et al.).

4.5 Downsizing of T Effector Lymphocytes

To date, there are no convincing data which would suggest that T cells entering the CNS ever leave this organ. It seems very likely that all undergo apoptosis and that apoptotic cells are phagocytozed by brain-resident microglia cells. So far, the death-triggering signal(s) are poorly defined in the virus-infected CNS.

CD95/FasL-mediated apoptosis might work in a way similar to that reported for the inflamed EAE brain (SUN et al. 1998). In EAE, glial as well as inflammatory $CD4^+$ T lymphocytes express both CD95 and FasL. This situation allows a two-way induction of apoptosis. Not only do T lymphocytes destroy glia cells, but glial cells can kill effector T lymphocytes. The regulatory events that influence the direction of killing is so far not entirely clear, but differential regulation of the density of death receptors on both cell populations could play a decisive role.

Theoretically, CD95/FasL-mediated killing might happen amongst T cells themselves. As discussed above, all activated T lymphocytes co-express CD95 and FasL. As long as the density of virus-infected target cells is high, it is unlikely that T cells meet and kill each other by CD95/FasL interaction. However, with increasing clearance of virus-infected cells from the tissue, the risk of contacting another CD95/FasL-expressing T lymphocyte increases, which could finally result in mutual killing of effector T lymphocytes. However, since expression of the CD95/FasL molecules is strictly dependent on continuous antigenic stimulation, these dangerous molecules will be down-regulated along with the decreasing number of virus-infected target cells, thereby probably allowing some T cells to survive.

Even more speculative than the CD95/FasL hypothesis is the idea that differential expression of T cell costimulatory molecules like B7 could contribute to T cell death in the brain. Engagement of the B7/CD 152 receptor/ligand pair principally causes abrogation of IL-2 synthesis in T cells. The consequences of this switch-off signal depend on the state of T cell differentiation. In resting non-activated T cells, B7/CD 152 interaction causes cell cycle arrest; in activated proliferating T lymphocytes, which depend on IL-2 for growth, apoptosis occurs after B7/CD 152 interaction (SCHEIPERS and REISER 1998). The death signal in the proliferating T cell is independent of CD95/FasL engagement. In Theiler's virus infection of the CNS, B7 is expressed in the virus-infected tissue on all MHC class II-positive cells including brain-infiltrating macrophages and T cells as well as brain-resident microglia (POPE et al. 1998).

In the model of EAE, BAUER et al. (1998) reported consequent elimination of inflammatory T lymphocytes in the parenchyma but not in connective tissue compartments of the brain. This process is totally independent from any antigen specificity. Since apoptosis of T cells in the brain occurs also in mice deficient in the CD95 gene (MALIPIERO et al. 1997), and apoptosis of T cells can be induced by microglial cells also in the absence of B7 expression (FORD et al. 1996), it is very likely that, besides CD95/FasL or B7/CD 152, other mechanisms of T cell elimination operate in the brain parenchyma with so far undefined signal pathways.

Whatever causes death of T lymphocytes in the brain, the CNS plays an active part in this process of elimination. With the dying T cells, the CNS-localized immune response terminates and the tissue returns to its relative immunological silence.

5 Consequences of T Cell Action

After having discussed how the T cell compartment is sensitized to viral antigens from brain, how these cells enter the CNS and home to virus-infected areas, and how they are eliminated, the consequences of T lymphocyte-mediated inflammation need to be examined. Many contradictory data have been published on this topic in the past, which roughly can be grouped into two opposite positions: (1) T cell action in the brain is protective and (2) T cell action in the brain is destructive and disease-inducing. Of course, careful evaluation of all published data reveal a much more complex and differentiated picture, which will be discussed in this chapter. Many routes were followed to make the important distinction between virus-induced and T lymphocyte-mediated destruction of CNS cells. The basis of most experimental approaches to this question was infection of animals that were genetically immunodeficient or that were immunocompromised by treatment with drugs or antibodies that interfere with immune system function.

The clinical consequences of T cell action in the virus-infected brain depend largely on: (a) What is the state of T cell differentiation in relation to the kinetics of viral spread in the tissue? (b) What type of non-T cell immune effector cell acts shortly before or concomitantly with T cells in tissue?

Both points are strongly influenced by the genetic background of the infected host. In general, it can be concluded that a rapid recruitment of a virus-specific T lymphocyte response to the CNS tissue, which is accompanied by virus-neutralizing antibody-secreting plasma cells, will cause much fewer clinical complications than a delayed and rather unspecific response. In the latter case it might even be less harmful to the host if no T cell response at all is recruited to the brain. A typical example demonstrating these rather complicated interrelationships is the experimental infection of the rat inbred strains Lewis and Brown Norway with coronavirus JHM (IMRICH et al. 1994; SCHWENDER et al. 1991). Brown Norway rats never showed any clinical signs of the infection, despite the fact that virus replicates in the brain. Infected rats showed a very early and specific T lymphocyte response in the CLNs and very rapidly recruited a strong virus-neutralizing antibody response to the brain. Consequently, extracellular viral spread was restricted efficiently, allowing a limited number of T lymphocytes to clear the virus from a few, small virus-infected foci in a subclinical manner. In contrast, Lewis rats responded in the early phase of the infection with a broad expansion of T cell clones in CLNs but a delayed and weak virus-neutralizing antibody response. This delay allowed the virus to spread throughout the CNS, including the spinal cord. The immune system responded with a strong infiltration of T lymphocytes ($CD4^+$ and $CD8^+$). $CD8^+$ T cells are MHC class I-restricted killers and, due to the widespread infection of the brain, $CD8^+$-mediated clearance of virus-infected cells contributes to neurological disease. These effects are enhanced by $CD4^+$ T lymphocytes, because they are of the TH1 type and thus attract inflammatory accessory cells like monocytes from the periphery. Many animals die within 7–10 days post-infection, but roughly 60% start to recover at about 12 days post-infection. In theses animals a weak but obviously sufficient virus-neutralizing antibody response can be detected in the CNS. From these data one would assume that the action of T lymphocytes in CNS tissue generally is dangerous for the host. Whether T cell action causes neurological disease depends largely on the quantity of recruited cells, which in turn is related to the extent of viral spread in the tissue. In other words, any mechanism which keeps the virus from spreading warrants a subclinical effector phase of T lymphocytes.

Recently, we have added further support to this idea by irradiation of Lewis rats followed by reconstitution of the immune system with naive spleen and lymph node cells, which were depleted from individual lymphocyte subsets (SCHWENDER et al. 1999). These partially deficient and immunologically naive animals were infected by JHMV and the development of neurological disease was monitored for the following 9 days. It turned out that, whenever B lymphocytes were absent from the transferred lymphocytes, the animals very rapidly became moribund within a few days. Infection of incompetent and not-reconstituted animals, however, resulted in a mild onset and progression of disease that in its severity was well below the level of T cell-reconstituted animals.

At first glance, these data contradict many reports which unequivocally showed that transfer of T lymphocytes protects from virus-induced neurological disease (ERLICH et al. 1989; KÖRNER et al. 1991; STOHLMAN et al. 1995; SUSSMAN et al. 1989; YAMAGUCHI et al. 1991). A more detailed look discloses that in all these cases primed T lymphocytes, T cell lines or T cell clones with specificity for the challenging virus were transferred before infection. As a result, the virus was faced with an overwhelming amount of highly differentiated T effector cells, which immediately interfere with viral spread. If, as in the case of a naive T cell compartment, the race between the virus and the defending T cell system starts with entry of the virus, the amount of time needed by the host to differentiate effector T cells determines the clinical outcome. This is also supported by experiments in which virus-specific T cell lines were transferred at different time points before or after viral infection. Virus-specific T lymphocytes are protective when given shortly before or concomitantly with the virus. Adoptive transfer of such cells 2 days after infection causes severe enhancement of neurological disease (YAMAGUCHI et al. 1991).

Taken together, T effector lymphocytes are a two-edged sword for the host. If they act soon after infection, they will most likely help to prevent disease. If their recruitment is delayed and a virus-neutralizing antibody response is absent, they can contribute significantly to disease.

6 Summary

T lymphocytes play a decisive role in the course and clinical outcome of viral CNS infection. Summarizing the information presented in this review, the following sequence of events might occur during acute virus infection: After invasion of the host and a few initial rounds of replication, the virus reaches the CNS in most cases by hematogeneous spread. After passage through the BBB, CNS cells are infected and replication of virus in brain cells causes activation of the surrounding microglia population. Moreover, local production of IFN-α/β induces expression of MHC antigens on CNS cells, and microglial cells start to phagocytose cellular debris, which accumulates as a result of virus-induced cytopathogenic effects. Upon phagocytosis, microglia becomes more activated; they up-regulate MHC molecules, acquire antigen presentation capabilities and secrete chemokines. This will initiate up-regulation of adhesion molecules on adjacent endothelial cells of the BBB. Transmigration of activated T lymphocytes through the BBB is followed by interaction with APC, presenting the appropriate peptides in the context of MHC antigens. It appears that $CD8^+$ T lymphocytes are amongst the first mononuclear cells to arrive at the infected tissue. Without a doubt, their induction and attraction is deeply influenced by natural killer cells, which, after virus infection, secrete IFN-γ, a cytokine that stimulates $CD8^+$ T cells and diverts the immune response to a TH1-type $CD4^+$ T cell-dominated response. Following the $CD8^+$ T lymphocytes, tissue-penetrating, TH1 $CD4^+$ T cells contact local APC. This results in a tremendous up-regulation of MHC molecules and secretion of more chemotactic and toxic substances. Consequently an increasing number of inflammatory cells, including macrophages/microglia and finally antibody-secreting plasma cells, are attracted to the site of virus infection.

All trapped cells are mainly terminally differentiated cells that are going to enter apoptosis during or shortly after exerting their effector functions. The clinical consequences and the influence of the effector phase on the further course of the infection depends on the balance and fine-tuning of the contributing lymphoid cell populations. Generally, any delay in the recruitment of effector lymphocytes to the tissue or an unbalanced combination of lymphocyte subsets allows the virus to spread in the CNS, which in turn will cause severe immune-mediated tissue effects as well as disease. If either too late or partially deficient, the immune system response may contribute to a lethal outcome or cause autosensitization to brain-specific antigens by epitope spreading to the antigen-presenting system in peripheral lymphoid tissue. This could form the basis for subsequent booster reactions of autosensitized $CD4^+$ T cells – a process that finally will end in an inflammatory autoimmune reaction, which in humans we call multiple sclerosis. In contrast, a rapid and specific local response in the brain tissue will result in efficient limitation of viral spread and thereby a subclinical immune system-mediated termination of the infection.

After clearance of virus-infected cells, downsizing of the local response probably occurs via self-elimination of the contributing T cell populations and/or by so far unidentified signal pathways. However, much of this is highly speculative, and more data have to be collected to make decisive conclusions regarding this matter.

Several strategies have been developed by viruses to escape T cell-mediated eradication, including interference with the MHC class I presentation pathway of the host cell or "hiding" in cells which lack MHC class I expression. This may result in life-long persistence of the virus in the brain, a state which probably is actively controlled by T lymphocytes. Under severe immunosuppression, however, reactivation of viral replication can occur, which is a lethal threat to the host.

References

Akaike T, Weihe E, Schaefer M, Fu ZF, Zheng YM, Vogel W, Schmidt H, Koprowski H, Dietzschold B (1995) Effect of neurotropic virus infection on neuronal and inducible nitric oxide synthase activity in rat brain. J Neurovirol 1:118–125
Aloisi F, Penna G, Cerase J, Menendez Iglesias B, Adorini L (1997) IL-12 production by central nervous system microglia is inhibited by astrocytes. J Immunol 159:1604–1612
Aloisi F, Ria F, Penna G, Adorini L (1998) Microglia are more efficient than astrocytes in antigen processing and in Th1 but not Th2 cell activation. J Immunol 160:4671–4680
Asensio VC, Campbell IL (1997) Chemokine gene expression in the brains of mice with lymphocytic choriomeningitis. J Virol 71:7832–7840
Bauer J, Bradl M, Hickley WF, Forss Petter S, Breitschopf H, Linington C, Wekerle H, Lassmann H (1998) T-cell apoptosis in inflammatory brain lesions: destruction of T cells does not depend on antigen recognition [see comments]. Am J Pathol 153:715–724
Becher B, Barker PA, Owens T, Antel JP (1998) CD95-CD95L: can the brain learn from the immune system? Trends Neurosci 21:114–117
Bell EB, Sparshott SM, Bunce C (1998) $CD4^+$ T-cell memory, CD45R subsets and the persistence of antigen-a unifying concept. Immunol Today 19:60–64
Bi Z, Barna M, Komatsu T, Reiss CS (1995) Vesicular stomatitis virus infection of the central nervous system activates both innate and acquired immunity. J Virol 69:6466–6472
Bilzer T, Stitz L (1994) Immune-mediated brain atrophy. $CD8^+$ T cells contribute to tissue destruction during borna disease. J Immunol 153:818–823
Brankin B, Hart MN, Cosby SL, Fabry Z, Allen IV (1995) Adhesion molecule expression and lymphocyte adhesion to cerebral endothelium: effects of measles virus and herpes simplex 1 virus. J Neuroimmunol 56:1–8
Butz EA, Hostager BS, Southern PJ (1994) Macrophages in mice acutely infected with lymphocytic choriomeningitis virus are primed for nitric oxide synthesis. Microb Pathog 16:283–295
Choe W, Stoica G, Lynn W, Wong PK (1998) Neurodegeneration induced by MoMuLV-ts1 and increased expression of Fas and TNF-alpha in the central nervous system. Brain Res 779:1–8
Christensen JP, Andersson EC, Scheynius A, Marker O, Thomsen AR (1995) Alpha 4 integrin directs virus-activated $CD8^+$ T cells to sites of infection. J Immunol 154:5293–5301
Christian AY, Barna M, Bi Z, Reiss CS (1996) Host immune response to vesicular stomatitis virus infection of the central nervous system in C57BL/6 mice. Viral Immunol 9:195–205

Cosby SL, Brankin B (1995) Measles virus infection of cerebral endothelial cells and effect on their adhesive properties. Vet Microbiol 44:135–139

Cserr HF, DePasquale M, Harling Berg CJ, Park JT, Knopf PM (1992a) Afferent and efferent arms of the humoral immune response to CSF-administered albumins in a rat model with normal blood–brain barrier permeability. J Neuroimmunol 41:195–202

Cserr HF, Harling Berg CJ, Knopf PM (1992b) Drainage of brain extracellular fluid into blood and deep cervical lymph and its immunological significance. Brain Pathol 2:269–276

Cserr HF, Knopf PM (1992) Cervical lymphatics, the blood–brain barrier and the immunoreactivity of the brain: a new view. Immunol Today 13:507–512

Czub M, Czub S, Rappold M, Mazgareanu S, Schwender S, Demuth M, Hein A, Dörries R (1995) Murine leukemia virus-induced neurodegeneration of rats: enhancement of neuropathogenicity correlates with enhanced viral tropism for macrophages, microglia, and brain vascular cells. Virology 214:239–244

del Pozo MA, Sánchez-Mateos P, Sánchez-Madrid F (1996) Cellular polarization induced by chemokines: a mechanism for leukocyte recruitment? Immunol Today 17:127–130

Dhib Jalbut SS, Xia Q, Drew PD, Swoveland PT (1995) Differential up-regulation of HLA class I molecules on neuronal and glial cell lines by virus infection correlates with differential induction of IFN-beta. J Immunol 155:2096–2108

Doherty PC, Allan JE (1986) Role of the major histocompatibility complex in targeting effector T cells into a site of virus infection [published erratum appears in Eur J Immunol 1986 Dec;16(12):1646]. Eur J Immunol 16:1237–1242

Doherty PC, Allan JE, Lynch F, Ceredig R (1990) Dissection of an inflammatory process induced by $CD8^+$ T cells. Immunol Today 11:55–59

Dörries R, Schwender S, Imrich H, Harms H (1991) Population dynamics of lymphocyte subsets in the central nervous system of rats with different susceptibility to coronavirus-induced demyelinating encephalitis. Immunology 74:539–545

Emmett CJ, Lawrence JM, Raisman G, Seeley PJ (1991) Cultured epithelioid astrocytes migrate after transplantation into the adult rat brain. J Comp Neurol 311:330–341

Erlich SS, Matsushima GK, Stohlman SA (1989) Studies on the mechanism of protection from acute viral encephalomyelitis by delayed-type hypersensitivity inducer T cell clones. J Neurol Sci 90:203–216

Fabry Z, Raine CS, Hart MN (1994) Nervous tissue as an immune compartment: the dialect of the immune response in the CNS. Immunol Today 15:218–224

Fierz W, Endler B, Reske K, Wekerle H, Fontana A (1985) Astrocytes as antigen-presenting cells. I. Induction of Ia antigen expression on astrocytes by T cells via immune interferon and its effect on antigen presentation. J Immunol 134:3785–3793

Fontana A, Fierz W, Wekerle H (1984) Astrocytes present myelin basic protein to encephalitogenic T-cell lines. Nature 307:273–276

Ford AL, Foulcher E, Lemckert FA, Sedgwick JD (1996) Microglia induce CD4T lymphocyte final effector function and death. J Exp Med 184:1737–1745

Ford AL, Goodsall AL, Hickey WF, Sedgwick JD (1995) Normal adult ramified microglia separated from other central nervous system macrophages by flow cytometric sorting. Phenotypic differences defined and direct ex vivo antigen presentation to myelin basic protein-reactive $CD4^+$ T cells compared. J Immunol 154:4309–4321

Frei K, Siepl C, Groscurth P, Bodmer S, Schwerdel C, Fontana A (1987) Antigen presentation and tumor cytotoxicity by interferon-gamma-treated microglial cells. Eur J Immunol 17:1271–1278

Froelich CJ, Dixit VM, Yang X (1998) Lymphocyte granule-mediated apoptosis: matters of viral mimicry and deadly proteases. Immunol Today 19:30–36

Gairin JE, Joly E, Oldstone MB (1991) Persistent infection with lymphocytic choriomeningitis virus enhances expression of MHC class I glycoprotein on cultured mouse brain endothelial cells. J Immunol 146:3953–3957

Georgsson G (1994) Neuropathologic aspects of lentiviral infections. Ann NY Acad Sci 724:50–67

Gombold JL, Weiss SR (1992) Mouse hepatitis virus A59 increases steady-state levels of MHC mRNAs in primary glial cell cultures and in the murine central nervous system. Microb Pathog 13:493–505

Grau V, Herbst B, van der Meide PH, Steiniger B (1997) Activation of microglial and endothelial cells in the rat brain after treatment with interferon-gamma in vivo. Glia 19:181–189

Greenwood J, Wang Y, Calder VL (1995) Lymphocyte adhesion and transendothelial migration in the central nervous system: the role of LFA-1, ICAM-1, VLA-4 and VCAM-1. Immunology 86:408–415

Griffith TS, Ferguson TA (1997) The role of FasL-induced apoptosis in immune privilege [published erratum appears in Imunol Today 1997 Jul; 18(7):361]. Immunol Today 18:240–244

Hafler DA, Kent SC, Pietrusewicz MJ, Khoury SJ, Weiner HL, Fukaura H (1997) Oral administration of myelin induces antigen-specific TGF-beta 1 secreting T cells in patients with multiple sclerosis. Ann NY Acad Sci 835:120–131

Harling Berg CJ, Knopf PM, Cserr HF (1991) Myelin basic protein infused into cerebrospinal fluid suppresses experimental autoimmune encephalomyelitis. J Neuroimmunol 35:45–51

Hayashi M, Luo Y, Laning J, Strieter RM, Dorf ME (1995) Production and function of monocyte chemoattractant protein-1 and other beta-chemokines in murine glial cells. J Neuroimmunol 60: 143–150

Hein A, Schwender S, Imrich H, Sopper S, Czub M, Dorries R (1995) Phenotypic and functional characterization of $CD8^+$ T lymphocytes from the central nervous system of rats with coronavirus JHM induced demyelinating encephalomyelitis [see comments]. J Neurovirol 1:340–348

Hickey WF, Hsu BL, Kimura H (1991) T-lymphocyte entry into the central nervous system. J Neurosci Res 28:254–260

Hickey WF, Kimura H (1988) Perivascular microglial cells of the CNS are bone marrow-derived and present antigen in vivo. Science 239:290–292

Hooper DC, Ohnishi ST, Kean R, Numagami Y, Dietzschold B, Koprowski H (1995) Local nitric oxide production in viral and autoimmune diseases of the central nervous system. Proc Natl Acad Sci USA 92:5312–5316

Imrich H, Schwender S, Hein A, Dorries R (1994) Cervical lymphoid tissue but not the central nervous system supports proliferation of virus-specific T lymphocytes during coronavirus-induced encephalitis in rats. J Neuroimmunol 53:73–81

Irani DN, Lin KI, Griffin DE (1997) Regulation of brain-derived T cells during acute central nervous system inflammation. J Immunol 158:2318–2326

Ishihara S, Sawada M, Chang L, Kim JM, Brightman M (1993) Brain vessels near muscle autografts are sites for entry of isogeneic macrophages into brain. Exp Neurol 124:219–230

Joly E, Mucke L, Oldstone MB (1991) Viral persistence in neurons explained by lack of major histocompatibility class I expression. Science 253:1283–1285

June CH, Bluestone JA, Nadler LM, Thompson CB (1994) The B7 and CD28 receptor families. Immunol Today 15:321–331

Kolb H, Kolb-Bachofen V (1998) Nitric oxide in autoimmune disease: cytotoxic or regulatory mediator. Immunology Today 19:556–561

Körner H, Schliephake A, Winter J, Zimprich F, Lassmann H, Sedgwick J, Siddell S, Wege H (1991) Nucleocapsid or spike protein-specific $CD4^+$ T lymphocytes protect against coronavirus-induced encephalomyelitis in the absence of $CD8^+$ T cells. J Immunol 147:2317–2323

Krakowka S, Cork LC, Winkelstein JA, Axthelm MK (1987) Establishment of central nervous system infection by canine distemper virus: breach of the blood-brain barrier and facilitation by antiviral antibody. Vet Immunol Immunopathol 17:471–482

Kulkarni AB, Holmes KL, Fredrickson TN, Hartley JW, Morse HCr (1997) Characteristics of a murine gammaherpesvirus infection immunocompromised mice. In Vivo 11:281–291

Kündig TM, Hengartner H, Zinkernagel RM (1993) T cell-dependent IFN-gamma exerts an antiviral effect in the central nervous system but not in peripheral solid organs. J Immunol 150:2316–2321

Lampson LA (1998) Beyond inflammation: site-directed immunotherapy. Immunol Today 19:17–22

Lane TE, Buchmeier MJ, Watry DD, Fox HS (1996) Expression of inflammatory cytokines and inducible nitric oxide synthase in brains of SIV-infected rhesus monkeys: applications to HIV-induced central nervous system disease. Mol Med 2:27–37

Lynch DH, Ramsdell F, Alderson MR (1995) Fas and FasL in the homeostatic regulation of immune responses [see comments]. Immunol Today 16:569–574

Male DK, Pryce G, Hughes CC (1987) Antigen presentation in brain: MHC induction on brain endothelium and astrocytes compared. Immunology 60:453–459

Malipiero U, Frei K, Spanaus KS, Agresti C, Lassmann H, Hahne M, Tschopp J, Eugster HP, Fontana A (1997) Myelin oligodendrocyte glycoprotein-induced autoimmune encephalomyelitis is chronic/relapsing in perforin knockout mice, but monophasic in Fas- and Fas ligand-deficient lpr and gld mice. Eur J Immunol 27:3151–3160

Marker O, Scheynius A, Christensen JP, Thomsen AR (1995) Virus-activated T cells regulate expression of adhesion molecules on endothelial cells in sites of infection. J Neuroimmunol 62:35–42

Massa P, Dörries R, ter Meulen V (1986) Viral particles induce Ia antigen expression on astrocytes. Nature 320:543–546

Menendez Iglesias B, Cerase J, Ceracchini C, Levi G, Aloisi F (1997) Analysis of B7-1 and B7-2 costimulatory ligands in cultured mouse microglia: up-regulation by interferon-gamma and lipopolysaccharide and downregulation by interleukin-10, prostaglandin E2 and cyclic AMP-elevating agents. J Neuroimmunol 72:83–93

Mercer JA, Albanesi JP, Brady ST (1994) Molecular motors and cell motility in the brain. Brain Pathol 4:167–179

Mills CD (1991) Molecular basis of "suppressor" macrophages. Arginine metabolism via the nitric oxide synthetase pathway. J Immunol 146:2719–2723

Mokhtarian F, Wesselingh SL, Choi S, Maeda A, Griffin DE, Sobel RA, Grob D (1996) Production and role of cytokines in the CNS of mice with acute viral encephalomyelitis. J Neuroimmunol 66: 11–22

Morimoto K, Hooper DC, Bornhorst A, Corisdeo S, Bette M, Fu ZF, Schafer MK, Koprowski H, Weihe E, Dietzschold B (1996) Intrinsic responses to Borna disease virus infection of the central nervous system. Proc Natl Acad Sci USA 93:13345–13350

Morris MM, Dyson H, Baker D, Harbige LS, Fazakerley JK, Amor S (1997) Characterization of the cellular and cytokine response in the central nervous system following Semliki Forest virus infection. J Neuroimmunol 74:185–197

Mucke L, Oldstone MB (1992) The expression of major histocompatibility complex (MHC) class I antigens in the brain differs markedly in acute and persistent infections with lymphocytic choriomeningitis virus (LCMV). J Neuroimmunol 36:193–198

Nansen A, Christensen JP, Ropke C, Marker O, Scheynius A, Thomsen AR (1998) Role of interferon-gamma in the pathogenesis of LCMV-induced meningitis: unimpaired leucocyte recruitment, but deficient macrophage activation in interferon-gamma knock-out mice. J Neuroimmunol 86:202–212

Neumann H, Boucraut J, Hahnel C, Misgeld T, Wekerle H (1996) Neuronal control of MHC class II inducibility in rat astrocytes and microglia. Eur J Neurosci 8:2582–2590

Neumann H, Cavalie A, Jenne DE, Wekerle H (1995) Induction of MHC class I genes in neurons [see comments]. Science 269:549–552

Neumann H, Misgeld T, Matsumuro K, Wekerle H (1998) Neurotrophins inhibit major histocompatibility class II inducibility of microglia: involvement of the p75 neurotrophin receptor. Proc Natl Acad Sci USA 95:5779–5784

Neumann H, Schmidt H, Cavalie A, Jenne D, Wekerle H (1997) Major histocompatibility complex (MHC) class I gene expression in single neurons of the central nervous system: differential regulation by interferon (IFN)-gamma and tumor necrosis factor (TNF)-alpha. J Exp Med 185:305–316

Neumann H, Wekerle H (1998) Neuronal control of the immune response in the central nervous system: linking brain immunity to neurodegeneration. J Neuropathol Exp Neurol 57:1–9

Nguyen VT, Walker WS, Benveniste EN (1998) Post-transcriptional inhibition of CD40 gene expression in microglia by transforming growth factor-beta. Eur J Immunol 28:2537–2548

Njenga MK, Pease LR, Wettstein P, Mak T, Rodriguez M (1997) Interferon alpha/beta mediates early virus-induced expression of H-2D and H-2K in the central nervous system. Lab Invest 77:71–84

Nottet HSLM, Gendelman HE (1995) Unraveling the neuroimmune mechanisms for the HIV-1-associated cognitive/motor complex. Immunol Today 16:441–448

Ohmori K, Hong Y, Fujiwara M, Matsumoto Y (1992) In situ demonstration of proliferating cells in the rat central nervous system during experimental autoimmune encephalomyelitis. Evidence suggesting that most infiltrating T cells do not proliferate in the target organ. Lab Invest 66:54–62

Oldstone MB, Blount P, Southern PJ, Lampert PW (1986) Cytoimmunotherapy for persistent virus infection reveals a unique clearance pattern from the central nervous system. Nature 321:239–243

Oleszak EL, Katsetos CD, Kuzmak J, Varadhachary A (1997) Inducible nitric oxide synthase in Theiler's murine encephalomyelitis virus infection. J Virol 71:3228–3235

Pardridge WM, Yang J, Buciak J, Tourtellotte WW (1989) Human brain microvascular DR-antigen. J Neurosci Res 23:337–341

Pearce BD, Hobbs MV, McGraw TS, Buchmeier MJ (1994) Cytokine induction during T-cell-mediated clearance of mouse hepatitis virus from neurons in vivo. J Virol 68:5483–5495

Pertile TL, Karaca K, Sharma JM, Walser MM (1996) An antiviral effect of nitric oxide: inhibition of reovirus replication. Avian Dis 40:342–348

Planz O, Bilzer T, Sobbe M, Stitz L (1993) Lysis of major histocompatibility complex class I-bearing cells in Borna disease virus-induced degenerative encephalopathy. J Exp Med 178:163–174

Planz O, Bilzer T, Stitz L (1995) Immunopathogenic role of T-cell subsets in Borna disease virus-induced progressive encephalitis. J Virol 69:896–903 issn: 0022–0538x

Pope JG, Vanderlugt CL, Rahbe SM, Lipton HL, Miller SD (1998) Characterization and Functional Antigen Presentation by Central Nervous System Mononuclear Cells from Mice Infected with Theiler's Murine Encephalomyelitis Virus. J Virol 72:7762–7771

Rall GF, Mucke L, Oldstone MB (1995) Consequences of cytotoxic T lymphocyte interaction with major histocompatibility complex class I-expressing neurons in vivo. J Exp Med 182:1201–1212

Rensing Ehl A, Malipiero U, Irmler M, Tschopp J, Constam D, Fontana A (1996) Neurons induced to express major histocompatibility complex class I antigen are killed via the perforin and not the Fas (APO-1/CD95) pathway. Eur J Immunol 26:2271–2274

Rossi CP, McAllister A, Tanguy M, Kagi D, Brahic M (1998) Theiler's virus infection of perforin-deficient mice. J Virol 72:4515–4519

Sandberg K, Kemper P, Stalder A, Zhang J, Hobbs MV, Whitton JL, Campbell IL (1994) Altered tissue distribution of viral replication and T cell spreading is pivotal in the protection against fatal lymphocytic choriomeningitis in mice after neutralization of IFN-alpha/beta. J Immunol 153:220–231

Sarawar SR, Sparshott SM, Sutton P, Yang CP, Hutchinson IV, Bell EB (1993) Rapid re-expression of CD45RC on rat CD4T cells in vitro correlates with a change in function. Eur J Immunol 23: 103–109

Scheipers P, Reiser H (1998) Fas-independent death of activated $CD4^+$ T lymphocytes induced by CTLA-4 crosslinking. Proc Natl Acad Sci USA 95:10083–10088

Schwender S, Hein A, Imrich H, Czub S, Dörries R (1999) Modulation of acute coronavirus-induced encephalomyelitis in g-irradiated rats by transfer of naive lymphocyte subsets before infection. J Neuro Virol 5:249–257

Schwender S, Imrich H, Dörries R (1991) The pathogenic role of virus-specific antibody-secreting cells in the central nervous system of rats with different susceptibility to coronavirus-induced demyelinating encephalitis. Immunology 74:533–538

Sedgwick JD, Ford AL, Foulcher E, Airriess R (1998) Central nervous system microglial cell activation and proliferation follows direct interaction with tissue-infiltrating T cell blasts. J Immunol 160: 5320–5330

Sedgwick JD, Mössner R, Schwender S, ter Meulen V (1991a) Major histocompatibility complex-expressing nonhematopoietic astroglial cells prime only $CD8^+$ T lymphocytes: astroglial cells as perpetuators but not initiators of $CD4^+$ T cell responses in the central nervous system. J Exp Med 173:1235–1246

Sedgwick JD, Schwender S, Gregersen R, Dorries R, ter Meulen V (1993) Resident macrophages (ramified microglia) of the adult Brown Norway rat central nervous system are constitutively major histocompatibility complex class II positive. J Exp Med 177:1145–1152

Sedgwick JD, Schwender S, Imrich H, Dörries R, Butcher GW, ter-Meulen V (1991b) Isolation and direct characterization of resident microglial cells from the normal and inflamed central nervous system. Proc Natl Acad Sci USA 88:7438–7442

Soilu Hanninen M, Roytta M, Salmi AA, Salonen R (1997) Semliki Forest virus infection leads to increased expression of adhesion molecules on splenic T-cells and on brain vascular endothelium. J Neurovirol 3:350–360

Springer TA (1994) Traffic signals for lymphocyte recirculation and leukocyte. Emigration: The multistep paradigm. Cell 76:301–314

Stohlman SA, Bergmann CC, Lin MT, Cua DJ, Hinton DR (1998) CTL effector function within the central nervous system requires $CD4^+$ T cells. J Immunol 160:2896–2904

Stohlman SA, Bergmann CC, van der Veen RC, Hinton DR (1995) Mouse hepatitis virus-specific cytotoxic T lymphocytes protect from lethal infection without eliminating virus from the central nervous system. J Virol 69:684–694

Stout RD, Suttles J (1996) The many roles of CD40 in cell-mediated inflammatory responses. Immunology Today 17:487–492

Subak Sharpe I, Dyson H, Fazakerley J (1993) In vivo depletion of $CD8^+$ T cells prevents lesions of demyelination in Semliki Forest virus infection. J Virol 67:7629–7633

Sun D, Whitaker JN, Cao L, Han Q, Sun S, Coleclough C, Mountz J, Zhou T (1998) Cell death mediated by Fas-FasL interaction between glial cells and MBP-reactive T cells. J Neurosci Res 52:458–467

Sussman MA, Shubin RA, Kyuwa S, Stohlman SA (1989) T-cell-mediated clearance of mouse hepatitis virus strain JHM from the central nervous system. J Virol 63:3051–3056

Tishon A, Eddleston M, de la Torre JC, Oldstone MB (1993) Cytotoxic T lymphocytes cleanse viral gene products from individually infected neurons and lymphocytes in mice persistently infected with lymphocytic choriomeningitis virus. Virology 197:463–467

Tontsch U, Rott O (1993) Cortical neurons selectively inhibit MHC class II induction in astrocytes but not in microglial cells. Int Immunol 5:249–254

Ward LA, Massa PT (1995) Neuron-specific regulation of major histocompatibility complex class I, interferon-beta, and anti-viral state genes. J Neuroimmunol 58:145–155

Watanabe R, Wege H, ter Meulen V (1983) Adoptive transfer of EAE-like lesions from rats with coronavirus-induced demyelinating encephalomyelitis. Nature 305:150–153

Watanabe R, Wege H, ter Meulen V (1987) Comparative analysis of coronavirus JHM-induced demyelinating encephalomyelitis in Lewis and Brown Norway rats. Lab Invest 57:375–384

Wekerle H, Engelhardt B, Risau W, Meyermann R (1991) Interaction of T lymphocytes with cerebral endothelial cells in vitro. Brain Pathol 1:107–114

Weller RO, Engelhardt B, Phillips MJ (1996) Lymphocyte targeting of the central nervous system: a review of afferent and efferent CNS-immune pathways. Brain Pathol 6:275–288

Wesselingh SL, Levine B, Fox RJ, Choi S, Griffin DE (1994) Intracerebral cytokine mRNA expression during fatal and nonfatal alphavirus encephalitis suggests a predominant type 2T cell response. J Immunol 152:1289–1297

Williams KC, Hickey WF (1995) Traffic of hematogenous cells through the central nervous system. Curr Top Microbiol Immunol 202:221–245

Wong GH, Bartlett PF, Clark Lewis I, Battye F, Schrader JW (1984) Inducible expression of H-2 and Ia antigens on brain cells. Nature 310:688–691

Yamaguchi K, Goto N, Kyuwa S, Hayami M, Toyoda Y (1991) Protection of mice from a lethal coronavirus infection in the central nervous system by adoptive transfer of virus-specific T cell clones. J Neuroimmunol 32:1–9

Virus-Induced Autoimmune Reactions in the CNS

P.J. Talbot[1], D. Arnold[2], and J.P. Antel[2]

1 Introduction.	247
2 Viruses and Multiple Sclerosis	249
3 Coronaviruses and Multiple Sclerosis	251
4 Murine Coronavirus-Induced Immune-Mediated CNS Disease	253
5 Systemic Virus Infection and Autoimmunity: Virus-Initiated Neuropathogenesis by Activation of Self-Reactive Lymphocytes (Molecular Mimicry)	255
6 Systemic Virus Infections: Induction of Non-Antigen-Specific Immune Activation	257
7 Virus-Induced Immune Activity within the CNS: Antigen-Specific Responses	258
8 Virus-Induced Immune Reactivity within the CNS: Effects on Glial Cells	259
9 Axonal Injury in Multiple Sclerosis.	259
References	263

1 Introduction

Neurologic diseases are diverse and often not well-understood, despite the tremendous health care problems they pose, such as the estimated 22 million persons worldwide who suffer from dementia, a characteristic loss of mental capacities. Virus infections are an established contributor to development of an array of diseases of the central nervous system (CNS) and are implicated in a further spectrum of disorders in which the etiology is not yet formally established. For example, 60% of acquired immune deficiency syndrome (AIDS) patients suffer neurologic sequelae presumably caused by human immunodeficiency virus (HIV). Viruses can contribute to the development of neurologic disease via an array of direct and indirect mechanisms, as summarized in Table 1 and described in detail in this chapter. Direct mechanisms refer to neural cell injury arising as a

---

[1] Centre de recherche en santé humaine, INRS-Institut Armand-Frappier 531, boulevard des Prairies, Laval, Québec, Canada, H7V 1B7
e-mail: Pierre.Talbot@inrs-iaf.uquebec.ca
[2] Montreal Neurologic Institute, McGill University, 3801 University Street, Montreal, Québec, Canada, H3A 2B4

Table 1. Virus involvement in autoimmune disease of the CNS

	Effects on systemic immune response	Effects within CNS compartment

Induction or maintenance of antigen-restricted responses	Cross-reactivity of virus and CNS antigen – molecular mimicry
	Response to CNS antigens transported to regional lymph nodes
		Response to viral antigens expressed in neural cells
		Increased expression or release of neural antigens by virus infected cells:
		  Sensitization to autoantigens – break in tolerance
		  Expansion of antigenic targets – epitope or determinant spreading
		Antigen presentation by APCs within CNS to pre-sensitized lymphocytes

Non-antigen-restricted enhancement of immune response	Activation of APCs
	Changes in immune regulatory cells and cytokines
	Induction of molecules involved in lymphocyte trafficking
		Blood–brain barrier:
		  Induction of chemoattractant and adhesion molecules
		  Activation of perivascular APCs
		Parenchyma:
		  Activation of APC and immune regulatory capacity of microglia
		  Enhance effector functions of glial cells:
		    Effector cytokines, proteases, NO
		    Antibody-dependent cell cytotoxicity

APCs, antibody-presenting cells; *NO*, nitric oxide.

consequence of the cells becoming infected with virus per se. Such infections can result in actual cell lysis, as in the case of poliomyelitis or herpes simplex encephalitis, or loss of cell function without cell death, a phenomenon termed loss of luxury function (OLDSTONE et al. 1982). Subacute sclerosing panencephalitis (SSPE) is an example of a chronic neurologic disease initially considered to be neurodegenerative but which is now established to be a result of a direct and persistent measles virus infection.

Indirect mechanisms of virus-induced neurologic disease would include those in which disease development is dependent on participation of non-viral mechanisms, which in the context of this chapter will be related to products of the immune system. Such indirect effector mechanisms could result from infection either outside or within the CNS. Infection of systemic lymphoid organs with virus could promote development of specific CNS antigen-directed immune responses by the process of molecular mimicry which implies that the neural-directed immune response arises because of antigenic homologies between viral and neural antigens (OLDSTONE 1987, 1998). The entity of acute disseminated encephalomyelitis, which follows immunization with neural tissue-containing vaccine such as the original Pasteur rabies vaccine, establishes the precedent that immune-mediated CNS demyelinating disease can occur, although in this condition the immune response is induced by non-viral antigens contained in the vaccine. Systemic virus infection can also alter immune regulatory activity so as to favor immune activation, which in turn would favor migration of the immune response into the CNS.

Virus infection within the CNS could trigger an immune response by an array of mechanisms. Infection of neural cells even with defective virus, such as with genetically engineered adenoviruses used in gene transfer paradigms, can evoke a viral antigen-directed immune response with harmful effects on host tissue. The Theiler's murine encephalomyelitis virus (TMEV) model of chronic immune-mediated CNS demyelination indicates that an immune response directed at myelin antigens can evolve over a prolonged time period, presumably as a result of a break in immune tolerance as such antigens become exposed or released consequent to the initial infection; this effect has been referred to as epitope or determinant spreading and is not dependent on molecular mimicry (LEHMANN et al. 1992; MILLER et al. 1997). Tissue injury can also result from release of effector molecules including cytokines from glial cells, particularly microglia, which become infected with virus or are activated by its presence; such an effect has been referred to as bystander injury (HORWITZ et al. 1998). This mechanism has been invoked to account for the neuronal injury associated with HIV encephalopathy (CONANT et al. 1998; KOLSON et al. 1998).

In this chapter, we will consider whether and how the direct and indirect consequences of viral infection can contribute to development of the human disease multiple sclerosis (MS), which is regarded as the prototype of a cellular immune-mediated CNS disorder. Whether an initial unique viral infection initiates the disease process remains speculative (TALBOT 1995). Since demyelination is the hallmark of MS, throughout this chapter, we will illustrate how virus-related immunopathogenic mechanisms can contribute to CNS demyelination. We will focus on coronaviruses to illustrate how experimental models have been used to characterize the immunopathogenesis of virally mediated CNS demyelinating disorders and how one can undertake a search for disease-related viruses in the human disease state. Although the immune system is regarded to be directed primarily at the myelin and its cell of origin, the oligodendrocyte in MS, there is increasing recognition that there is substantial axonal injury as well. We will consider in some detail the basis whereby such injury can result either as a primary or secondary event in the disease course and how it contributes to the extent of clinical neurologic disability.

Multiple sclerosis is a chronic debilitating neurologic disease that affects 0.1%–0.2% of the population in high-risk areas such as Canada and northern parts of the United States of America and of Europe. Both genes and environment have been brought to the "Supreme Court" of scientific suspicions, although a final judgment is still awaited and may be a long-time in coming, since several genes and several environmental factors may contribute to neuropathogenesis.

2 Viruses and Multiple Sclerosis

Even though the etiology of MS remains elusive, a working hypothesis is that it is a polygenic disease (SADOVNICK et al. 1997) in which an infectious agent or agents play(s) a role, either in disease initiation and/or in its propagation and relapses. Interestingly, an infectious etiology was postulated when the disease was first described in the nineteenth century. Viruses would be the most likely infectious culprit in genetically predisposed individuals (KURTZKE 1993; COOK et al. 1995). Several viruses have been implicated in the last two decades, although not one has so far withstood the test of time or closer scrutiny, perhaps because several different viruses, acting through similar direct and/or indirect mechanisms, could be involved (TALBOT 1995). Neurotropic viruses recovered from patients include: measles, herpes simplex, parainfluenza, tick-borne encephalitis and rabies viruses, simian virus 5, coronaviruses and cytomegalovirus (JOHNSON 1985), and more recently a retrovirus (PERRON et al. 1997). A series of studies have used the findings of increased serum titers and/or local antibody synthesis against various viral agents within the CNS of MS patients as evidence for their participation in the MS disease process (SALMI et al. 1981). Measles, herpes simplex, and coronaviruses (SALMI et al. 1982) were again implicated, as were mumps, varicella, influenza C, rubella and vaccinia viruses. Findings regarding simian virus 5 are inconsistent (GOSWAMI et al. 1987; VANDVIK and NORRBY 1989). Epstein-Barr (BRAY et al. 1992; KINNUNEN et al. 1990) and human T lymphotropic (SHIRAZIAN et al. 1993) viruses are further examples. Analysis of viral antibodies in MS affected and non-affected twin pairs could only confirm elevated cerebrospinal fluid antibodies to rubella and vaccinia viruses (WOYCIECHOWSKA et al. 1985). In situ hybridization detected measles genomes in only a few MS patients (HAASE et al. 1981; COSBY et al. 1989). This technique was also used to detect murine-related coronaviruses in brains of 12 of 22 MS patients, with antigen detected by immunohistochemistry in two patients (MURRAY et al. 1992a). The development of the very sensitive polymerase chain reaction (PCR) technique led to the detection of human T lymphotropic virus genetic information in MS patients (REDDY et al. 1989), although other groups have failed to confirm this result (e.g., DEKABAN and RICE 1990), which has unfortunately given a bad reputation to the PCR-based search for MS-associated viruses, even when studies are very carefully performed.

A recent study attempting to link human herpes virus-6 with MS described an IgM response, suggestive of viral reactivation, but only in patients with the relapsing-remitting form of MS. PCR-based viral DNA detection in blood cells was positive in less than a third of MS patients tested (SOLDAN et al. 1997), which prompted words of caution (STEINMAN and OLDSTONE 1997). A further cautionary note is that detection of virus and disease causality are not necessarily linked. One need consider whether there is a normal viral flora in both the CNS and in the lymphoid system which may be more readily expressed in response to inflammatory conditions. Thus, a viral etiology for MS remains speculative, as do the pathogenic mechanisms that would lead from initial viral exposure in a genetically susceptible individual to chronic disease.

3 Coronaviruses and Multiple Sclerosis

Coronaviruses appear in the long list of viruses occasionally associated with MS. Human coronaviruses (HCoV) were first isolated in the mid-1960s from patients with upper respiratory tract disease (TYRRELL and BYNOE 1965; HAMRE and PROCKNOW 1966; MCINTOSH et al. 1967). Their characteristic morphological appearance led to the creation of a new virus family (TYRRELL et al. 1968). Serological studies distinguish HCoV into two groups, named 229E and OC43 after the initial isolates. Viral proteins have slowly been identified, either directly or through the cloning, sequencing, and expression of viral RNAs, and resemble proteins of other coronavirus strains (ARPIN and TALBOT 1990; MOUNIR and TALBOT 1992, 1993a,b; MOUNIR et al. 1994; LABONTÉ et al. 1995). These viruses are recognized respiratory pathogens involved in up to one-third of common colds and to which close to 100% of the population has been exposed, as shown by seroconversion (MYINT 1994).

The possible involvement of these viruses in neurologic disease such as MS remains an intriguing possibility that is sustained by several lines of evidence. Direct relevance of coronaviruses to MS was initially suggested by isolation of two coronaviruses from the CNS of two MS patients (BURKS et al. 1980). Even though a murine origin of these isolates was suggested from genomic (WEISS 1983) and antigenic (FLEMING et al. 1988) assessments, recent reports strengthen the possible importance of coronaviruses in MS. First, there was a preferential association of murine-like coronavirus genes with MS brain tissue, shown by in situ hybridization (MURRAY et al. 1992a). Second, susceptibility of primates to murine coronavirus-induced demyelinating disease, after intracerebral (MURRAY et al. 1992b, 1997), or even peripheral inoculation (CABIRAC et al. 1994) was demonstrated. This suggests that previous MS isolates may indeed have derived from infection by murine-like viruses. As in rodents, astrocytes were described as a site of viral persistence in primates (MURRAY et al. 1997) and virus could find its way to the CNS after a peripheral inoculation (CABIRAC et al. 1994). Other findings implicating coronaviruses in MS include intrathecal antibody synthesis (SALMI et al. 1982) and controversial ultrastructural observations (TANAKA et al. 1976). However, serological assessments failed to show increased levels of anti-HCoV antibodies in MS serum and/or cerebrospinal fluid (FLEMING et al. 1988; HOVANEC and FLANAGAN 1983; JOHNSON-LUSSENBURG and ZHENG 1987). One study did show a significant association of colds with MS exacerbations and an equally significant association of HCoV-229E infection in MS patients compared to controls (JOHNSON-LUSSENBURG and ZHENG 1987). Interestingly, in the first report on the association of viral infections and MS, it was commented that seasonal HCoV infection patterns do indeed fit the observed occurrence of MS exacerbations (SIBLEY et al. 1985).

More recently, molecular biologic approaches have been applied to resolving the relation between coronavirus infection and MS. An in situ hybridization study with cDNA synthesized from the RNA of HCoV-OC43 failed to detect this viral genome in four autopsy and one biopsy samples (SORENSEN et al. 1986). However, these authors speculated that more sensitive methods of detection may be required in view of the probably low viral titers and the small number of infected CNS cells (SORENSEN et al. 1986). Thus, we initiated a search for HCoV in human brains a few years ago with a more sensitive assay. Our initial pilot study did reveal the presence of HCoV-229E RNA in MS autopsy brain tissue (STEWART et al. 1992). Extremely stringent RT-PCR RNA amplifications followed by Southern blot hybridization and nucleotide sequencing were performed on coded tissue. After this was completed, codes were broken and we found the following from 90 donors: both strains of HCoV were surprisingly prevalent (44% for HCoV-779E and 23% for HCoV-OC43), without signs of preferential histological expression sites. A statistically significant higher prevalence of HCoV-OC43 was found in MS patients than in either neurologically normal controls or controls with other neurologic diseases (ARBOUR et al. 2000). These results are consistent with the neurotropic potential of HCoV but, as in any other such studies, relevance to MS cannot be directly inferred. The next phase of our HCoV detection study involves in situ detection of viral RNA and the identification of sites of persistence by double-labeling of virus and cell markers. We have obtained initial evidence for the expression of HCoV-OC43 in neural tissue, after demonstrating the relative abundance of viral material by Northern blotting (ARBOUR et al. 2000). Our studies are consistent with persistence of HCoV in vivo, which is indicative that these viruses could be part of a CNS viral flora that only become pathogenically significant under conditions that remain to be determined but may for example include the expression of susceptibility genes.

Persistent HCoV infections in vitro have been reported, as for example HCoV-229E in the L132 human fibroblast cell line (CHALONER-LARSSON and JOHNSON-LUSSENBURG 1981) and HCoV-OC43 in human glioblastoma and rhabdomyosarcoma cells (COLLINS and SORENSEN 1986). Our own work has demonstrated that cell lines of neural origin, including oligodendrocytes, are susceptible to persistent infection by both strains of HCoV (ARBOUR et al. 1999a,b, 1998). We further showed that CD13 is the cellular receptor for HCoV-229E on susceptible neural cell lines (TALBOT et al. 1994; LACHANCE et al. 1998). Primary cultures of mouse CNS cells can also be infected with HCoV-OC43 (PEARSON and MIMS 1985). Neuronal cells produced infectious virus whereas astrocytes only produced viral antigens. Oligodendrocytes were apparently not infected in that study. Human fetal brain cells were also susceptible to HCoV-OC43 infection, although no infectious virus was detected, indicative of an abortive viral infection. Such abortive or incomplete viral replication cycles in neural cells could also have pathological consequences, for example by altering the "luxury" cellular functions or attracting an immune response to the CNS. Our own studies have also demonstrated the susceptibility of primary human microglia and astrocyte cultures to infection by both HCoV strains (BONAVIA et al. 1997), and more recently oligodendrocytes (Viau et al. unpublished data). Moreover, it is known that murine and human coronaviruses can infect human brain endothelial cells (CABIRAC et al. 1995), a potential route of entry into the CNS from peripheral blood, after viremia (MYINT 1994), or infection of blood lymphocytes or macrophages (BANG and WARWICK 1960; LAMONTAGNE et al. 1989). Importantly, as shown in studies with murine coronaviruses, replication in neuronal and glial cell cultures has correlated with in vivo susceptibility to infection and disease development (e.g. PASICK and DALES 1991; LAMONTAGNE et al. 1989), which adds credence to the possible in vivo relevance of in vitro studies. Thus, coronaviruses may provide more than animal models for demyelinating disorders, as described in the next section: they may be active participants in human disease (HOUTMAN and FLEMING 1996).

4 Murine Coronavirus-Induced Immune-Mediated CNS Disease

Like HCoV, murine coronaviruses are respiratory pathogens. However, they have also attracted attention as causal agents of liver infections, hence their name MHV, for mouse hepatitis virus. Neurotropism of murine coronaviruses was first reported in 1949, when the JHM strain of MHV was isolated from mice with disseminated encephalomyelitis with extensive demyelination (CHEEVER et al. 1949; BAILEY et al. 1949). The hemagglutinating encephalomyelitis virus of pigs represents another example of neurologic involvement in a coronavirus infection (WEGE 1995). Infection of rodents with MHV provides an excellent animal model for virus-induced demyelinating disease, as originally described for the JHM strain (WEINER 1973; LAMPERT et al. 1973). Infection of adult mice culminates in encephalomyelitis with infection of both glial and neuronal cells and over 95% mortality. The few survivors develop chronic white matter pathology characterized by focal CNS demyelinating lesions, with subsequent remyelination and recurrent demyelination. Demyelination was long thought to be a primary effect of virus infection and destruction of myelin-synthesizing oligodendrocytes. However, an immunopathological mechanism, as is observed in MS, was subsequently documented (WANG et al. 1990; HOUTMAN and FLEMING 1996). Genetic susceptibility to MHV-mediated CNS disease in mice correlates, both in vivo and in vitro, with the infection of neurons and macrophages and is controlled by a single autosomal dominant gene on chromosome 7 (KNOBLER et al. 1981).

Murine coronaviruses can access the CNS through anteroneuronal transport via the olfactory nerve (LAVI et al. 1988; BARNETT and PERLMAN 1993), although the possibility remains that infected macrophages (BANG and WARWICK 1960; KNOBLER et al. 1981) may carry virus to the CNS, as is observed with HIV. Coronavirus CNS persistence appears to involve mainly oligodendrocytes (KNOBLER et al. 1982) and astrocytes (PERLMAN and RIES 1987), with possible neuronal involvement (PASICK and DALES 1991). Possible infection of microglial cells has not been seriously investigated, which is surprising given the long-known macrophage susceptibility (BANG and WARWICK 1960) and the above-mentioned genetically determined susceptibility of neurons and macrophages to infection (KNOBLER et al. 1981).

The underlying mechanisms explaining why neurons are involved in the acute phase of MHV (and TMEV) disease but appear spared in the chronic infection are not known but may involve resistance of glial cells to immune clearance (BUCHMEIER et al. 1984; STOHLMAN et al. 1995) or the appearance of putative glial cell-specific viral variants that are resistant to the action of cytotoxic T cells (PEWE et al. 1996) or to antibody neutralization (DALZIEL et al. 1986; FLEMING et al. 1986). Virus mutants, either thermosensitive (HASPEL et al. 1978), small plaque (ERLICH et al. 1987) or neutralizing monoclonal antibody (mAb)-selected (DALZIEL et al. 1986; FLEMING et al. 1986), show modified pathology with sparing of neurons and infection of oligodendrocytes. Passive transfer of neutralizing mAbs modulated disease from fatal encephalomyelitis to chronic recurrent demyelination (BUCHMEIER et al. 1984). Selective cell susceptibilities to infection could also relate to initial virus-cell interactions that may differ according to the cell type that is infected. A member of the carcinoembryogenic (CEA) family of antigens has been identified as a cellular receptor for MHV (WILLIAMS et al. 1991), and a CNS receptor from the same CEA family has also been identified (CHEN et al. 1995a). A variety of co-factors are also required for viral penetration after receptor binding (YOKOMORI et al. 1993). It is conceivable to imagine that these early events in viral infection are less prone to immune intervention when glial cells are the viral target.

Observations applicable to possible viral-associated immune mechanisms underlying development of MS have been made with the MHV model. Immunosuppression prevents virus-induced demyelination. The disease can be adoptively transferred by spleen cells from virus-infected donors (WANG et al. 1990). Cellular sensitization to myelin basic protein (MBP) and adoptive transfer of encephalomyelitis with such cells, as is done in the classic experimental allergic encephalomyelitis (EAE) models, has been reported in rats infected with MHV-JHM (WATANABE et al. 1983). MHV-JHM infection of mice also induces self-reactive T cells (KYUWA et al. 1991). Interestingly, MHV infection can enhance the course of EAE (CROSS et al. 1987).

Various immunopathogenically relevant alterations are shown to occur within the CNS after MHV infection. Viral particles induce expression of class II major histocompatibility (MHC) antigens on astrocytes (MASSA et al. 1986) and class I antigens on astrocytes and oligodendrocytes (SUZUMURA et al. 1986), which could presumably activate local immune responses within the CNS. Actual infection of astrocytes is essential for MHC class I induction (GILMORE et al. 1994). Infection of oligodendrocytes results in an initial down-regulation of mRNA for proteolipid protein (PLP) followed by necrotic death; demyelination continues even in the absence of detectable viral antigen with apoptosis becoming the prevalent mode of cell death (BARAC-LATAS et al. 1997). Infection of activated glial cells, astrocytes and microglia, releases various inflammatory mediators as also observed in MS, including nitric oxide (NO) synthase and NO (GRZYBICKI et al. 1997; EDWARDS et al. 2000), interleukin (IL)-6, tumor necrosis factor-(TNF)-α, IL-1β (SUN et al. 1995; EDWARDS et al. 2000), as well as macrophage chemoattractant protein-1 and matrix metalloproteinases (EDWARDS et al. 2000). Each of these could contribute to tissue injury.

5 Systemic Virus Infection and Autoimmunity: Virus-Initiated Neuropathogenesis by Activation of Self-Reactive Lymphocytes (Molecular Mimicry)

Myelin basic protein and PLP constitute the most abundant proteins in the CNS (HASHIM 1978; MIKOSHIBA et al. 1991). It is believed that MBP contributes to the compactness of myelin lamellae by fusing the cytoplasmic surfaces of oligodendrocytes into major dense lines. PLP is the primary constituent protein of myelin, and it represents 50% of the myelin membrane proteins. It bears covalently linked lipids, shows five strongly hydrophobic transmembrane domains and surface-exposed regions and is thought to play a crucial role in myelination in the CNS, probably by promoting the apposition of extracellular surfaces of the myelin lamellae (DIEHL et al. 1986). Myelin-associated glycoprotein (MAG) and myelin-oligodendrocyte glycoprotein (MOG) are minor protein constituents of myelin (SATO et al. 1989; MIKOL et al. 1990). We know that MAG is a heavily glycosylated membrane protein with a long extracellular domain and a structure characteristic of the immunoglobulin superfamily. Although it is a quantitatively minor component of myelin (about 1% of total myelin protein), it is believed to play a role in axon-myelinating cell interactions. Similarly, MOG is also a minor glycoprotein of myelin, detectable in white matter tracts. It is anchored in the outer leaflet of the oligodendrocyte membrane through a glycosylphosphatidylinositol lipid intermediate. The presence on MOG of a series of leucine-rich repeats and the association of the so-called HNK-1 carbohydrate adhesion marker suggests a function of the protein in adhesion and thus in myelination. It was also shown to be a member of the immunoglobulin superfamily (PHAM-DINH et al. 1993).

The amino acid sequences of MBP and PLP are well-conserved among several species (DIEHL et al. 1986; FRITZ and McFARLIN 1989) and their injection, either whole or as specific peptides, into various animals causes EAE, a T cell-mediated neuro-autoimmune disease that shares many clinical and histopathological features with MS (SWANBORG 1995). Various encephalitogenic determinants of MBP have been described (FRITZ and McFARLIN 1989; LENNON et al. 1970; HASHIM et al. 1991). Encephalitogenic determinants on PLP were also reported (LININGTON et al. 1990; GREER et al. 1996; TUOHY et al. 1995). Immunization with MOG also induces EAE, and this particular model demonstrates that antibody can also be a major determinant of the extent of demyelination that develops, presumably through an antibody-dependent cell-mediated cytotoxic mechanism (SCHLUESENER et al. 1987).

Several studies have indicated autoimmune reactivity to various myelin antigens in MS, although such reactivity is sometimes also found in normal individuals, albeit often at lower levels. The best-studied target myelin antigen is MBP (reviewed in WUCHERPFENNIG et al. 1991). Several target epitopes for HLA-DR-restricted T lymphocytes of MS and/or normal patients were identified on human MBP (JINGWU et al. 1990; PETTE et al. 1990; BURNS et al. 1991; LIBLAU et al. 1991; SALVETTI et al. 1993; MEINL et al. 1993; CHOU et al. 1994). Recent evidence suggests that the response to human MBP is dominated in at least some subjects by expanded clones that may persist in vivo for long periods of time (WUCHERPFENNIG et al. 1994). A study of the reactivity of 15,824 short-term T cell lines from MS patients and controls indicated that epitope 84-102 of human MBP is immunodominant in DR2+ MS patients, suggesting that this may be the encephalitogenic antigenic determinant in these individuals (OTA et al. 1990). This epitope may also be immunodominant for B cell responses (WARREN et al. 1995). Importantly, circulating in vivo-activated T cells reactive to MBP were detected in MS patients (ALLEGRETTA et al. 1990; LODGE et al. 1996). Numerous studies have identified other myelin antigens as possible autoimmune targets in MS patients (e.g. CHOU et al. 1992; SUN et al. 1991; BAIG et al. 1991; DEROSBO et al. 1993; MARKOVICPLESE et al. 1995; CORREALE et al. 1995). We have also recently reported on the possible importance of antigen-specific humoral responses in the disease pathogenesis (QIN et al. 1998). Non-myelin autoantigens, such as members of the heat shock protein family, are also candidate autoantigens in MS (VAN NOORT et al. 1995).

Among the various mechanisms proposed for virus induction of autoimmune diseases, an intriguing hypothesis which has yet to be definitively associated with human pathology is molecular mimicry (OLDSTONE 1987, 1998). It is well known that viral proteins can share antigenic determinants with host cell proteins. Upon viral infection, these common epitopes could, at least theoretically, induce cross-reacting immune responses leading to autoimmune disease. The biological relevance of such a pathogenic mechanism has already been demonstrated in an animal model (FUJINAMI and OLDSTONE 1985). Computer searches showed the conservation of a six-amino acid stretch between a site of MBP known to be encephalitogenic in rabbits and the hepatitis B viral polymerase. The corresponding peptide was synthesized chemically and injected into rabbits, which produced both anti-virus and anti-MBP antibodies, generated cellular reactivity to the CNS antigen, and developed an EAE-like disease. In the TMEV model, a virus-directed monoclonal antibody was shown to react with the oligodendroglial galactocerebroside and enhance virus-induced demyelination (YAMADA et al. 1990).

Other striking homologies observed between viral proteins and CNS antigens include conserved domains between MBP and human T lymphotropic virus (LIQUORI 1991), cytomegalovirus (ROOTBERNSTEIN 1995), influenza virus and adenovirus (JAHNKE et al. 1985) and between PLP and adenovirus, polyoma virus, Epstein-Barr virus, influenza virus and human T lymphotropic virus (JAHNKE et al. 1985; SHAW et al. 1986). Some viruses, previously listed as having been implicated in the etiology of MS, do share common determinants with the myelin antigens MBP and PLP, although the significance of such observations remains to be elucidated, since molecular mimicry is not automatically involved in disease (OLDSTONE 1987; RICHTER et al. 1994).

It has now been shown that peptides containing only four native residues of MBP can stimulate MBP-specific T cells and that this sequence is found in a number of viruses, including coronaviruses (GAUTAM et al. 1992). Certainly the best evidence for the possible relevance of molecular mimicry in the pathogenesis of autoimmune diseases such as MS was the recent demonstration of activation of several MBP-specific T cell clones established from MS patients to seven viral and one bacterial peptide which contained the predicted motifs for binding to HLA-DR2 (WUCHERPFENNIG and STROMINGER 1995). This publication revived interest in pathogen-myelin molecular mimicry as a possible trigger of MS pathology (STEINMAN 1996). Our own contribution describing HCoV and MBP cross-reactive T cells in MS patients was timely in showing possible clinical relevance (TALBOT et al. 1996). Indeed, a recent review article, which quoted the Wucherpfennig and Strominger study (WUCHERPFENNIG and STROMINGER 1995) commented that no data are available to suggest the reverse mechanism, which is actually required to explain the development of MS, i.e. recognition of CNS protein sequences by T cells originally activated by a naturally processed viral or bacterial pathogen (STEINMAN 1996), an observation that we did provide for eight of 16 MS patients studied (TALBOT et al. 1996). Our published studies have now been extended to show cross-reactivity at the T cell clone level between both strains of HCoV (229E and OC43) and MBP as well as with PLP (BOUCHER et al. 1998). The next phase of our study involves the elucidation of the molecular basis, MS specificity and CNS relevance of this T cell cross-reactivity between a common respiratory pathogen and major myelin antigens against which the immune system appears to be activated in MS.

Another important observation from the WUCHERPFENNIG and STROMINGER (1995) study was that, with one exception, activating peptides could not have been predicted by simple sequence alignment. Indeed, the concept has now emerged that apparently very dissimilar peptides can activate the same T cells. This was evident in the memory T cell response to antigenically different arenaviruses (SELIN et al. 1994) and suggests that this may be explained by the concept of "molecular shape mimicry" i.e. recognition by the T cell receptor of a three-dimensional structure formed by the MHC–peptide complex (BHARDWAJ et al. 1993; QUARATINO et al. 1995; GARZA and TUNG 1995; KERSH and ALLEN 1996).

6 Systemic Virus Infections: Induction of Non-Antigen-Specific Immune Activation

The development of target-directed cellular immune mediated disorders is dependent on migration of lymphocytes from the systemic circulation into the involved target site. The extent of migration of T cells into the CNS is determined by a range of factors acting either upon the immune cells or the target tissue, both of which can be influenced by viral infection. Under basal conditions there does seem to be a level of ongoing immune surveillance within the CNS carried out by T cells which enter and then rapidly depart or die. As will be discussed later, the endogenous neural cells contribute to migration by providing chemoattractant signals, by serving as partners in the process of cell adhesion, and by providing immune accessory signals. The lymphocyte migration process is dependent on the state of activation of the immune cells in the systemic circulation. We have shown that the migration rate of T cells derived from MS patients in vitro through a fibronectin barrier is increased compared to controls (UHM et al. 1997; PRAT et al. 1998). This process is dependent on production by the T cells of matrix metalloproteinases which increases when the cells are activated (STUVE et al. 1996). T cells, when activated, also up-regulate adhesion molecules, which will promote interaction with neural cells at the blood–brain barrier (BBB). Systemic viral infections are one set of stimuli that can induce immune activation including production of cytokines, which in turn can up-regulate expression of adhesion molecules and production of proteases (reviewed in ANTEL and BECHER 1998). Perhaps this could explain why exacerbations of MS are linked to seasonal occurrence of viral infections (SIBLEY et al. 1985; PANITCH 1994).

7 Virus-Induced Immune Activity within the CNS: Antigen-Specific Responses

Within the CNS compartment, a number of endogenous cell types have varying capacity to serve as antigen-presenting cells (APCs). At the BBB, the perivascular microglia are shown to be competent APCs (reviewed in ANTEL and BECHER 1998). In chimeric animals, histocompatibility between these cells and encephalitogenic myelin-reactive T cells is required for development of EAE. Within the parenchyma of the CNS, the microglia are the cell type that most convincingly demonstrate the properties of competent APCs. A central issue in understanding the basis for development of chronic immune-mediated demyelinating disease of the CNS, including after initial viral infection, is whether there is ongoing immune sensitization to antigen release during the initial process of tissue injury. In both the TMEV and coronavirus infection models and in the chronic EAE model, there are data that indicate that determinant spreading has occurred and that myelin-reactive T cells can be recovered from the infected animals. In the TMEV model, molecular mimicry was apparently not involved (MILLER et al. 1997). In the case of MBP peptide-induced EAE, T cells reactive to other MBP peptide sequences and other myelin antigens (PLP) can be recovered over time (MILLER et al. 1995). One need consider that the determinant spreading immune responses could be generated either in the CNS or in the regional lymph nodes. As regards the latter, it is well-demonstrated that antigens released into the CNS will reach these structures.

Determinant spreading to cryptic determinants on the same antigen or to other myelin antigens after an original attack on a specific determinant on a myelin antigen, as shown in the previously discussed experimental models, and which may in some instances involve molecular mimicry, could explain the multitude of complementary or often contradictory studies on immune reactivity to myelin antigens that have been reported in MS (McRae et al. 1995). The observations of myelin-autoreactive lymphocytes in apparently healthy individuals have caused difficulties in inferring a pathological relevance for these cells in MS patients. Evidence was provided that both MBP- and PLP-reactive T cells are in an activated state, both in the periphery and the CNS compartment, preferentially in MS patients and not in healthy donors or patients with other neurologic disorders (Zhang et al. 1994). This study strengthens the proposed involvement of myelin-reactive T cells in the pathogenesis of MS (reviewed in Hohlfeld et al. 1995), and gives hope for treatment of disease by induction of tolerance to the target antigen or specific elimination of autoreactive T cells (Chen et al. 1995b; Brocke et al. 1996; Steinman 1996). However, it does not address the mechanism of induction of these autoreactive T cells in the initiation and development of MS.

8 Virus-Induced Immune Reactivity within the CNS: Effects on Glial Cells

Activation of CNS glial cells, namely astrocytes and especially microglia, is becoming recognized as a hallmark of various neurologic disorders (Coyle 1996; Ludwin 1997; Benveniste 1997; Zielasek and Hartung 1996; Chao et al. 1996; Brown and Kretzschmar 1997), including MS (Sriram and Rodriguez 1997) and Alzheimer's disease (Barger and Harmon 1997). Viruses that enter the CNS and target these cells, such as HCoV (Bonavia et al. 1997; Arbour et al. 2000) and various retroviruses (Gravel et al. 1993; Dhibjalbut et al. 1994; Sharpless et al. 1992), are therefore prime candidate mediators of at least some of this neuropathologically relevant activation (Bilzer and Stitz 1996). The role of microglial cells in CNS biology has only recently become recognized, since it has been difficult to differentiate these brain-resident cells from infiltrating macrophages (Ulvestad et al. 1994). Activation of glial cells results in increased expression of the immune accessory molecules involved in the process of antigen presentation. As previously mentioned, activation is further characterized by the release of various pro-inflammatory molecules such as cytokines, chemokines, NO, and reactive oxygen intermediates (Zielasek and Hartung 1996). These mediators can contribute both to regulating the activity of lymphocytes which enter this compartment and to actually effecting target directed injury.

9 Axonal Injury in Multiple Sclerosis

Although MS is regarded as primarily a myelin- and oligodendrocyte-directed autoimmune disorder, there is increasing recognition that the extent of axonal injury and loss may account for at least a component of the neurologic dysfunction (reviewed in MATTHEWS et al. 1998). Conversely, one need consider whether axonal compensatory responses to demyelination may be sufficient to permit continued, relatively unimpaired neurologic function. The early pathological description of MS lesions by Marie and Charcot does indicate an appreciation for the axonal loss component, although emphasis was placed on the relative preservation of axons compared to their myelin sheaths. More recent pathological studies indicate axonal disruption even in early active lesions (DE STEFANO et al. 1995). Clinical brain-imaging correlative studies indicate that axonal involvement can contribute both to reversible and irreversible neurologic deficits, which characterize the MS disease process. The advent of magnetic resonance spectroscopy (MRS), which is an MR technique that can suppress the dominant water signal that forms the basis for conventional MR imaging (MRI), permits quantitation of multiple other molecules in the CNS, including N-acetyl aspartate (NAA), an amino acid exclusively expressed in neurons and their axons in the mature CNS. NAA is the dominant peak seen using long-sequence ^{3}H-based MRS. NAA production is dependent on intact mitochondrial function. Although MRS does not have the spatial resolution of MRI because of the increased voxel sizes needed, whole-brain images can now be reconstructed with MRS imaging (MRSI), permitting analysis of individual lesions and normal-appearing white matter. An example of such spectra derived from a chronic MS lesion and a conventional MRI-defined normal-appearing white matter is presented in Fig. 1.

In initial prospective studies of MS patients, using selected large volumes of interest, we and others demonstrated that there was a significant, albeit imperfect, correlation between progressive loss of NAA and clinical neurologic disability scores (EDSS). Such correlations were less robust between conventional MRI defined lesion volumes and EDSS score (DE STEFANO et al. 1995; FU et al. 1998). In the EAE model, animals with the chronic form of the disease, with its associated, persistent neurologic sequel, are also recognized as having a significant extent of axonal injury (TAUPIN et al. 1997).

Using the MRSI technique, we have had the opportunity to examine NAA values in MS patients presenting with acute neurologic deficits (DE STEFANO et al. 1998). In a series of such patients, we documented initially reduced NAA values in MRI-defined lesions corresponding to the anatomic site that coincided with the effected neurologic function, namely motor function in our patients. In these cases, we observed gradual recovery of the NAA over time, with a strong overall correlation between NAA values and clinical function (DE STEFANO et al. 1998). The basis for this reversible pattern of NAA depression could reflect several non-exclusive mechanisms. The inflammatory mediators present in the acute MS lesion could directly act upon the neuron/axon, perhaps having greater access to those whose myelin covering is damaged. As mentioned, NAA production is dependent on intact mitochondrial function. In vitro, one can demonstrate that neuron expression of NAA can be reversibly down-regulated by manipulation of culture conditions, such as by serum deprivation. The clinical worsening that occurred in MS patients receiving the anti-CD4 CAMPATH 1H mAb, was shown to be

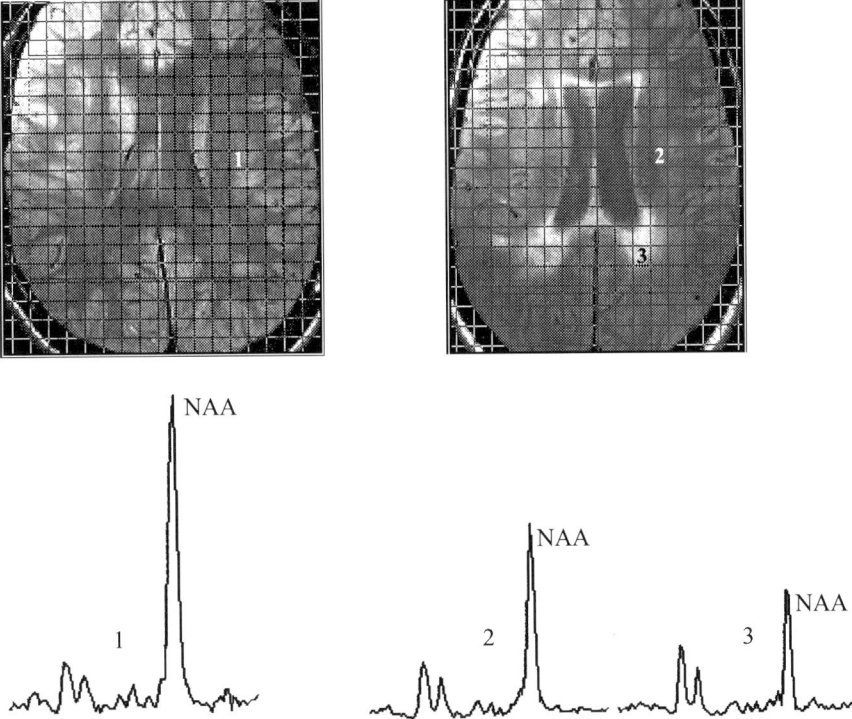

Fig. 1. Magnetic resonance spectroscopy profiles showing decrease in the peak of N-acetyl aspartate (*NAA*) in a chronic MS lesion and normal appearing white matter

dependent on axonal conduction block induced by NO, which in the active MS lesion would be produced by activated glial cells and probably infiltrating macrophages (WING et al. 1996).

The neuronal dysfunction in MS could also reflect an indirect effect of demyelination. Oligodendrocytes and astrocytes are likely important sources of trophic support for axons. The size of axons is known to be proportional to the extent of myelination. Depletion of myelin results in neurofilaments undergoing a change in their phosphorylation state, with subsequent effects on the entire axonal cytoskeleton. Axons in vitro show a propensity to grow on astrocyte substrates. Similarly, microglia can support axonal regrowth in selected injury paradigms (DAVID et al. 1990). One speculates whether the trophic vs destructive potential effect of these glial cells is shifted when such cells are activated in the inflammatory environment that characterizes MS.

Recent MRI pathologic correlative studies, have found that some MRI-defined lesions identified on post-mortem tissue specimens are not seen by gross visual inspection. Furthermore, studies using more specialized MRI techniques, such as magnetization transfer imaging and MRS have demonstrated abnormalities in what by usual MRI criteria is apparently normal tissue. These observations suggest

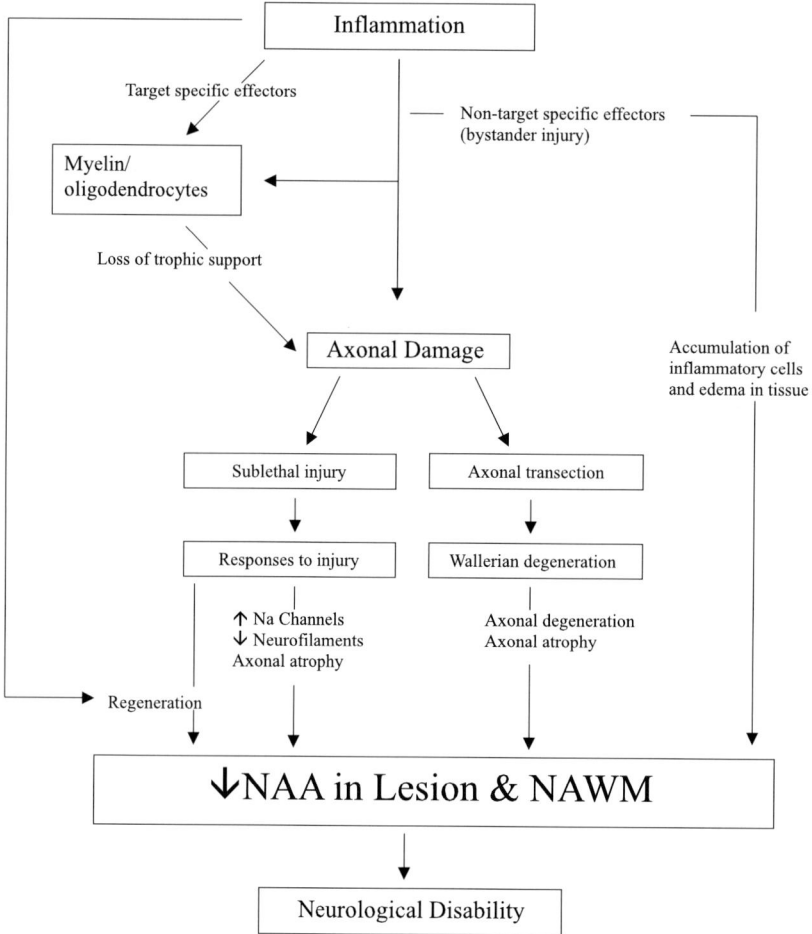

Fig. 2. Outline of basis of axonal injury and recovery in multiple sclerosis (*MS*), with resultant effect on *N*-acetyl aspartate (*NAA*) expression in lesions and normal appearing white matter (*NAWM*)

more widespread tissue abnormalities in MS than previously suspected. These could include dying back of axons transected at the remote lesion sites or diffuse tissue responses to pro-inflammatory cytokines. In this regard, focal stab-wound injury models in animals are shown to result in release of an array of cytokines which diffuse and induce reactive glial changes over a considerable distance (MOUMDJIAN et al. 1991). Such changes could make the normal appearing white matter more susceptible to subsequent injury.

Serial MRI studies of MS patients have documented that white matter lesions without apparent corresponding clinical deficits can also arise. These patients do not show the profound decreases of NAA within lesions in contrast to a symptomatic cohort. Short-echo-time MRS studies have also identified areas of myelin destruction, as defined by free lipid release without evidence of NAA loss or clinical deficits. In the TMEV demyelination model induced in CD8 T cell-deficient animals, Rodriguez and colleagues demonstrated that extensive demyelination can occur without apparent functional neurologic deficits (RIVERA-QUINONES et al. 1998). The axons of these animals are shown to have compensated for their myelin loss by rearrangement of expression of Na^+ channels, so that such channels are now expressed throughout the course of the axon and are not restricted to the perinodal region. In this way, the animal can overcome the conduction block otherwise expected if only saltatory conduction over a relatively large distance (node-node) is the only means to maintain electrical conduction. Analysis of axons contained within active MS lesions indicates that such compensation also occurs in the human disease.

Additional compensatory mechanisms are now recognized to occur even in the mature CNS. Axonal sprouting would provide a means whereby an intact axon might replace the function of an adjacent injured one. Results from functional MRI studies suggest that functional reorganization within the CNS does occur following acute MS lesions. One raises the concern that over time these compensatory mechanisms may fail, either through ageing or additional lesions, and may account for the evolution of MS into a more chronic disorder as occurs in more than 50% of cases. The above considerations indicate that the axon is a central participant in the evolution of the MS disease process, both with regard to it being directly (immune-mediated) or indirectly (trophic support withdrawal) injured or, conversely, by providing a means to compensate for the primary myelin injury (see Fig. 2).

References

Allegretta M, Nicklas JA, Sriram S, Albertini RJ (1990) T-cells responsive to myelin basic protein in patients with multiple sclerosis. Science 247:718–721

Antel J, Becher B (1998) Central nervous system-immune interactions: contribution to neurologic disease and recovery. In: Antel J, Birnbaum G, Hartung H-P (eds) Clinical neuroimmunology. Blackwell Science pp 26–39

Arbour N, Day R, Newcombe J, Talbot PJ (2000) Neuroinvasion by human respiratory coronaviruses. J Virol 74:8913–8971

Arbour N, Côté G, Lachance C, Tardieu M, Cashman NR, Talbot PJ (1999a) Acute and persistent infection of human neural cell lines by human coronavirus OC43. J Virol 73:3338–3350

Arbour N, Ékandé S, Côté G, Lachance C, Chagnon F, Cashman NR, Talbot PJ (1999b) Persistent infection of human oligodendrocytic and neuronal cell lines by human coronavirus 229E. J Virol 73:3326–3337

Arbour N, Talbot PJ (1998) Persistent infection of neural cell lines by human coronaviruses. Adv Exp Med Biol 440:575–581

Arpin N, Talbot PJ (1990) Molecular characterization of the 229E strain of human coronavirus. Adv Exp Med Biol 276:73–80

Baig S, Olsson T, Yuping J, Hojeberg B, Cruz M, Link H (1991) Multiple sclerosis – cells secreting antibodies against myelin-associated glycoprotein are present in cerebrospinal fluid. Scand J Immunol 33:73–79

Bailey OT, Pappenheimer AM, Cheever FS, Daniels JB (1949) A murine virus (JHM) causing disseminated encephalomyelitis with extensive destruction of myelin. II. Pathology. J Exp Med 90: 195–212

Bang FB, Warwick A (1960) Mouse macrophages as host cells for the mouse hepatitis virus and the genetic basis of their susceptibility. Proc Natl Acad Sci USA 46:1065–1075

Barac-Latas V, Suchanek G, Breitschopf H, Stuehler A, Wege H, Lassmann H (1997) Patterns of oligodendrocyte pathology in coronavirus-induced subacute demyelinating encephalomyelitis in the Lewis rat. Glia 19:1–12

Barger SW, Harmon AD (1997) Microglial activation by Alzheimer amyloid precursor protein and modulation by apolipoprotein E. Nature 388:878–881

Barnett EM, Perlman S (1993) The olfactory nerve and not the trigeminal nerve is the major site of CNS entry for mouse hepatitis virus strain JHM. Virology 194:185–191

Benveniste EN (1997) Role of macrophages/microglia in multiple sclerosis and experimental allergic encephalomyelitis. J Mol Med 75:165–173

Bhardwaj V, Kumar V, Geysen HM, Sercarz EE (1993) Degenerate recognition of a dissimilar antigenic peptide by myelin basic protein-reactive T cells. J Immunol 151:5000–5010

Bilzer T, Stitz L (1996) Immunopathogenesis of virus diseases affecting the central nervous system. Crit Rev Immunol 16:145–222

Bonavia A, Arbour N, Wee Yong V, Talbot PJ (1997) Infection of primary cultures of human neural cells by human coronaviruses 229E and OC43. J Virol 71:800–806

Boucher A, Mercier G, Duquette P, Talbot PJ (1998) Clonal T cell cross-reactivity between myelin antigens MBP and PLP and human respiratory coronavirus in multiple sclerosis. J Neuroimmunol 90:33 (abstract)

Bray PF, Luka J, Bray PF, Culp KW, Schlight JP (1992) Antibodies against Epstein-Barr nuclear antigen (EBNA) in multiple sclerosis CSF and two pentapeptide sequence identities between EBNA and myelin basic protein. Neurology 42:1798–1804

Brocke S, Gijbels K, Allegretta M, Ferber I, Piercy C, Blankenstein T, Martin R, Utz U, Karin N, Mitchell D, Veromaa T, Waisman A, Gaur A, Conlon P, Ling N, Fairchild PJ, Wraith DC, Ogarra A, Fathman CG, Steinman L (1996) Treatment of experimental encephalomyelitis with a peptide analogue of myelin basic protein. Nature 379:343–346

Brown DR, Kretzschmar HA (1997) Microglia and prion disease: a review. Histol Histopathol 12: 883–892

Buchmeier MJ, Lewicki HA, Talbot PJ, Knobler RL (1984) Murine hepatitis virus-4 (strain JHM) – induced neurologic disease is modulated in vivo by monoclonal antibody. Virology 132:261–270

Burks JS, DeVald BL, Jankovsky LD, Gerdes JC (1980) Two coronaviruses isolated from central nervous system tissue of two multiple sclerosis patients. Science 209:933–934

Burns J, Littlefield K, Gomez C, Kumar V (1991) Assessment of antigenic determinants for the human T-cell response against myelin basic protein using overlapping synthetic peptides. J Neuroimmunol 31:105–113

Cabirac GF, Murray RS, McLaughlin LB, Skolnick DM, Hogue B, Dorovini-Zis K, Didier PJ (1995) In vitro interaction of coronaviruses with primate and human brain microvascular endothelial cells. Adv Exp Med Biol 380:79–88

Cabirac GF, Soike KF, Zhang JY, Hoel K, Butunoi C, Cai GY, Johnson S, Murray RS (1994) Entry of coronavirus into primate CNS following peripheral infection. Microbial Path 16:349–357

Chaloner-Larsson G, Johnson-Lussenburg CM (1981) Establishment and maintenance of a persistent infection of L132 cells by human coronavirus 229E. Arch Virol 69:117–129

Chao CC, Hu SX, Peterson PK (1996) Glia: the not so innocent bystanders. J Neurovirol 2:234–239

Cheever FS, Daniels JB, Pappenheimer AM, Bailey OT (1949) A murine virus (JHM) causing dissemi- nated encephalomyelitis with extensive destruction of myelin. I. Isolation and biological properties of the virus. J Exp Med 90:181–194

Chen YH, Inobe J, Marks R, Gonnella P, Kuchroo VK, Weiner HL (1995a) Peripheral deletion of antigen-reactive T cells in oral tolerance. Nature 376:177–180

Chen DS, Asanaka M, Yokomori K, Wang FI, Hwang SB, Li HP, Lai MMC (1995b) A pregnancy-specific glycoprotein is expressed in the brain and serves as a receptor for mouse hepatitis virus. Proc Natl Acad Sci USA 92:12095–12099

Chou YK, Bourdette DN, Offner H, Whitham R, Wang RY, Hashim GA, Vandenbark AA (1992) Frequency of T cells specific for myelin basic protein and myelin proteolipid protein in blood and cerebrospinal fluid in multiple sclerosis. J Neuroimmunol 38:105–113

Chou YK, Jones RE, Bourdette D, Whitham R, Hashim G, Atherton J, Offner H, Vandenbark AA (1994) Human myelin basic protein (MBP) epitopes recognized by mouse MBP-selected T cell lines from multiple sclerosis patients. J Neuroimmunol 49:45–50

Collins AR, Sorensen O (1986) Regulation of viral persistence in human glioblastoma and rhabdomyosarcoma cells infected with coronavirus OC43. Microbial Path 1:573–582

Conant K, Garzino-Demo A, Nath A, McArthur JC, Halliday W, Power C, Gallo RC, Major EO (1998) Induction of monocyte chemoattractant proteon-1 in HIV-1 Tat-stimulated astrocytes and elevation in AIDS dementia. Proc Natl Acad Sci USA 95:3117–3121

Cook SD, Rohowskykochan C, Bansil S, Dowling PC (1995) Evidence for multiple sclerosis as an infectious disease. Acta Neurol Scand 91:34–42

Correale J, McMillan M, McCarthy K, Le T, Weiner LP (1995) Isolation and characterization of autoreactive proteolipid protein-peptide specific T-cell clones from multiple sclerosis patients. Neurology 45:1370–1378

Cosby SL, McQuaid S, Taylor MJ, Bailey M, Rima BK, Martin SJ, Allen IV (1989) Examination of 8 cases of multiple sclerosis and 56 neurological and non-neurological controls for genomic sequences of measles virus, canine distemper virus, simian virus-5 and rubella virus. J Gen Virol 70:2027–2036

Coyle PK (1996) The neuroimmunology of multiple sclerosis. Adv Neuroimmunol 6:143–154

Cross AH, McCarron R, McFarlin DE, Raine CS (1987) Adoptively transferred acute and chronic relapsing autoimmune encephalomyelitis in the PL/J mouse and observations on altered pathology by intercurrent virus infection. Lab Invest 57:499–512

Dalziel RG, Lampert PW, Talbot PJ, Buchmeier MJ (1986) Site-specific alteration of murine hepatitis virus type 4 peplomer glycoprotein E2 results in reduced neurovirulence. J Virol 59:463–471

David S, Bouchard C, Tsatas O, Giftochristos N (1990) Macrophages can modify the nonpermissive nature of the adult mammalian central nervous system. Neuron 5:463–469

Dekaban GA, Rice GPA (1990) Retroviruses and multiple sclerosis. 2. Failure of gene amplification techniques to detect viral sequences unique to the disease. Neurology 40:1254–1258

Derosbo NK, Milo R, Lees MB, Burger D, Bernard CCA, Ben-Nun A (1993) Reactivity to myelin antigens in multiple sclerosis – Peripheral blood lymphocytes respond predominantly to myelin oligodendrocyte glycoprotein. J Clin Invest 92:2602–2608

De Stefano N, Matthews PM, Antel JP, Preul M, Francis G, Arnold DL (1995) Chemical pathology of acute demyelinating lesions and its correlation with disability. Ann Neurol 38:901–909

De Stefano N, Matthews PM, Fu L, Narayanan S, Stanley J, Francis GS, Antel JP, Arnold DL (1998) Axonal damage correlates with disability in patients with relapsing-remitting multiple sclerosis. Results of a longitudinal magnetic resonance spectroscopy study. Brain 121:1469–1477

Dhibjalbut S, Hoffman PM, Yamabe T, Sun D, Xia J, Eisenberg H, Bergey G, Ruscetti FW (1994) Extracellular human T-cell lymphotropic virus type I tax protein induces cytokine production in adult human microglial cells. Ann Neurol 36:787–790

Diehl HJ, Scharch M, Budzinski RM, Stoffel W (1986) Individual exons encode the integral membrane domains of human myelin proteolipid protein. Proc Natl Acad Sci USA 83:9807–9811

Edwards J, Denis F, Talbot PJ (2000) Activation of glial cells by human coronavirus OC43. J Neuroimmunol 108:73–81

Erlich SS, Fleming JO, Stohlman SA (1987) Experimental neuropathology of chronic demyelination by a JHM virus variant. Arch Neurol 44:839–842

Fleming JO, El Zaatari AK, Gilmore W, Berne JD, Burks JS, Stohlman SA, Tourtelotte WW, Weiner LP (1988) Antigenic assessment of coronaviruses isolated from patients with multiple sclerosis. Arch Neurol 45:629–633

Fleming JO, Trousdale MD, El-Zaatari F, Stohlman SA, Weiner LP (1986) Pathogenicity of antigenic variants of murine coronavirus JHM selected with monoclonal antibodies. J Virol 58:869–875

Fritz RB, Mc Farlin DE (1989) Encephalitogenic epitopes of myelin basic protein. In: Sercarz E (ed) Antigenic determinants and immune regulation chem immunol. Karger, Basel pp 101–125

Fu L, Matthews PM, De Stefano N, Worseley KJ, Narayanan S, Francis GS, Antel JP, Wolfson C, Arnold DL (1998) Imaging axonal damage of normal-appearing white matter in multiple sclerosis. Brain 121:103–113

Fujinami RS, Oldstone MBA (1985) Amino acid homology between the encephalitogenic site of myelin basic protein and virus: mechanism for autoimmunity. Science 230:1043–1045

Garza KM, Tung KSK (1995) Frequency of molecular mimicry among T cell peptides as the basis for autoimmune disease and autoantibody induction. J Immunol 155:5444–5448

Gautam AM, Pearson CI, Smilek DE, Steinman L, McDevitt HO (1992) A polyalanine peptide with only 5 native myelin basic protein residues induces autoimmune encephalomyelitis. J Exp Med 176: 605–609

Gilmore W, Correale J, Weiner LP (1994) Coronavirus induction of class I major histocompatibility complex expression in murine astrocytes is virus strain specific. J Exp Med 180:1013–1023

Goswami KKA, Randall RE, Lange LS, Russell WC (1987) Antibodies against the paramyxovirus SV5 in the cerebrospinal fluids of some multiple sclerosis patients. Nature 327:244–247

Gravel C, Kay DG, Jolicoeur P (1993) Identification of the infected target cell type in spongiform myeloencephalopathy induced by the neurotropic Cas-Br-E murine leukemia virus. J Virol 67: 6648–6658.

Greer JM, Sobel RA, Sette A, Southwood S, Lees MB, Kuchroo VK (1996) Immunogenic and encephalitogenic epitope clusters of myelin proteolipid protein. J Immunol 156:371–379

Grzybicki DM, Kwack KB, Perlman S, Murphy SP (1997) Nitric oxide synthase type II expression by different cell types in MHV-JHM encephalitis suggests distinct roles for nitric oxide in acute versus persistent virus infection. J Neuroimmunol 73:15–27

Haase AT, Ventura P, Gibbs CJ, Tourtellotte WW (1981) Measles virus nucleotide sequences: detection by hybridization in situ. Science 212:672–674

Hamre D, Procknow JJ (1966) A new virus isolated from the human respiratory tract. Proc Soc Exp Biol Med 121:190–193

Hashim G (1978) Myelin basic protein: structure, function and antigenic determinants. Immunol Rev 39:60–107

Hashim G, Vandenbark AA, Gold DP, Diamanduros T, Offner H (1991) T-cell lines specific for an immunodominant epitope of human basic protein define an encephalitogenic determinant for experimental autoimmune encephalomyelitis-resistant LOU/M rats. J Immunol 146:515–520

Haspel MV, Lampert PW, Oldstone MBA (1978) Temperature-sensitive mutants of mouse hepatitis virus produce a high incidence of demyelination. Proc Natl Acad Sci USA 75:4033–4036

Hohlfeld R, Meinl E, Weber F, Zipp F, Schmidt S, Sotgiu S, Goebels N, Voltz R, Spuler S, Iglesias A, Wekerle H (1995) The role of autoimmune T lymphocytes in the pathogenesis of multiple sclerosis. Neurology 45:S33–S38

Horwitz MS, Bradley LM, Harbertson J, Krahl T, Lee J, Sarvetnick N (1998) Diabetes induced by Coxsackie virus: initiation by bystander damage and not molecular mimicry. Nature Med 4:781–785

Houtman JJ, Fleming JO (1996) Pathogenesis of mouse hepatitis virus-induced demyelination. J Neurovirol 2:361–376

Houtman JJ, Fleming JO (1996) Dissociation of demyelination and viral clearance in congenitally immunodeficient mice infected with murine coronavirus JHM. J Neurovirol 2:101–110

Hovanec DL, Flanagan TD (1983) Detection of antibodies to human coronaviruses 229E and OC43 in the sera of multiple sclerosis patients and normal subjects. Infect Immun 41:426–429

Jahnke U, Fischer EH, Alvord EC (1985) Sequence homology between certain viral proteins and proteins related to encephalomyelitis and neuritis. Science 229:282–284

Jingwu Z, Chou CHJ, Hashim G, Medaer R, Raus JCM (1990) Preferential peptide specificity and HLA restriction of myelin basic protein-specific T-cell clones derived from MS patients. Cell Immunol 129:189–198

Johnson RT (1985) Viral aspects of multiple sclerosis In: Koetsier JC (ed) Handbook of Clinical Neurology Demyelinating Diseases. Elsevier, Amsterdam pp 319–336

Johnson-Lussenburg CM, Zheng Q (1987) Coronavirus and multiple sclerosis: results of a case/control longitudinal serological study. Adv Exp Med Biol 218:421–429

Kersh GJ, Allen PM (1996) Essential flexibility in the T-cell recognition of antigen. Nature 380:495–498

Kinnunen E, Valle M, Piirainen L, Kleemola M, Kantanen ML, Juntunen J, Klockars M, Koskenvuo M (1990) Viral antibodies in multiple sclerosis – a nationwide co-twin study. Arch Neurol 47:743–746

Knobler RL, Haspel MV, Oldstone MBA (1981) Mouse hepatitis virus type 4 (JHM strain) – induced fatal central nervous system disease. I. Genetic control and the murine neuron as the susceptible site of disease. J Exp Med 153:32–843

Knobler RL, Lampert PW, Oldstone MBA (1982) Virus persistence and recurring demyelination produced by a temperature-sensitive mutant of MHV-4. Nature 298:289–281

Kolson DL, Lavi E, Gonzalez-Scarano F (1998) The effects of human immunodeficiency virus in the central nervous system. Adv Virus Res 50:1–47

Kurtzke JF (1993) Epidemiologic evidence for multiple sclerosis as an infection. Clin Microbiol Rev 6:382–427

Kyuwa S, Yamaguchi K, Toyoda Y, Fujiwara K (1991) Induction of self-reactive T-cells after murine coronavirus infection. J Virol 65:1789–1795

Labonté P, Mounir S, Talbot PJ (1995) Sequence and expression of the ns2 protein gene of human coronavirus OC43. J Gen Virol 76:431–435

Lachance C, Arbour N, Cashman NR, Talbot PJ (1998) Involvement of aminopeptidase N (CD13) in infection of human neural cells by human coronavirus 229E. J Virol 72:6511–6519

Lamontagne L, Descôteaux JP, Jolicoeur P (1989) Mouse hepatitis virus 3 replication in T and B lymphocytes correlate with viral pathogenicity. J Immunol 142:4458–4465

Lampert PW, Sims JK, Kniazeff AJ (1973) Mechanism of demyelination in JHM virus encephalomyelitis. Acta Neuropathol 24:76–85

Lavi E, Fishman PS, Highkin MK, Weiss SR (1988) Limbic encephalitis after inhalation of a murine coronavirus. Lab Invest 58:31–36

Lehmann PV, Forsthuber T, Miller A, Sercarz EE (1992) Spreading of T cell autoimmunity to cryptic determinants of an autoantigen. Nature 358:155–157

Lennon VA, Wilks AV, Carnegie PR (1970) Immunologic properties of the main encephalitogenic peptide from the basic protein of human myelin. J Immunol 105:1223–1230

Liblau R, Tournierlasserve E, Maciazek J, Dumas G, Siffert O, Hashim G, Bach MA (1991) T-cell response to myelin basic protein epitopes in multiple sclerosis patients and healthy subjects. Eur J Immunol 21:1391–1395

Linington C, Gunn CA, Lassmann H (1990) Identification of an encephalitogenic determinant of myelin proteolipid protein for the rabbit. J Neuroimmunol 30:135–144

Liquori AM (1991) Myelin basic protein (MBP) displays significant homologies with GAG core proteins of HTLV retroviruses. J Theor Biol 148:279–281

Lodge PA, Johnson C, Sriram S (1996) Frequency of MBP and MBP peptide-reactive T cells in the HPRT mutant T-cell population of MS patients. Neurology 46:1410–1415

Ludwin SK (1997) The pathobiology of the oligodendrocyte. J Neuropathol Exp Neurol 56:111–124

Markovicplese S, Fukaura H, Zhang JW, Alsabbagh A, Southwood S, Sette A, Kuchroo VK, Hafler DA (1995) T cell recognition of immunodominant and cryptic proteolipid protein epitopes in humans. J Immunol 155:982–992

Massa PT, Dörries R, ter Meulen V (1986) Viral particles induce Ia antigen expression on astrocytes. Nature 320:543–546

Matthews PM, De Stefano N, Narayanan S, Francis GS, Wolinsky JS, Antel JP, Arnold DL (1998) Putting MRS studies in context: axonal damage and disability in multiple sclerosis. Semin Neurol 18:327–336

McIntosh K, Becker WB, Chanock RM (1967) Growth in suckling mouse brain of "IBV-like" viruses from patients with upper respiratory tract disease. Proc Natl Acad Sci USA 58:2268–2273

McRae BL, Vanderlugt CL, DalCanto MC, Miller SD (1995) Functional evidence for epitope spreading in the relapsing pathology of experimental autoimmune encephalomyelitis. J Exp Med 182:75–85

Meinl E, Weber F, Drexler K, Morelle C, Ott M, Saruhandireskeneli G, Goebels N, Ertl B, Jechart G, Giegerich G, Schonbeck S, Bannwarth W, Wekerle H, Hohlfeld R (1993) Myelin basic protein specific T lymphocyte repertoire in multiple sclerosis – complexity of the response and dominance of nested epitopes due to recruitment of multiple T cell clones. J Clin Invest 92:2633–2643

Mikol DD, Gulcher JR, Stefansson K (1990) The oligodendrocyte-myelin glycoprotein belongs to a distinct family of proteins and contains the HNK-1 carbohydrate. J Cell Biol 110:471–479

Mikoshiba K, Okano H, Tamura T, Ikenaka K (1991) Structure and function of myelin protein genes. Annu Rev Neurosci 14:201–217

Miller SD, Mcrae BL, Vanderlugt CL, Nikcevich KM, Pope JG, Pope L, Karpus WJ (1995) Evolution of the T-cell repertoire during the course of experimental immune-mediated demyelinating diseases. Immunol Rev 144:225–244

Miller SD, Vanderlugt CL, Begolka WS, Pao W, Yauch RL, Neville KL, Katz-Levy Y, Carrizosa A, Kim BS (1997) Persistent infection with Theiler's virus leads to CNS autoimmunity via epitope spreading. Nature Med 3:1133–1136

Moumdjian RA, Antel JP, Yong VW (1991) Origin of contralateral reactive gliosis in surgically injured rat cerebral cortex. Brain Res 547:223–228

Mounir S, Talbot PJ (1992) Sequence analysis of the membrane protein gene of human coronavirus OC43 and evidence for O-glycosylation. J Gen Virol 73:2731–2736

Mounir S, Talbot PJ (1993a) Molecular characterization of the S protein gene of human coronavirus OC43. J Gen Virol 74:1981–1987

Mounir S, Talbot PJ (1993b) Human coronavirus OC43 RNA 4 lacks 2 open reading frames located downstream of the S gene of bovine coronavirus. Virology 192:355–360

Mounir S, Labonté P, Talbot PJ (1994) Characterization of the nonstructural and spike proteins of the human respiratory coronavirus OC43: comparison with bovint enteric coronavirus. Coronaviruses and their diseases. Adv Exp Biol Med 342:61–68

Murray RS, Brown B, Brian D, Cabirac GF (1992a) Detection of coronavirus RNA antigen in multiple sclerosis brain. Ann Neurol 31:525–533

Murray RS, Cai GY, Hoel K, Zhang JY, Soike KF, Cabirac GF (1992b) Coronavirus infects and causes demyelination in primate central nervous system. Virology 188:274–284

Murray RS, Cai GY, Soike KF, Cabirac GF (1997) Further observations on coronavirus infection of primate CNS. J Neurovirol 3:71–75

Myint SH (1994) Human coronaviruses – a brief review. Rev Med Virol 4:35–46

Oldstone MB, Sinha YN, Blount P, Tishon A, Rodriguez M, von Wedel R, Lampert PW (1982) Virus-induced alterations in homeostasis: alteration in differentiated functions of infected cells in vivo. Science 218:1125–1127

Oldstone MBA (1987) Molecular mimicry and autoimmune disease. Cell 50:819–820

Oldstone MBA (1998) Molecular mimicry and immune-mediated diseases. FASEB J 12:1255–1265.

Ota K, Matsui M, Milford EL, Mackin GA, Weiner HL, Hafler DA (1990) T-cell recognition of an immunodominant myelin basic protein epitope in multiple sclerosis. Nature 346:183–187

Panitch HS (1994) Influence of infection on exacerbations of multiple sclerosis. Ann Neurol 36: S25–S28

Pasick JMM, Dales S (1991) Infection by coronavirus JHM of rat neurons and oligodendrocyte-type-2 astrocyte lineage cells during distinct developmental stages. J Virol 65:5013–5028

Pearson J, Mims CA (1985) Differential susceptibility of cultured neural cells to the human coronavirus OC43. J Virol 53:1016–1019

Perlman S, Ries D (1987) The astrocyte is a target cell in mice persistently infected with mouse hepatitis virus strain JHM. Microbial Path 3:309–314

Perron H, Garson JA, Bedin F, Beseme F, Paranhos-Baccala G, Komurian-Pradel F, Mallet F, Tuke PW, Voisset C, Blond JL, Lalande B, Seigneurin JM, Mandrand B (1997) Molecular identification of a novel retrovirus repeatedly isolated from patients with multiple sclerosis. Proc Natl Acad Sci USA 94:7583–7588

Pette M, Fujita K, Wilkinson D, Altmann DM, Trowsdale J, Giegerich G, Hinkkanen A, Epplen JT, Kappos L, Wekerle H (1990) Myelin autoreactivity in multiple sclerosis – recognition of myelin basic protein in the context of HLA-DR2 products by lymphocytes-T of multiple-sclerosis patients and healthy donors. Proc Natl Acad Sci USA 87:7968–7972

Pewe L, Wu GF, Barnett EM, Castro RF, Perlman S (1996) Cytotoxic T cell-resistant variants are selected in a virus-induced demyelinating disease. Immunity 5:253–262

Pham-Dinh D, Mattei MG, Nussbaum JL, Roussel G, Pontarotti P, Roeckel N, Mather IH, Artzt K, Lindahl KF, Dautigny A (1993) Myelin/oligodendrocyte glycoprotein is a member of a subset of the immunoglobulin superfamily encoded within the major histocompatibility complex. Proc Natl Acad Sci USA 90:7990–7994

Prat A, Weinrib L, Becher B, Duquette P, Couture R, Antel J (1998) Expression of brabykinin B1 receptor on lymphocytes from MS patients. Neurology 50:A151

Qin Y, Duquette P, Zhang Y, Talbot PJ, Poole R, Antel J (1998) Clonal expansion and somatic hypermutation of VH genes of B cells from cerebrospinal fluid in multiple sclerosis. J Clin Invest 102:1045–1050

Quaratino S, Thorpe CJ, Travers PJ, Londei M (1995) Similar antigenic surfaces, rather than sequence homology, dictate T-cell epitope molecular mimicry. Proc Natl Acad Sci USA 92:10398–10402

Reddy EP, Sandberg-Wollheim M, Mettus RV, Ray PE, DeFreitas E, Koprowski H (1989) Amplification and molecular cloning of HTLV-1 sequences from DNA of multiple sclerosis patients. Science 243:529–533

Richter W, Mertens T, Schoel B, Muir P, Ritzkowsky A, Scherbaum WA, Boehm BO (1994) Sequence homology of the diabetes-associated autoantigen glutamate decarboxylase with coxsackie B4-2C protein and heat shock protein 60 mediates no molecular mimicry of autoantibodies. J Exp Med 180:721–726

Rivera-Quinones C, McGavern D, Schmelzer JD, Hunter SF, Low PA, Rodriguez M (1998) Absence of neurological deficits following extensive demyelination in a class I-deficient murine model of multiple sclerosis. Nature Med 4:187–193

Rootbernstein RS (1995) Preliminary evidence for idiotype antiidiotype immune complexes cross-reactive with lymphocyte antigens in AIDS and lupus. Med Hypoth 44:20–27

Sadovnick AD, Dyment D, Ebers GC (1997) Genetic epidemiology of multiple sclerosis. Epidemiol Rev 19:99–106

Salmi A, Reunanen M, Ilonen J (1981) Possible viral etiology of multiple sclerosis. In: Katsuki S, Tsubaki T, Toyokura Y (eds) International Congress Series Neurology. Excerpta Medica, Amsterdam pp 416–431

Salmi A, Ziola B, Hovi T, Reunanen M (1982) Antibodies to coronaviruses OC43 and 229E in multiple sclerosis patients. Neurology 32:292–295

Salvetti M, Ristori G, Damato M, Buttinelli C, Falcone M, Fieschi C, Wekerle H, Pozzilli, C (1993) Predominant and stable T cell responses to regions of myelin basic protein can be detected in individual patients with multiple sclerosis. Eur J Immunol 23:1232–1239

Sato S, Fujita N, Kurihara T, Kuwano R, Sakimura K, Takahashi Y, Miyatake T (1989) cDNA cloning and amino acid sequence for human myelin-associated glycoprotein. Biochem Biophys Res Comm 163:1473–1480

Schluesener HJ, Sobel RA, Linington C, Weiner HL (1987) A monoclonal antibody against a myelin oligodendrocyte glycoprotein induces relapses and demyelination in central nervous system autoimmune disease. J Immunol 139:4016–4021

Selin LK, Nahill SR, Welsh RM (1994) Cross-reactivities in memory cytotoxic T lymphocyte recognition of heterologous viruses. J Exp Med 179:1933–1943

Sharpless NE, Obrien WA, Verdin E, Kufta CV, Chen ISY, Dubois-Dalcq M (1992) Human immunodeficiency virus type-1 tropism for brain microglial cells is determined by a region of the env glycoprotein that also controls macrophage tropism. J Virol 66:2588–2593

Shaw SY, Laursen RA, Lees MB (1986) Analogous amino acid sequences in myelin proteolipid and viral proteins. FEBS Lett 207:266–270

Shirazian D, Mokhtarian F, Herzlich BC, Miller AE, Grob D (1993) Presence of cross-reactive antibodies to HTLV-1 and absence of antigens in patients with multiple sclerosis. J Lab Clin Med 122:252–259

Sibley WA, Bamford CR, Clark K (1985) Clinical viral infections and multiple sclerosis. Lancet 1:1313–1315

Soldan SS, Berti R, Salem N, Seccherio P, Flamand L, Calabresi PA, Brennan MB, Maloni HW, McFarland HF, Lin H-C, Patnaik M, Jacobson S (1997) Association of human herpes virus 6 (HHV-6) with multiple sclerosis: increased IgM response to HHV-6 early antigen and detection of serum HHV-6 DNA. Nature Med 3:1394–1397

Sorensen O, Collins AR, Flintoff W, Ebers G, Dales S (1986) Probing for the human coronavirus OC43 in multiple sclerosis. Neurology 36:1604–1606

Sriram S, Rodriguez M (1997) Indictment of the microglia as the villain in multiple sclerosis. Neurology 48:464–470

Steinman L (1996) A few autoreactive cells in an autoimmune infiltrate control a vast population of nonspecific cells: a tale of smart bombs and the infantry. Proc Natl Acad Sci USA 93:2253–2256

Steinman L, Oldstone MBA (1997) More mayhem from molecular mimics. Nature Med 3:1321–1322

Steinman L (1996) Multiple sclerosis: a coordinated immunological attack against myelin in the central nervous system. Cell 85:299–302

Stewart JN, Mounir S, Talbot PJ (1992) Human coronavirus gene expression in the brains of multiple sclerosis patients. Virology 191:502–505

Stohlman SA, Bergmann CC, van der Veen RC, Hinton DR (1995) Mouse hepatitis virus-specific cytotoxic T lymphocytes protect from lethal infection without eliminating virus from the central nervous system. J Virol 69:684–694

Stuve O, Dooley NP, Uhm JH, Antel JP, Francis GS, Williams G, Yong VW (1996) Interferon beta-1b decreases the migration of T lymphocytes in vitro: effects on matrix metalloproteinase-9. Ann Neurol 40:853–863

Sun JP, Olsson T, Wang WZ, Xiao BG, Kostulas V, Fredrikson S, Ekre HP, Link H (1991) Autoreactive T-cell and B-cell responding to myelin proteolipid protein in multiple sclerosis and controls. Eur J Immunol 21:1461–1468

Sun N, Grzybicki D, Castro RF, Murphy S, Perlman S (1995) Activation of astrocytes in the spinal cord of mice chronically infected with a neurotropic coronavirus. Virology 213:482–493

Suzumura A, Lavi E, Weiss SR, Silberberg DH (1986) Coronavirus infection induces H-2 antigen expression on oligodendrocytes and astrocytes. Science 232:991–993

Swanborg RH (1995) Animal models of human disease: experimental autoimmune encephalomyelitis in rodents as a model for human demyelinating disease. Clin Immunol Immunopathol 77:4–13

Talbot P (1995) Implication of viruses in multiple sclerosis. Med Sci 11:837–843
Talbot P, Jouvenne P (1992) Neurotropic potential of coronaviruses. Med Sci 8:119–125
Talbot PJ, Ékandé S, Cashman NR, Mounir S, Stewart JN (1994) Neurotropism of human coronavirus 229E. Adv Exp Med Biol 342:339–346
Talbot PJ, Paquette JS, Ciurli C, Antel JP, Ouellet F (1996) Myelin basic protein and human coronavirus 229E cross-reactive T cells in multiple sclerosis. Ann Neurol 39:233–240
Tanaka R, Iwasaki Y, Koprowski HJ (1976) Intracisternal virus-like particles in the brain of a multiple sclerosis patient. J Neurosci Res 28:121–126
Taupin V, Renno T, Bourbonnière L, Peterson AC, Rodriguez M, Owens T (1997) Increased severity of experimental autoimmune encephalomyelitis, chronic macrophage/microglial reactivity, and demyelination in transgenic mice producing tumor necrosis factor-alpha in the central nervous system. Eur J Immunol 27:905–913
Tuohy VK, Thomas DM, Haqqi T, Yu M, Johnson JM (1995) Determinant-regulated onset of experimental autoimmune encephalomyelitis: distinct epitopes of myelin proteolipid protein mediate either acute or delayed disease in SJL/J mice. Autoimmunity 21:203–213
Tyrrell DAJ, Almeida JD, Berry DM, McIntosh K (1968) Coronaviruses. Nature 220:650
Tyrrell DAJ, Bynoe ML (1965) Cultivation of a novel type of common-cold virus in organ cultures. Brit Med J 1:1467–1470
Uhm, Joon H, Dooley, Nora P, Stuve O, Francis G, Duquette P, Antel JP, Yong VW (1997) Migratory behaviour of T lymphocytes isolated from MS patients undergoing treatment with β-interferon (IFN-β1b). Neurology 48:A80
Ulvestad E, Williams K, Mork S, Antel J, Nyland H (1994) Phenotypic differences between human monocytes/macrophages and microglial cells studied in situ and in vitro. J Neuropathol Exp Neurol 53:492–501
van Noort JM, van Sechel AC, Bajramovic JJ, Elouagmiri M, Polman CH, Lassmann H, Ravid R (1995) The small heat-shock protein alpha B-crystallin as candidate autoantigen in multiple sclerosis. Nature 375:798–801
Vandvik B, Norrby E (1989) Paramyxovirus SV5 and multiple sclerosis. Nature 338:769–771
Wang FI, Stohlman SA, Fleming JO (1990) Demyelination induced by murine hepatitis virus JHM strain (MHV-4) is immunologically mediated. J Neuroimmunol 30:31–41
Warren KG, Catz I, Steinman L (1995) Fine specificity of the antibody response to myelin basic protein in the central nervous system in multiple sclerosis: the minimal B-cell epitope and a model of its features. Proc Natl Acad Sci USA 92:11061–11065
Watanabe R, Wege H, ter Meulen V (1983) Adoptive transfer of EAE-like lesions from rats with coronavirus-induced demyelinating encephalomyelitis. Nature 305:150–153
Wege H (1995) Immunopathological aspects of coronavirus infections. Springer Sem Immunopathol 17:133–148
Weiner LP (1973) Pathogenesis of demyelination induced by mouse hepatitis virus (JHM virus). Arch Neurol 28:298–303
Weiss SR (1983) Coronaviruses SD and SK share extensive nucleotide homology with murine coronavirus MHV-A59 more than that shared between human murine coronaviruses. Virology 126:669–677
Williams RK, Jiang GS, Holmes KV (1991) Receptor for mouse hepatitis virus is a member of the carcinoembryonic antigen family of glycoproteins. Proc Natl Acad Sci USA 88:5533–5536
Wing MC, Moreau T, Greenwood J, Smith RM, Hale G, Isaacs J, Waldmann H, Lachmann PJ, Compston A (1996) Mechanism of first-dose cytokine-release syndrome by CAMPATH 1-H: involvement of CD16 (FcgammaRIII) and CD11a/CD18 (LFA-1) on NK cells. J Clin Invest 98:2819–2826
Woyciechowska JL, Dambrozia J, Leinikki P, Shekarchi C, Wallen W, Sever J, McFarland H, McFarlin D (1985) Viral antibodies in twins with multiple sclerosis. Neurology 35:1176–1180
Wucherpfennig KW, Strominger JL (1995) Molecular mimicry in T cell-mediated autoimmunity: viral peptides activate human T cell clones specific for myelin basic protein. Cell 80:695–705
Wucherpfennig KW, Weiner HL, Hafler DA (1991) T-cell recognition of myelin basic protein. Immunol Today 12:277–282
Wucherpfennig KW, Zhang JW, Witek C, Matsui M, Modabber Y, Ota K, Hafler DA (1994) Clonal expansion and persistence of human T cells specific for an immunodominant myelin basic protein peptide. J Immunol 152:5581–5592

Yamada M, Zurbriggen A, Fujinami RS (1990) Monoclonal antibody to Theilers murine encephalomyelitis virus defines a determinant on myelin and oligodendrocytes, and augments demyelination in experimental allergic encephalomyelitis. J Exp Med 171:1893–1907

Yokomori K, Asanaka M, Stohlman SA, Lai MMC (1993) A spike protein-dependent cellular factor other than the viral receptor is required for mouse hepatitis virus entry. Virology 196:45–56

Zhang JW, Markovicplese S, Lacet B, Raus J, Weiner HL, Hafler DA (1994) Increased frequency of interleukin 2-responsive T cells specific for myelin basic protein and proteolipid protein in peripheral blood and cerebrospinal fluid of patients with multiple sclerosis. J Exp Med 179:973–984

Zielasek, Hartung HP (1996) Molecular mechanisms of microglial activation. Adv Neuroimmunol 6:191–222

Subject Index

A
N-acetyl aspartate (NAA) 260
acetylcholine receptors 9, 10
N-acetylcysteine (NAC) 134
ACh receptors 123, 127
α-actinin-2 125
activated microglia 233
acute disseminated encephalomyelitis 4
acute viral encephalitis 16–30
– mechanisms of cell death 22
adenovirus 249, 256
adhesion 227
– molecules 228, 258
adoptive transfer 238, 254
adrenalectomy 135
adrenergic neurons 123
age-related virulence 108
AIDS 11, 137, 247
Alphaherpesvirinae 63
altruistic cell suicide 97, 114
Alzheimer's disease 105
American equine encephalitides 6
D-2-amino-5-phosphonovalerate (APV) 136
α-amino-3-hydroxy-5-methylisoxazole propionic acid (AMPA) 123, 133
– receptor 126, 131
β-adrenergic receptors 10
animal herpesvirus 77–86
antibody/antibodies 152
– anti-idiotypic 9, 127
– neutralization 254
antigen-presenting 229
– cells 258
– macrophages 229
anti-idiotypic antibodies 9, 127
antioxidant enzyme 211
antisense 71
apoptosis (*see also* cell suicide) 22–24, 95, 98, 133, 138, 148, 173, 205, 236
– mediators 172
apoptotic cell death 138
APV 137, 138
arachidonic acid 137

arachinodate metabolites 136
aspartate 129
astrocytes 127, 128, 131, 193, 251
astroglia 136
ataxia 81
autoimmune aggression 52
autoreceptor 134
axonal
– injury 260
– transport
– – anterograde 125, 126
– – retrograde 125, 126
axons 261

B
B cell 86
B7-molecules 224
BDV receptor 131
blood lymphocytes 252
blood-brain barrier 187, 189, 191, 258
bone marrow 78
Borna disease 2, 5, 8
– virus 10, 145, 157–173
– – adult infected rat 160
– – neonatal rats 161
bovine
– herpesviruses 83
– spongiform encephalopathy (BSE) 203
α-bungarotoxin 122

C
calcium 208
caspases 102
CD4+
– and CD8+ T cell interactions 234
– regulator T cells 232
CD8+ T lymphocytes 234
CD40 molecule 233
CD45RC 231
CD95 225
CD95/FasL 234
CD95/FasL-mediated killing 236
CEA 254

cell
- culture 207
- death (*see also* apoptosis) 98, 100
- suicide 96
- - altruistic 97, 114
cellular receptor 252
cerebellar granule cells 166
cerebellum 165, 210
cerebrospinal fluid 27
cervical lymph nodes 225
chemoattractant signals 257
chemokine 149, 188, 189, 193
- receptor 185
cleavage pattern 65
clinical complications 237
CNS parenchyma 27
cofactor 75, 83
common colds 251
copper 211
coronaviruses 249–251
Creutzfeldt-Jakob disease 203
cross-protection 79
CSF 77
cuprizone 211
CVS strain 125, 127
CXCR4 193, 194
cytokines 25, 42, 72, 136, 187, 192, 208, 249, 258
cytomegalovirus 250
cytotoxic T cells 97, 114, 254
cytotoxicity 207
cytotropism 7

D
death
- receptors 99
- regulators 99
dementia, subcortical 192
demyelination 184, 188, 249
dentate gyrus 124–126, 129, 131, 134, 135, 165
destruction of CNS cells, T lymphocyte-mediated 237
dizocilpine (MK-801) 124, 136–138
dopamine system 135

E
EAE (*see* experimental allergic encephalitis)
ectodermoses neurotropes 2
electron microscopy 7
encephalitogenic determinants 255
endonuclease 65
endothelial cells 80, 252
engagement of the B7/CD152 236
entorhinal cortex 126, 129
epidemic encephalitis (*von Economo*) 5
epitopes 249, 255
Epstein-Barr virus 256

exacerbations 258
exanthema subitum 77
excitatory amino acids 10, 133
excitotoxic damage 137, 138
excitotoxicity 133
experimental allergic
- encephalomyelitis 254
- encephalitis (EAE) 5, 51
extravasation 227

F
FasL 225
FasL/CD95 pathway 234
focal retinopathy 81
free oxygen radicals 133

G
GABA 127, 131
GABAergic 125–127
Gammaherpesvirinae 83
ganglia 85
gangliocytotropism 5
gangliosides 72
Gardella gel electrophoresis 86
general neurotropism 5
glial cells 127, 128
gliosis 203
GluR1 131
glutamate 129, 137, 186
- receptors 123–125, 131, 135
glutathione 208
glycoprotein 79
glycosylphosphatidylinositol 210
gp120 136–138
Guinea pigs 67

H
HAM/TSP 188
hemagglutinating encephalomyelitis virus of pigs 253
herpes
- simplex 250
- - encephalitis 17–22, 27, 248
- - - neonatal infection 20
- - virus 6, 110, 250
- - - bovine 83
high temperature 68
hippocampal formation 124, 128, 131
hippocampus 125, 129, 132, 134, 165
histones 69
HIV 109, 135, 136, 247
- encephalitis 190
HIV-1 11
- encephalitis 110
- envelope protein gp120 135
HLA-DR2 257

homing patterns 229
homologies 256
human T lymphotropic virus 250
Huntington's disease 105

I
ICAM-1 228
ICP0 71
IFN-α/β 230
IFN-γ 138, 230
IL (*see* interleukin)
immune
– complexes 81
– surveillance 257
– tolerance 249
immunohistochemistry 7, 8
immunopathological disease 36
immunosuppressive effects 232
in situ end-labeling 206
in situ hybridization 65, 251
inclusion bodies 2
indicator cells 64
inducible nitric oxide synthetase 232
infected monocytes 77
infections
– latent 39
– persistent 38
– reversibly non-productive 62
– slow virus infections 39
infiltrating T cells 231
inflammatory mediators 149
influenza 87, 250
– virus 256
interferon 68
interleukin
– IL-1β 136, 138, 210, 254
– IL-2r expression 231
– IL-6 73, 254
interneurons 125–127
intra-axonal virus spread 128
ischaemia 104
ischaemic
– injuries 81
– necrosis 28

J
Japanese encephalitis 6
– virus 110

K
kainate 123
– (KA-1) receptor 130, 131, 133

L
La Crosse virus 111
lameness 81

latency-associated protein 84
latent infections 39
LATs 71
LCMV (*see* lymphocytic choriomeningitis virus)
leukoencephalitis 5
liposomes 207
locus coeruleus 123
long-term culture 64
lymphocyte migration 258
lymphocytic choriomeningitis virus (LCMV) 48, 112
lymphoid tissue 78, 80
lymphotropic herpesvirus 76

M
macrophage chemoattractant protein-1 254
macrophages 252
Maedi/Visna virus 184
magnetic resonance spectroscopy 260
mamillitis 84
matrix metalloproteinases 254
MBP 255
measles 250
– virus 41, 248
memantine 138
N-methyl-D-aspartate (NMDA) 123
MHC antigens 42, 221
– class I
– – antigens 165
– – expression on neurons 223
– – molecules 222
– class II
– – antigens 165
– – expression on
– – – astrocytes 224
– – – microglia cells 224
– up-regulation 230
MHV 253
microglia 127, 193, 208, 210, 232, 249
microglial activation 24
MK-801 (dizocilpine) 124, 136–138
molecular
– mimicry 248
– shape mimicry 257
monoclonal antibodies 65
monocytes, infected 77
mononegavirales 128
mossy fiber system 126, 129
motor neuron disease 114
mouse hepatitis virus 253
MRI 260
MRS imaging 260
MS (*see* multiple sclerosis)
multinucleated giant cells 181, 184, 185, 187, 191

multiple sclerosis (MS)   114, 249, 251
– pathogenesis   258–263
mumps   250
murine leukemia virus   181–184
mutants   70
myelin   249
– basic protein   255
myelin-associated glycoprotein   255
myelin-oligodendrocyte glycoprotein   255

N

N2a cells   207
NAA (see N-acetyl aspartate)
NAC (N-acetylcysteine)   134
nervous tissue   78
neural
– development   103
– spread   1
neurodegeneration   208
neurodegenerative disorders   105
neurofilaments   261
neurologic disability scores (EDSS)   260
neuronal
– damage   145
– latency   83
– loss   203
– nucleus   69
neurons   76, 80, 254
neurotoxicity   136
neurotransmitter
– receptors   10, 122
– system   161
neurotrophic factors   170
neurotropic alphaherpesvirus   82
neurotropism   8, 122
– general   5
– special   5
neurovirulence   38
nifedipine   136
nitric oxide (NO)   25, 26, 136
– synthase   254
NMDA (N-methyl-D-aspartate)   123, 138
– receptor   125, 126, 128, 134, 135, 137
NO (nitric oxide)   136, 254
non-cytolytic clearance   235
norepinephrine   211
NR1   125–128
nucleoside
– analogue   75
– inhibitors   68

O

OC43   251
olfactory
– bulb   127
– pathways   107

oligodendrocyte   127, 128, 131, 249
oxidative stress   133, 208, 210, 211

P

PAF (see platelet activating factor)
panencephalitides   6
panencephalitis   7
papovaviruses   7, 86
parainfluenza   250
Parkinson's disease   105
parvalbumin   209
pathoclisis   8
pathogenesis   145
PC12 cells   207
perforant path   126, 129, 131
perforin pathway   234
perivascular space   229
peroxynitrite   136
persistent infections   38
physical state   79
picornavirus   87
platelet activating factor (PAF)   137
PLP   255
PML (see progressive multifocal leuko-
    encephalopathy)
polioencephalitis   5
poliomyelitis   1, 9, 248
poliovirus   37
– receptor   9, 10
polyoma virus   256
post-herpetic neuralgia   76
post-polio syndrome   114
prion   203
– diseases   203–212
– protein   204
programmed cell death   98
progressive multifocal leukoencephalopathy
    (PML)   7
pro-inflammatory substances   233
proliferation   231
– switch-off   231
prophylaxis   75
protein kinase C   134
PrP106-126   207
PrPC   204
PrPSc   204
Purkinje cells   109, 111, 166
pyramidal cells   125, 126

Q

quinolinate   136, 138
quinolinic acid (QUIN)   183, 188, 193

R

rabbits   67
rabies   1, 8, 28, 248, 250

Subject Index

– virus   9, 10, 109, 145
reactivation   62
– stimulus   64
reactive oxygen species   208
receptor   9, 10
– acetylcholine   9, 10
– ACh   123, 127
– β-adrenergic   10
– AMPA   126, 131
– autoreceptor   134
– BDV   131
– cellular   252
– chemokine   185
– death   99
– glutamate   123–125, 131, 135
– kainate   130, 131, 133
– neurotransmitter   10, 122
– NMDA   125, 126, 128, 134, 135, 137
– poliovirus   9, 10
– RV   124–127
– virus   122
recombinant viruses   65
reovirus   9, 10, 111
replication cycle   69
reservoir   85
retina   134, 135, 206
retrovirus   250
reversibly non-productive infection   62
riluzole   137
rostral migratory stream   113
RT-PCR   66, 252
rubella   250
– virus   109
RV receptor   124–127

S
satellite cells   76
Schaffer collaterals   129
scrapie   203
– strains   206
selective vulnerability   8
Semliki Forest virus   106
sensitive method   66
septum pellucidum   125, 126
shingles   76
simian virus   250
– SIV encephalitis   186
Sindbis virus   106
– infection   23
slow virus infections   39
Southern blot hybridisation   65
special neurotropism   5
spongiform
– degeneration   181
– encephalopathies, transmissible   105
SSPE (*see* subacute sclerosing panencephalitis)

St. Louis encephalitis   6
staggering disease   134
strain   205
stress   63
subacute sclerosing panencephalitis (SSPE)   6, 7
subcortical dementia   192
superinfection   68
superior cervical ganglion   123
superoxide   208
– anions   136
– dismutase   211

T
T cell   72, 86, 227, 229, 255
– action in the brain   237
– clone   257
– cytotoxic   97, 114, 254
– homing   228
– infiltrating   231
– responses   26, 27
– – in the CLNs   226
targeting   205
TH1-determined cytokine milieu   233
Theiler's murine encephalomyelitis virus   249
thymidine kinase   70
tick-borne encephalitis   250
time factor   6
tissue tropism   9
TNF-α (*see* tumor necrosis factor α)
transactivation factors   72
transcription factors   146
transmissible spongiform encephalopathies   105
transmission   80
tree shrews   133
tumor necrosis factor (TNF)-α   136, 138, 182, 187, 189, 190, 192, 193, 210, 254
tumour viruses   85
TUNEL   206
tyrosine kinase   208

U
ubiquinone   210
upper respiratory distress   81
UV light   73

V
vaccines   75, 85
vaccinia   250
varicella   250
vasculities   28
vasoactive intestinal polypeptide (VIP)   136
VCAM-1   228
Venezuelan equine encephalitis virus   108, 113

vesicular stomatitis virus 111
veterinary diseases 78
VIP (vasoactive intestinal polypeptide) 136
viraemia 80
viral
– attachment proteins 9, 122
– persistence 98

virus
– clearance 150
– infection 247
– receptors 122
virus-cell interactions 36–42
virus-cleansing property 235

X
xenotransplants 80

Current Topics in Microbiology and Immunology

Volumes published since 1989 (and still available)

Vol. 212: **Vainio, Olli; Imhof, Beat A. (Eds.):** Immunology and Developmental Biology of the Chicken. 1996. 43 figs. IX, 281 pp. ISBN 3-540-60585-1

Vol. 213/I: **Günthert, Ursula; Birchmeier, Walter (Eds.):** Attempts to Understand Metastasis Formation I. 1996. 35 figs. XV, 293 pp. ISBN 3-540-60680-7

Vol. 213/II: **Günthert, Ursula; Birchmeier, Walter (Eds.):** Attempts to Understand Metastasis Formation II. 1996. 33 figs. XV, 288 pp. ISBN 3-540-60681-5

Vol. 213/III: **Günthert, Ursula; Schlag, Peter M.; Birchmeier, Walter (Eds.):** Attempts to Understand Metastasis Formation III. 1996. 14 figs. XV, 262 pp. ISBN 3-540-60682-3

Vol. 214: **Kräusslich, Hans-Georg (Ed.):** Morphogenesis and Maturation of Retroviruses. 1996. 34 figs. XI, 344 pp. ISBN 3-540-60928-8

Vol. 215: **Shinnick, Thomas M. (Ed.):** Tuberculosis. 1996. 46 figs. XI, 307 pp. ISBN 3-540-60985-7

Vol. 216: **Rietschel, Ernst Th.; Wagner, Hermann (Eds.):** Pathology of Septic Shock. 1996. 34 figs. X, 321 pp. ISBN 3-540-61026-X

Vol. 217: **Jessberger, Rolf; Lieber, Michael R. (Eds.):** Molecular Analysis of DNA Rearrangements in the Immune System. 1996. 43 figs. IX, 224 pp. ISBN 3-540-61037-5

Vol. 218: **Berns, Kenneth I.; Giraud, Catherine (Eds.):** Adeno-Associated Virus (AAV) Vectors in Gene Therapy. 1996. 38 figs. IX,173 pp. ISBN 3-540-61076-6

Vol. 219: **Gross, Uwe (Ed.):** Toxoplasma gondii. 1996. 31 figs. XI, 274 pp. ISBN 3-540-61300-5

Vol. 220: **Rauscher, Frank J. III; Vogt, Peter K. (Eds.):** Chromosomal Translocations and Oncogenic Transcription Factors. 1997. 28 figs. XI, 166 pp. ISBN 3-540-61402-8

Vol. 221: **Kastan, Michael B. (Ed.):** Genetic Instability and Tumorigenesis. 1997. 12 figs.VII, 180 pp. ISBN 3-540-61518-0

Vol. 222: **Olding, Lars B. (Ed.):** Reproductive Immunology. 1997. 17 figs. XII, 219 pp. ISBN 3-540-61888-0

Vol. 223: **Tracy, S.; Chapman, N. M.; Mahy, B. W. J. (Eds.):** The Coxsackie B Viruses. 1997. 37 figs. VIII, 336 pp. ISBN 3-540-62390-6

Vol. 224: **Potter, Michael; Melchers, Fritz (Eds.):** C-Myc in B-Cell Neoplasia. 1997. 94 figs. XII, 291 pp. ISBN 3-540-62892-4

Vol. 225: **Vogt, Peter K.; Mahan, Michael J. (Eds.):** Bacterial Infection: Close Encounters at the Host Pathogen Interface. 1998. 15 figs. IX, 169 pp. ISBN 3-540-63260-3

Vol. 226: **Koprowski, Hilary; Weiner, David B. (Eds.):** DNA Vaccination/Genetic Vaccination. 1998. 31 figs. XVIII, 198 pp. ISBN 3-540-63392-8

Vol. 227: **Vogt, Peter K.; Reed, Steven I. (Eds.):** Cyclin Dependent Kinase (CDK) Inhibitors. 1998. 15 figs. XII, 169 pp. ISBN 3-540-63429-0

Vol. 228: **Pawson, Anthony I. (Ed.):** Protein Modules in Signal Transduction. 1998. 42 figs. IX, 368 pp. ISBN 3-540-63396-0

Vol. 229: **Kelsoe, Garnett; Flajnik, Martin (Eds.):** Somatic Diversification of Immune Responses. 1998. 38 figs. IX, 221 pp. ISBN 3-540-63608-0

Vol. 230: **Kärre, Klas; Colonna, Marco (Eds.):** Specificity, Function, and Development of NK Cells. 1998. 22 figs. IX, 248 pp. ISBN 3-540-63941-1

Vol. 231: **Holzmann, Bernhard; Wagner, Hermann (Eds.):** Leukocyte Integrins in the Immune System and Malignant Disease. 1998. 40 figs. XIII, 189 pp. ISBN 3-540-63609-9

Vol. 232: **Whitton, J. Lindsay (Ed.):** Antigen Presentation. 1998. 11 figs. IX, 244 pp. ISBN 3-540-63813-X

Vol. 233/I: **Tyler, Kenneth L.; Oldstone, Michael B. A. (Eds.):** Reoviruses I. 1998. 29 figs. XVIII, 223 pp. ISBN 3-540-63946-2

Vol. 233/II: **Tyler, Kenneth L.; Oldstone, Michael B. A. (Eds.):** Reoviruses II. 1998. 45 figs. XVI, 187 pp. ISBN 3-540-63947-0

Vol. 234: **Frankel, Arthur E. (Ed.):** Clinical Applications of Immunotoxins. 1999. 16 figs. IX, 122 pp. ISBN 3-540-64097-5

Vol. 235: **Klenk, Hans-Dieter (Ed.):** Marburg and Ebola Viruses. 1999. 34 figs. XI, 225 pp. ISBN 3-540-64729-5

Vol. 236: **Kraehenbuhl, Jean-Pierre; Neutra, Marian R. (Eds.):** Defense of Mucosal Surfaces: Pathogenesis, Immunity and Vaccines. 1999. 30 figs. IX, 296 pp. ISBN 3-540-64730-9

Vol. 237: **Claesson-Welsh, Lena (Ed.):** Vascular Growth Factors and Angiogenesis. 1999. 36 figs. X, 189 pp. ISBN 3-540-64731-7

Vol. 238: **Coffman, Robert L.; Romagnani, Sergio (Eds.):** Redirection of Th1 and Th2 Responses. 1999. 6 figs. IX, 148 pp. ISBN 3-540-65048-2

Vol. 239: **Vogt, Peter K.; Jackson, Andrew O. (Eds.):** Satellites and Defective Viral RNAs. 1999. 39 figs. XVI, 179 pp. ISBN 3-540-65049-0

Vol. 240: **Hammond, John; McGarvey, Peter; Yusibov, Vidadi (Eds.):** Plant Biotechnology. 1999. 12 figs. XII, 196 pp. ISBN 3-540-65104-7

Vol. 241: **Westblom, Tore U.; Czinn, Steven J.; Nedrud, John G. (Eds.):** Gastroduodenal Disease and Helicobacter pylori. 1999. 35 figs. XI, 313 pp. ISBN 3-540-65084-9

Vol. 242: **Hagedorn, Curt H.; Rice, Charles M. (Eds.):** The Hepatitis C Viruses. 2000. 47 figs. IX, 379 pp. ISBN 3-540-65358-9

Vol. 243: **Famulok, Michael; Winnacker, Ernst-L.; Wong, Chi-Huey (Eds.):** Combinatorial Chemistry in Biology. 1999. 48 figs. IX, 189 pp. ISBN 3-540-65704-5

Vol. 244: **Daëron, Marc; Vivier, Eric (Eds.):** Immunoreceptor Tyrosine-Based Inhibition Motifs. 1999. 20 figs. VIII, 179 pp. ISBN 3-540-65789-4

Vol. 245/I: **Justement, Louis B.; Siminovitch, Katherine A. (Eds.):** Signal Transduction and the Coordination of B Lymphocyte Development and Function I. 2000. 22 figs. XVI, 274 pp. ISBN 3-540-66002-X

Vol. 245/II: **Justement, Louis B.; Siminovitch, Katherine A. (Eds.):** Signal Transduction on the Coordination of B Lymphocyte Development and Function II. 2000. 13 figs. XV, 172 pp. ISBN 3-540-66003-8

Vol. 246: **Melchers, Fritz; Potter, Michael (Eds.):** Mechanisms of B Cell Neoplasia 1998. 1999. 111 figs. XXIX, 415 pp. ISBN 3-540-65759-2

Vol. 247: **Wagner, Hermann (Ed.):** Immunobiology of Bacterial CpG-DNA. 2000. 34 figs. IX, 246 pp. ISBN 3-540-66400-9

Vol. 248: **du Pasquier, Louis; Litman, Gary W. (Eds.):** Origin and Evolution of the Vertebrate Immune System. 2000. 81 figs. IX, 324 pp. ISBN 3-540-66414-9

Vol. 249: **Jones, Peter A.; Vogt, Peter K. (Eds.):** DNA Methylation and Cancer. 2000. 16 figs. IX, 169 pp. ISBN 3-540-66608-7

Vol. 250: **Aktories, Klaus; Wilkins, Tracy, D. (Eds.):** Clostridium difficile. 2000. 20 figs. IX, 143 pp. ISBN 3-540-67291-5

Vol. 251: **Melchers, Fritz (Ed.):** Lymphoid Organogenesis. 2000. 62 figs. XII, 215 pp. ISBN 3-540-67569-8

Vol. 252: **Potter, Michael; Melchers, Fritz (Eds.):** B1 Lymphocytes in B Cell Neoplasia. 2000. XIII, 326 pp. ISBN 3-540-67567-1

Printing: Saladruck, Berlin
Binding: H. Stürtz AG, Würzburg